Schriftenreihe

Innovative Betriebswirtschaftliche Forschung und Praxis

Band 536

ISSN 1437-787X (Print)

Verlag Dr. Kovač

Madita Amelie Pesch

„Standing on the Shoulders of Giants“: The Case of Negotiation Research

Structuring the Field of Negotiation Research and Aggregating Contradictory Empirical Findings of Emotions on Negotiation Outcomes

Verlag Dr. Kovač

Hamburg
2021

VERLAG DR. KOVAČ GMBH
FACHVERLAG FÜR WISSENSCHAFTLICHE LITERATUR

Leverkusenstr. 13 · 22761 Hamburg · Tel. 040 - 39 88 80-0 · Fax 040 - 39 88 80-55

E-Mail info@verlagdrkovac.de · Internet www.verlagdrkovac.de

D6

Bibliografische Information der Deutschen Nationalbibliothek
Die Deutsche Nationalbibliothek verzeichnet diese Publikation
in der Deutschen Nationalbibliografie;
detaillierte bibliografische Daten sind im Internet
über http://dnb.d-nb.de abrufbar.

ISSN: 1437-787X (Print)

ISBN: 978-3-339-12022-9
eISBN: 978-3-339-12023-6

Zugl.: Dissertation, Westfälische Wilhelms-Universität Münster, 2020

Printed in Germany

Gedruckt auf holz-, chlor- und säurefreiem, alterungsbeständigem Papier. Archivbeständig nach ANSI 3948 und ISO 9706.

Geleitwort

Zwerge, die auf den Schultern von Riesen stehen: Diese Metapher stammt von Bernhard von Chartres aus dem 12. Jahrhundert und zeigt wie Forschung offensichtlich stattfindet. Die Riesen stellen die große Anzahl von Wissenschaftlern vergangener Epochen dar und die Zwerge die aktuellen Forscher, die auf den Erkenntnissen ihrer Vorgänger aufbauen. Indem sich aktuelle Forscher auf die Ergebnisse ihrer Vorgänger beziehen, können sie weiter schauen als die Riesen, da sie von ihrer gigantischen Statur emporgehoben werden. Diese Metapher bildet den Leitrahmen für die vorliegende Arbeit, in der die Verfasserin die Verhandlungsforschungssituation beleuchtet.

Verhandlungen sind allgegenwärtig. Sie prägen sowohl den beruflichen Alltag, als auch das Privatleben. Neben ihrer hohen praktischen Relevanz oder gerade aufgrund ihrer großen Bedeutung in der Praxis, haben Verhandlungen bei einer Vielzahl von Forschern aus verschiedenen Forschungsbereichen hohe Beachtung gefunden (z.B. Psychologie, Wirtschaftswissenschaften oder Rechtswissenschaften). Die einzelnen Forschungsergebnisse stehen jedoch weitestgehend isoliert nebeneinander, selbst wenn gleichartige Themen behandelt werden. Darüber hinaus bestätigen sich manche Erkenntnisse, andere sind jedoch von Divergenzen gekennzeichnet.

Auf Basis der aktuellen Forschungssituation stellt die Verfasserin heraus, dass die vielen vorliegenden Einzelerkenntnisse der Verhandlungsforschung einer Synthese und Strukturierung bedürfen, um Verhandlungsforschern ein „geordnetes" Bild des Forschungsfeldes zu präsentieren.

Diesen Aspekt greift die Verfasserin auf und adressiert ihn im Rahmen ihrer Dissertation. In ihrer ersten Studie zeigt sie unter Einsatz von bibliometrischen Methoden die Struktur und die dynamische Entwicklung des Forschungsfeldes auf. Basierend auf der Annahme, dass Veröffentlichungen, die häufiger gemeinsam zitiert werden, inhaltlich näher beieinanderliegen, ermittelt die Verfasserin mit Hilfe einer Cluster-Analyse die inhaltliche Vernetzung der Publikationen. Dadurch erhält sie die Themenschwerpunkte der Verhandlungsforschung der vergangenen Jahrzehnte.

In einer zweiten Studie greift die Verfasserin die inkonsistenten Ergebnisse von Emotionsstudien auf. Mit Hilfe einer Meta-Analyse versucht sie, diese Inkonsistenzen in den Griff zu bekommen. Basierend auf dem Emotions as Social Information (EASI)-Modell, das die zwischenmenschlichen Auswirkungen von Emotionen in sozialen Interaktionen beschreibt, zeigt sie in regressions-analytischen Modellen, welche Ergebnisse stabil sind und welche nicht. Dabei wird unterstrichen, dass ein Ausdruck von Wut die ökonomischen Verhandlungsergebnisse des wütenden Verhandlers signifikant erhöht (im Vergleich zu Emotionsneutralität und Freude). Dieser positive Effekt hängt jedoch von einer Vielzahl moderierender Faktoren ab, die ihn verstärken oder abschwächen. Neben den positiven Auswirkungen von Wut in Verhandlungen wird allerdings auch gezeigt, dass sich Wut nachteilig auf die subjektiven Verhandlungsergebnisse auswirkt.

Die Arbeit adressiert unter Zuhilfenahme von quantitativ-empirischen Methoden bestehende Forschungsdefizite und liefert wertvolle Erkenntnisse für Verhandlungen in Forschung und Praxis. Im Rahmen der Arbeit werden sowohl die Schultern der Verhandlungsriesen gestärkt, als auch die Verfasserin der Dissertation als Zwerg auf diesen Schultern emporgehoben, indem sie auf dem meist untersuchten Thema Emotionen aufbaut und das Wissen über Wut und Freude in Verhandlungen erweitert. Ich wünsche der Arbeit daher die ihr zustehende weite Verbreitung in Wissenschaft und Praxis.

Münster, im September 2020 Prof. Dr. Dr. h.c. Klaus Backhaus

Vorwort

Die vorliegende Arbeit ist während meiner Zeit als wissenschaftliche Mitarbeiterin am Institut für Anlagen und Systemtechnologien (IAS) unter dem Dach des Marketing Center Münsters entstanden. Sie wurde im Juli 2020 von der Wirtschaftswissenschaftlichen Fakultät der Westfälischen Wilhelms-Universität als Dissertation angenommen. Auch wenn eine Dissertation als Individualleistung angefertigt wird, möchte ich an dieser Stelle hervorheben, dass mir einige Personen in meiner Promotionszeit zur Seite standen ohne deren Unterstützung und Beistand die Entstehung dieses Buches nicht möglich gewesen wäre. An dieser Stelle möchte ich mich von Herzen bei diesen Menschen bedanken.

Zunächst gilt mein Dank meinem akademischen Lehrer und Doktorvater, Prof. Dr. Dr. h.c. Klaus Backhaus. Bereits in meiner Zeit als studentische Mitarbeiterin habe ich Sie als warmherzigen und engagierten Leiter des IAS kennengelernt, der für seine Mitarbeiter stets ein offenes Ohr hat und ihnen ein hohes Maß an Vertrauen und Freiräumen entgegenbringt. Dieser Eindruck hat sich in meiner Zeit als wissenschaftliche Mitarbeiterin weiter gefestigt. Ich habe es sehr genossen einen Doktorvater zu haben, mit dem die Zusammenarbeit so viel Freude bereitet, der mir eine Promotionszeit ermöglicht hat, in welcher ich mich fachlich sowie persönlich weiterentwickeln konnte und der immer zu mir gestanden hat. Sie haben eine positive Atmosphäre durch Ihre liebevolle Art am IAS erschaffen, die meines Erachtens einmalig ist. Wer bei Ihnen promoviert, gehört zu einer Familie, die stets zusammenhält. Mein Dank gilt weiterhin Prof. Dr. Manfred Krafft für die Erstellung des Zweitgutachtens zu meiner Dissertation sowie Prof. Dr. Karl-Hans Hartwig für den Beisitz in meiner Disputation.

Des Weiteren möchte ich mich herzlich bei meinen Kollegen vom IAS bedanken. Buffi, Max, Stefan, Amir, Ulf, Thommy und Niklas. Ihr habt maßgeblich dazu beigetragen, dass meine Zeit am IAS so einzigartig und wunderschön geworden ist. Wer am IAS gearbeitet hat, weiß, dass das Dasein als Arbeitskollegen den Weg für gute Freundschaften ebnet, die einen durch das weitere Leben begleiten werden. Ich erinnere mich an unzählige tolle Erlebnisse mit euch. Seien es diverse Partys zu Karneval oder Weihnachten, „Klassenfahrten“ oder All-you-can-Eats im Le Feu – wir hatten eine super Zeit, die ich sehr vermissen werde. Selbst die kleinen Dinge wie das alltägliche Mittagessen, die Diskussionen über die Dissertation oder die vielen Gespräche über Gott und die Welt haben mich in manch stressiger Phase abgelenkt und mir viele neue Perspektiven aufgezeigt. Ich danke euch für eure Loyalität und fürs Zuhören und Ratschläge geben.

Insbesondere möchte ich mich an dieser Stelle bei Ulf, Thommy und Niklas bedanken. Mit euch dreien habe ich die längste Zeit zusammengearbeitet und euch dabei sehr schätzen gelernt. Ulfi, du warst mit mir an der Seite die Zuckerschnute des Instituts, mit der ich sehr gerne zu diversen Bäckereien und Eisdielen gegangen bin. Durch deine super offene, fröhliche und witzige Art habe ich mich immer gefreut in das Ulfi-Büro zu kommen und auf deiner super intakten Couch mit dir zu quatschen. Thommy, du hast mir ebenfalls viele Lachanfälle beschert, indem du durch deinen einzigartigen Musikgeschmack, deine super witzige und direkte Art und deine Laser-Auszeiten in meinem Büro Abwechslung in das Promotionsleben gebracht hast. Niklas, wir haben uns sogar noch als Hiwis am IAS kennengelernt und ich bin heilfroh, dass wir dich noch für das WMA-Team dazugewinnen konnten. Unser Zweier-Team am Ende der Laufzeit des Instituts hat wunderbar harmoniert und dich zeichnet deine sehr warmherzige und loyale Art aus.

Zudem gilt mein Dank den vielen weiteren IAS-Mitgliedern, die meine Zeit am Lehrstuhl bereichert und mich maßgeblich unterstützt haben. Hier seien zunächst Britta Jansen und Britta Weiguny genannt, die mir am IAS nicht nur organisatorisch zur Seite standen, sondern auch menschlich durch ihre positive Art sehr dazu beigetragen haben, das IAS zu diesem wunderbaren Ort zu machen. Ich danke euch für viele Gespräche und Ratschläge, die mir oft weitergeholfen haben. Außerdem möchte ich mich herzlich bei den studentischen Hilfskräften des Instituts bedanken. Ihr unterstützt uns bei so vielen Aufgaben und erleichtert unser Dasein als wissenschaftliche Mitarbeiter immens. Ausdrücklich möchte ich an dieser Stelle vier Hilfskräfte hervorheben, die mich besonders bei meiner Dissertation unterstützt haben. Viktoria Heldt, Joscha Hofmann, Rafael Kansy und Kaja Banach möchte ich für ihre ausdauernde und sehr genaue Arbeit danken.

Abseits der IAS-Familie gibt es einige Personen, die in der Promotionszeit für mich da waren und mir egal in welchem Gemütszustand ich mich befunden habe zur Seite standen. Dazu gehören zum einen meine Freunde aus Schul-und Studienzeiten. Katha, Anna, Clara, Theresa, Marie, Julia, Lea, Joschi, Steffi, Hiesti und Martina. Ihr begleitet mich nun schon seit kürzerer oder längerer Zeit durch mein Leben. Ich blicke auf unzählig viele schöne Momente mit euch zurück und bin dankbar so wundervolle Menschen in meinem Leben zu haben. Insbesondere möchte ich an dieser Stelle Lena und Carsten hervorheben. Ich habe selten Menschen in meinem Leben getroffen, die ich in so kurzer Zeit so schnell in mein Herz geschlossen habe. Ich liebe es mit dir, Lena, zu quatschen und beim gemeinsamen Yogieren die Seele baumeln zu lassen. Ihr zaubert mir immer ein Lächeln ins Gesicht bei unseren diversen Spiele-Abenden, Kochevents usw.

Neben meinen Freunden möchte ich mich von ganzem Herzen bei meinem wunderbaren Freund Timo bedanken. Timo, dir danke ich für deine bedingungslose Liebe in den letzten Jahren. Du hast mich aus manch einem Tal tiefer Verzweiflung geholt. Du hast mir immer zugehört, mit mir Diss-Themen ausdiskutiert und mich abgelenkt, wenn ich mal Abwechslung brauchte. Du bist für mich mein Fels in der Brandung, der mich zum Lachen bringt, der mich runterholt und der mich einfach in den Arm nimmt, wenn ich es brauche. Ebenfalls möchte ich an dieser Stelle auch deiner tollen Familie danken, die mich so herzlich und offen willkommen geheißen hat. Heike, Peter, Jasmin, Kai, Tabea, Jonas und Carlo: Ich danke euch, dass ich einen Platz in eurer Familie einnehmen darf.

Zu guter Letzt gilt mein Dank besonders meiner Familie. Es ist nicht selbstverständlich so große Fans in der Familie zu haben, die einen bedingungslos lieben und unterstützen. Insbesondere denke ich hier an meine Oma und meinen Opa sowie an meine tollen Eltern. Es fing bereits mit kleinen Pokalen in der Grundschulzeit und gemeinsam gemalten Kunstbildern an und reicht bis heute. Ihr habt mir immer zur Seite gestanden, mich bei allem, was ich mir vorgenommen habe unterstützt und mich immer begleitet. Insbesondere für die unendlich vielen Gespräche, Umarmungen, Ratschläge, Urlaube, Unternehmungen, ja, eigentlich für alles, möchte ich euch danken. Ich weiß, wie stolz euch meine Promotion macht und möchte euch nun meine Arbeit widmen, auch wenn ich dadurch nicht annähernd das zurückgeben kann, was ihr alles für mich getan habt.

Münster, im September 2020 Madita Amelie Pesch

Contents

List of Figures

List of Tables

List of Abbreviations

A	Article
ANOVA	Analysis of Variance
Avg.	Average
B	Behaviour
BATNA	Best Alternative to a Negotiated Agreement
BDR	Behavioural Decision Research
Brexit	Composite of Britain and Exit
CEO	Chief Executive Officer
CF	Cell Frequency
ci.lb	Confidence Interval Lower Bound
ci.ub	Confidence Interval Upper Bound
CoCit	Co-Citation
Comp	Competitive
Coop	Cooperative
CV	Citation Value
D	Demographic
df	Degrees of Freedom
DOI	Digital Object Identifier
DV	Dependent Variable
e-	Electronic
EASI	Emotions as Social Information
EBSCO	Elton B. Stephens Company
EFA	Exploratory Factor Analysis
e.g.	Exempli Gratia (for example)

EI	Emotional Intelligence
EO	Economic Outcome
et al.	Et Alii (and others)
etc.	Et Cetera (and so on)
FTF	Face-to-Face
H	Hypothesis
HCA	Hierarchical Cluster Analysis
ICC	Intra-Class Correlation
i.e.	Id Est (in other words)
IT	Information Technology
IV	Independent Variable
MDS	Multidimensional Scaling
ns.	Not Stated
NSS	Negotiation Support System
O	Organizational
P	Psychographic
Psyc	Psychological
R	Reference
S	Situative
SC	Study Characteristics
SCIE	Science Citation Index Expanded
SNA	Social Network Analysis
SO	Subjective Outcome
SSCI	Social Science Citation Index
VIF	Variance Inflation Factor
vs.	Versus
VW	Volkswagen
WOS	Web of Science

List of Symbols

°	Statistically Significant at the Level of 10%
*	Statistically Significant at the Level of 5%
**	Statistically Significant at the Level of 1%
***	Statistically Significant at the Level of 0.1%
a	Document within Reference List
α	Intercept of Meta-Regression Model
b	Document within Reference List
β	Estimated Coefficient of a Moderator in Random-Effects Model
C	Citing Document
δ	True Effect Size
e	Fraction of Edges Linking Nodes of one Subgroup
ε	Sampling Error
ES	Standardized Mean Difference (Effect Size)
$\overline{ES}$	Weighted Mean Effect Size
F	Value of F-Statistic
g	Consecutive Number of Moderators (Within-Study Level)
h	True Variation of Effect Size at Within-Study Level
i	Consecutive Number of Effect Sizes
I^2	Degree of True Heterogeneity between Effect Sizes
j	Subgroup in a Cluster Network
k	Consecutive Number of Moderators in Random-Effects Model
κ	Cohen's Kappa
l	Consecutive Number of Moderators (Between-Study Level)
L_0	Second Estimator of Missing Effect Sizes of the Trim and Fill Method

λ	Estimated Coefficient of a Moderator at Between-Study Level
m	Consecutive Number of Effect Sizes per Study
M	Moderator in Random-Effects Model
μ_{0s}	Intercept of Studies in Multilevel Model
n	Sample Size of Group
N	Overall Number of Effect Sizes
p	Value of Statistical Significance
Pr(o)	Observed Percentage Agreement
Pr(x)	Expected Percentage Agreement
q	Expected Value of Fraction of Edges Linking Nodes of one Subgroup
Q	Q-value of Cochran's Q-Test
r	Pearson's Correlation Coefficient
R_0	First Estimator of Missing Effect Sizes of the Trim and Fill Method
R^2	Coefficient of Determination
ρ	Estimated Coefficient of a Moderator at Within-Study Level
s	Consecutive Number of Studies
sd	Standard Deviation
s.e.	Standard Error
σ_{BS}^2	Variance at the Between-Study Level
σ_{WS}^2	Variance at the Within-Study Level
Σ	Sum
t	Value of t-Statistic
T	Sum of Ranks of Effect Sizes Greater than the Overall Mean
τ	Standard Deviation of True Variation in Random-Effects Model
ϑ	True Variation of Effect Size in Random-Effects Model
u	Error Term of Study Level in Multilevel Model
v	Variance of Standardized Mean Difference
V	Moderator at Within-Study Level

ϖ	Standard Deviation of Sampling Error
W	Moderator at Between-Study Level
$\overline{X}$	Mean Value of Group Sample
y*	Rightmost Series of Consecutive Ranks (Trim and Fill Method)
Y	Modularity
z	Value of z-Statistic
ζ_0	Level Three Intercept of Multilevel Model

List of Appendices

1 Introduction

1.1 Negotiation Research: Paradigm of a Research Process

In 2008, Porsche planned the takeover of Volkswagen (VW) after having steadily increased their shares in the company (here and in the following Topham 2012).[1] Nevertheless, this undertaking failed fast as Porsche ruined itself by speculating with bank credits at the peak of the financial crisis. Thereafter, VW acquired nearly 50% of the shares of Porsche in 2009 and took over the entire company in 2012. All these developments were accompanied by a multitude of negotiations between both parties. Porsche and VW reverted to their best management experts, strategists and lawyers, in order to bargain about the future of the two companies (Polke-Majewski 2009). Besides, the negotiations attracted a great deal of interest in the media, both because of the outcome's huge economic impact and due to the fact that some of the most powerful of German decision makers gathered to take part in them, including both companies' chief executive officers (CEOs) and the chief minister of Lower Saxony (Schultz 2009). Over the course of the negotiations, however, the relationships among the negotiation leaders became damaged, as VW patriarch Piech withdrew his trust in Porsche's CEO Wiedeking for becoming – in Piech's eyes – overly self-aware, which is one reason why the takeover by Porsche failed (Hawranek 2009).

The example of Porsche and VW is just one of many instances in which negotiations constitute the root of high-impact and long-lasting changes. Next to such merger-and-acquisition negotiations, the most common kind of business negotiations are price negotiations (Backhaus and Voeth 2014; Voeth and Herbst 2015a). Imagine the following:

[1] A citation which is supplemented by the amendment "here and in the following" is valid until a new citation is introduced.

VW might need automotive components and identifies Hella as a relevant automotive supplier. In order to reach a consensus between both parties concerning the conditions of a potential cooperation, the companies come together to negotiate. These business price negotiations are characterized by the inclusion of multiple decision makers on each side; as a rule by high investment volumes and multiple negotiation objects, whereby the latter encompass the price itself, and commonly further factors such as delivery dates or purchase quantities (Backhaus and Voeth 2014; Voeth and Herbst 2015a).

Nevertheless, negotiations do not occur exclusively in business settings; indeed, in politics they also play a major role in such situations as coalition negotiations between political parties (Dupont 1996). A further, rather recent political application is the negotiations between the British Parliament and the European Union over the withdrawal of the United Kingdom from the European Union (*Brexit*) (here and in the following Mueller 2019; Schüller, Imöhl, and Berndt 2019). As of this writing, the Brexit negotiations have been going on for several years, and have involved a great deal of public interest, many postponed decisions and rejected proposals, and changes in the negotiators: the British Prime Minister Theresa May resigned in May 2019 and Boris Johnson assumed her position. Yet another context in which negotiations take place is the field of legal affairs and involving trials such as the recent proceedings concerning the VW diesel emissions scandal, or the trials following Porsche's failed takeover of VW (Hägler 2016; Wahlquist 2019). Lastly, we all know negotiations from our own daily lives all too well. Whether it is the debate with one's partner on the next travel destination or the negotiation with a car dealer about the purchase price of an automobile – in either cases we are negotiators. Consequently, one can conclude that every individual becomes a negotiator multiple times in his or her life, be it in professional or in private settings (Fisher and Ury 1981).

Next to their high practical relevance, negotiations have been addressed extensively by researchers. Due to the fact that negotiations play an important role in many fields, a multitude of research areas, including political and legal sciences, (business) economics, psychology and sociology, deal with the topic of negotiations (Backhaus and König 2019; Bichler, Kersten, and Strecker 2003; Voeth and Herbst 2015b). Negotiation research dates back to early 1940s when the mathematicians von Neumann and Morgenstern (1944) developed mathematical models in order to explain decision-making for rational individuals. From there onwards, the focus in negotiation research shifted: In particular, the 1960s and 1970s were characterized by social-psychological studies focusing on personal factors in negotiations (Rubin and Brown 1975). Another turning point occurred during the 1980s and 1990s, when negotiation research sought to explain the deviations by bargaining parties from rational behaviour (Bazerman et al. 2000; Bazerman and Neale 1991). However, since the mid 1990s, the focus of negotiation research has shifted back to social-psychological studies, and now emphasizes such factors as social relationships among and the emotions of negotiators (Bazerman et al. 2000).

Within the last few decades, the already extensive body of negotiation research that had emerged from this long history and broad coverage of research areas has increased to ever higher amounts (Buelens et al. 2008; Thompson 2006). However, the bulk of this scientific research is poorly structured. Much and detailed knowledge exists, but the individual conclusions of negotiation publications are discrete from one another; hence it becomes rather difficult to capture the field of negotiations and to distil its key topics and findings (Agndal, Åge, and Eklinder-Frick 2017).

To be sure, one can find a number of structuring approaches in the literature, whose aim it is to organize the research field of negotiations; however, all these approaches are restricted in multiple respects: Nearly all existing structuring studies are simple qualitative reviews which aim to describe the research field (see here and in the following chapter 3.1.1). Unfortunately, qualitative approaches are prone to subjective biases, owing to scholars' subjective review on the research, which in turn merely reflects their own intuitive beliefs in one way or another (Backhaus, Lügger, and Koch 2011). Qualitative approaches are also limited with respect to the number of articles reviewed, as the human processing capacity is finite (Tranfield, Denyer, and Smart 2003). Besides, nearly all published structuring studies focus either on very specific negotiation topics (exempli gratia (e.g.), the role of gender in negotiations), journals (e.g., Group Decision and Negotiation) or research areas (e.g., psychology), so that a more comprehensive perspective is lacking (Kray and Thompson 2005; Laengle et al. 2018; Tsay and Bazerman 2009). As a consequence, to date major confusion – if not an utter lack of structured knowledge! – prevails. We are in most cases stymied when it comes to the question: "How can we integrate the knowledge garnered from negotiation research?" Further research can hardly build upon a state of affairs in which no reliable knowledge exists either of the structure of the field or of the contributions of single publications arising from the large amounts.

By far the greatest portion of this mass of negotiation research consists of empirical studies (Buelens et al. 2008; Druckman and Robinson 1998), of which especially the field of *personal factors* has produced a plethora of empirical findings (Agndal 2007; Voeth and Herbst 2015b). This is hardly astonishing in light of the fact that negotiations' basic characteristic is the involvement of individuals who are in turn influenced by many personal factors, such as their personality, their demographics, their organizational roles or their emotions (Lax and Sebenius 1986; Sharma, Bottom, and Elfenbein 2013; Thompson, Wang, and Gunia 2010; Voeth and Herbst 2015b). That being said, the empirical research findings of personal factors exhibit multiple contradictions (Bowles, Babcock, and McGinn 2005; Stuhlmacher and Gillespie 2006; Stuhlmacher and Walters 1999). To name one example, different findings can be deduced from the literature pertaining to the influence of organizational roles on conflict styles in negotiations. On the one hand, studies find that superiors are more forcing with subordinates while the latter prefer smoothing when dealing with their superiors as they might be afraid of losing their job (e.g., Phillips and Cheston 1979; Rahim 1986). On the other hand, other studies reach quite opposite results, for instance, that superiors relate to smoothing and subordinates draw on forcing (e.g., Rahim 1983; Renwick 1975).

Clearly, the negotiation research community longs for and deserves clarification on these contradictions. To meet this purpose, though, qualitative review studies would have to be able to deliver an aggregation of these contradictory findings, which they cannot do, as they provide only verbal descriptions of the inconsistencies among negotiation studies. The quantitative method of a meta-analysis, however, determines one overall effect that is both reliable and valid, and can in this way enable the aggregation of the results of multiple empirical studies dealing with a specific topic (Bijmolt and Pieters 2001; Borenstein et al. 2009). A few meta-analyses have already been conducted regarding the influence of personal factors in negotiations (see here and in the following chapter 3.2.1).

Nonetheless, one factor has been neglected up until now, despite having been repeatedly identified as of particular importance in recent times: *emotions*.

As the beginning of negotiation research presupposed rational bargainers who seek to maximize their negotiation outcomes, emotions were seen as sources of irrationality, which in turn lead to the advice that they be suppressed in a negotiation setting (Olekalns and Druckman 2014). Based on this belief, research on emotions declined at first (Neale and Northcraft 1991; Pruitt and Carnevale 1993). Nevertheless, since the beginning of the century, a renaissance of research on emotions in negotiations has taken place (Martinovski 2010), and since then emotions have moved to the front stage of investigation and have attracted ever more scientific attention (Barry 2008; Barry, Fulmer, and van Kleef 2004). Negotiation scholars of the new research front argue that bargainers can hardly blank out or ignore neither their own emotions nor the emotional expressions of their counterparts, for which reason negotiations are a highly emotional setting (Bazerman and Chugh 2006; Elfenbein et al. 2007; Voeth and Herbst 2015b).

Due to the divergent views on emotions arising from their scientific history, their concrete functions and effects in negotiations remain unclear: Should negotiators try to suppress emotions, or is it advisable to display certain emotions in specific situations in order to influence the counterpart in favour of the own negotiation outcomes? Besides, the inconsistency of the findings on emotions in negotiations is amplified by the various different contextual manipulations in which they are examined (e.g., the power positions of the bargainers or the competitiveness of the situation) (Barry 2008; Hine et al. 2009; Olekalns and Druckman 2014). These inconsistencies hold especially true for the effect that emotional expressions have on the negotiating counterpart as compared to the perception of one's own emotions (Kopelman, Rosette, and Thompson 2006; Sinaceur and Tiedens 2006). Furthermore, the inconsistencies arise in particular for the most frequently examined emotions of *anger* and *happiness* (e.g., Druckman and Olekalns 2008; Olekalns and Druckman 2014; Pietroni et al. 2008) (see chapter 3.2.1). In this manner, studies find on the one hand that anger leads the opposing party to give in, so that higher outcomes can be achieved than with a display of happiness (e.g., van Kleef et al. 2006; van Kleef, De Dreu, and Manstead 2004a; Sinaceur and Tiedens 2006). On the other hand, it has been shown that the display of positive rather than negative emotions leads to a greater chance of agreement and better outcomes (e.g., Hine et al. 2009; Kopelman, Rosette, and Thompson 2006). Thus, it remains rather unclear which emotional expression is superior.

1.2 How Research Works: The Story of Giants and Dwarfs

Research can be described as a process in which information is collected and analysed in order to increase the knowledge and understanding of the topic of investigation (here and in the following Creswell 2012). It aims at detecting new findings, closing existing research gaps, validating or refuting previous research results, and supporting existing theories or developing new ones. A quite memorable and intuitive metaphor which explains the way in which research is conducted is the image of dwarfs standing on the shoulders of giants (Clandinin and Hamilton 2011). This image was applied for the first time by the scholar and theologian Bernard of Chartres in the 12th century (Stock 1979). From then on, it reached considerable popularity when Isaac Newton (1676) used the image and claimed "If I have seen farther it is by standing on the shoulders of giants" (Clandinin and Hamilton 2011; Keith et al. 2016). This metaphor sets the latest research in relation to the achievements of previous generations, as it assumes that truth can be discovered and progress made by recent scholars only by building on the work of preceding scientific pioneers (Keith et al. 2016), whereby the scientific predecessors of past epochs are seen as the giants and the current researchers as the dwarfs standing on their shoulders and profiting from their knowledge and findings (Stock 1979). One quite important rationale of this image is that the dwarfs sitting on the shoulders of giants see more and farther than the giants themselves, but not because they are wiser or keener, but rather because they are raised by the gigantic stature of the previous researchers (McGarry 1955; von Salisbury 1159). Thus, the metaphor is frequently used in order to honour the work of preceding scholars (vom Brocke et al. 2015).

Transferring the metaphor to the case of negotiation research, one can recognize that the research field is characterized by the large shoulders of negotiation giants who have accumulated enormous and ever-increasing amounts of negotiation publications over the last decades. However, as underlined in chapter 1.1, the bulk of scientific research is poorly structured, which makes it difficult to grasp its key topics and findings. But without structured negotiation research results, it becomes difficult for negotiation researchers to identify key findings and act upon them. As illustrated by the metaphor of giants and dwarfs, scholars build on insights from prior scholarly work and seek to detect and fill research gaps. Consequently, as the shoulders at our disposal have no adequate structuring so far, proceeding researchers feel insecure about its resilience since, robust though it may seem to be, it only *appears* to be a stable construct, which in reality it is not, thus making it dangerous to build upon. Accordingly, the revelation of the intellectual structure of the field of negotiation research and its development over time are of the utmost importance for academia.

Chapter 1.1 has furthermore shown that within the large set of negotiation publications a variety of empirical studies focusing on personal influence factors exists which lead to contradictory results – among others for the factor emotions. Far from being able to deduce a reliable effect from these findings, researchers and practitioners are much rather confused about which results to follow. Thus, an aggregation of the existent study findings seems to be quite important, in that it can help negotiation researchers and professionals along. The attempt to synthesize inconsistent empirical results lifts researchers up as dwarfs – relying as they do on the previous results of the giants – and enables them to see further thanks to an overall reliable effect.

1.3 Structure and Contribution of the Thesis

The thesis at hand proceeds analogously to the structure illustrated in Figure 1-1. This structure will be explained in more detail in the following.

Subsequent to the introduction, *chapter 2* provides the conceptual foundations for this dissertation, as it represents the basis on which the following chapters are built on. It therefore introduces the reader to the characteristics of negotiations by defining the term, and by deducing not only its main features, but also its considerable importance in business settings, business settings being as they are the background of this thesis. Thereupon the historical development both of the field of negotiation research and of its defining epochs are summarized briefly, after which follows a more precise examination of negotiations, as the process of negotiations and its multiple influencing factors are presented. The chapter closes with the observation that negotiation research is characterized by a multitude of different topics and findings which lead to a vast set of details.

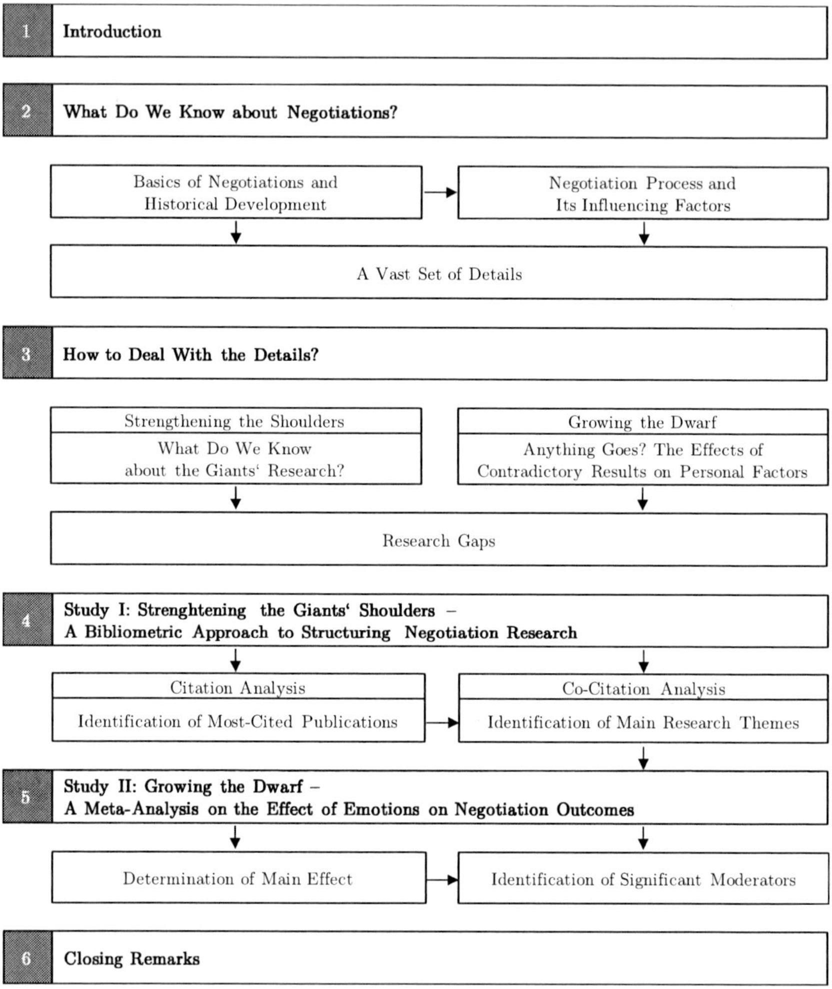

Figure 1-1: Structure of the Thesis
Source: Author's own illustration.

Chapter 3 provides the reader with two thorough overviews pertaining to the question of how the diverse findings of negotiation research should be structured and aggregated. The chapter proceeds in two successive layers. The first layer considers the giants' shoulders and offers an overview of literature reviews whose aim it is to provide an appraisal of the negotiation research field. The goal of the first layer is to identify structuring studies that depict the wide-ranging spectrum of negotiation research, that deliver an objective perspective on negotiation research, and thus that strengthen the shoulders of the giants. With this end in mind, existing structuring approaches of negotiation research are identified and classified by predefined criteria, which are then discussed in more detail in order to ascertain shortcomings and potential for further research. As a second layer, the dwarf perspective is taken into account by offering an overview of meta-analyses which aim at distilling an overall effect from empirical studies on specific personal factors influencing negotiators' behaviours and outcomes. The meta-analyses are then categorized into predefined classes and less covered categories are identified. The scientific shortcomings ascertained by both literature reviews are translated into research gaps, which are then taken up by the subsequent empirical studies. While the first study aims at approaching the negotiation research field from the perspective of the giants, the second study covers the perspective of a dwarf.

The first empirical study is reported in *chapter 4*. *Study I* quantitatively maps the structure and development of the field of negotiations by yielding key publications and topics as well as their interconnections and evolution over the last 60 years. In order to do so, citation and co-citation analysis are implemented as quantitative bibliometric methods that permit the processing and analyzing of the huge underlying dataset of 66,294 references. First, the citation analysis reveals the most-cited and therefore most influential publications (Zupic and Čater 2015). Then follows the co-citation analysis, which goes on processing a fraction of the most-cited publications identified by the citation analysis. One major outcome of Study I is that the topic of *emotions* represents the most highly studied research theme over the last two decades yielding a plethora of scientific publications, which in turn are characterized by several contradictory empirical findings. This underscores the importance of a meta-analysis that would quantitatively aggregate the contradictory study results of emotions on negotiation outcomes, for the sake of deducing reliable and valid insights.

The aim of Study I is to empower the negotiation giants' shoulders by generating topical networks that allow for organising the manifold existing negotiation research results and to detect their interconnections. Researchers can rely on these networks in order to assess the field, work with the identified defining research topics, and identify remaining gaps for further research. In addition, practitioners who want to broaden their knowledge of the field of negotiations stand to benefit from the first study, as they receive a "guided tour" (including key publications and topics) they can rely on in order to improve negotiations in their own daily businesses. However, the claim of Study I is not (nor can it be!) to account for every publication or topic which has shaped negotiation research in the last decades. On the contrary, the aim of the study is to work out the most defining ones for negotiation research. Thereby, it ensures a quite broad coverage of the field, as the study by far exceeds, with 66,294 references examined, the existing structuring studies reviewed in chapter 3.1.1. Consequently, the determination of defining research takes place on the basis of an enormous database that allows for the inclusion of various potentially relevant publications and themes.

Chapter 5 comprises the second empirical study of this thesis. In its implementation of a meta-analysis on the effect of emotions on economic and subjective negotiation outcomes, one specificity of *Study II* lies in the fact that it builds on the key research topic identified in the co-citation analysis of Study I. In doing so, Study II focuses on the by far most examined emotion of anger, and compares it to the second most frequently investigated emotion happiness and to emotional neutrality (see here and in the following chapter 3.2.1).[2] In this way, the meta-analysis aggregates the empirical findings of 94 empirical studies, which account for 16,795 participants and 267 effect sizes in all. Thereby, it stands out from other meta-analyses in the field with respect to its topic and extent. In addition to analyzing the main effects in the meta-analysis, several moderators are also investigated in Study II that may influence the main effect on economic outcomes. These moderators, which are conceptually deduced from the Emotions as Social Information model (van Kleef 2009), are the competitiveness of the negotiation setting, the power level of the emotion receiver, the negotiation medium used, the emotion direction, the authenticity of the anger expression, the degree of time pressure and study characteristics such as the publication status.

The aim of Study II is to shed new light on the mass of empirical (contradictory) research results on the effect of emotions on negotiation outcomes in general and the effect of

[2] Certainly, emotions can be divided into more subforms – such as sadness, fear, disgust or surprise – than anger and happiness (Russell and Barrett 1999). Since, however, these emotions are investigated too infrequently by negotiation researchers (see chapter 3.2.1), the determination of an overall effect is particularly important for anger and happiness.

anger compared to happiness and emotional neutrality on outcomes in particular. For this purpose, the diverging empirical results are quantitatively aggregated in order to derive one overall effect and significant moderating effects are identified that explain the deviations in the main effect. Thereupon, both researchers and practitioners can rely. In this manner, specific and concrete recommendations for real life negotiations can be deduced for the enhancement of negotiation success. By doing so, the researcher of this thesis builds on the strengthened shoulders of negotiation giants and is thereby raised up as dwarf herself.

The thesis closes with *chapter 6*, which summarizes the results of both empirical studies, provides implications for negotiation researchers and practitioners, and indicates limitations of the thesis and directions for future research.

To sum up, negotiations are well-researched constructs for which detailed scientific knowledge is available. Nevertheless, the research field is characterized by a lack of clarity due to its unstructured complexity and contradictory findings. Consequently, a great deal of primary research exists in the negotiation research field for which adequate secondary research is lacking. The thesis at hand aims to draw on this and uses the story of giants and dwarfs as a guideline to cover both roles for purposes of negotiation research. For both the giants' perspective and the dwarf's perspective this thesis yields new knowledge and so advances negotiation research from each side. In the process, both studies make use of quantitative methods for structuring and aggregating the huge body of (empirical) negotiation studies. Thus, the central added value of this dissertation lies in boiling down the overwhelming number and contradictory nature of negotiation studies in a quantitative way. By so doing, both researchers and practitioners may find an unbiased view of the state of the art of negotiation research in general and the role of emotions in particular.

2 What Do We Know about Negotiations?

2.1 Characteristics and Peculiarities of Negotiations

The term *negotiation* is not consistently defined in the literature (here and in the following Voeth and Herbst 2015b). This is on the one hand caused by terminological clutter: in addition to the term *negotiation*, the term *bargaining* can also be found in negotiation studies. While these terms can be used as synonyms (e.g., Kristensen and Gärling 1997; Thompson and Hrebec 1996; Wilken, Jacob, and Prime 2013), they can also demonstrate different constructs. In the latter case, bargaining represents a form of haggling as opposed to negotiation, which reflects a formal process of finding solutions to problems (Lewicki and Hiam 2006). Nevertheless, the thesis at hand follows the other approach, taken by numerous studies, and uses the terms "bargaining" and "negotiation" interchangeably. On the other hand, there are multiple research disciplines, such as economics, psychology, computer science, political sciences and legal sciences, that deal with negotiations, hence negotiation research is influenced by quite different domains (Voeth and Herbst 2015b). Thus the divergence in the definitions for the term "negotiation" can be explained as stemming from terminological clutter and the influence of divergent research areas.

This thesis relies on the definition given by Lax and Sebenius (1986), which best comprises the most relevant aspects of negotiations. According to these two scholars, negotiation is a process of interaction between two or more parties that exhibit an apparent conflict and that seek to do better through a jointly made decision than they could do otherwise (Lax and Sebenius 1986).[3] Although a multitude of different definitions exist,

[3] For other definitions that basically refer to similar characteristics of negotiations, see for instance Pruitt and Carnevale (1993) or Curhan, Elfenbein, and Xu (2006).

there are specific characteristics that recur in the literature which can be identified as quite defining for negotiations (Fisher and Ury 1981). According to Voeth and Herbst (2015b), these characteristics encompass

- the presence of two or more parties (e.g., persons/organizations),
- a congruency between the parties of the goals pursued, as all parties exhibit an interest for agreement on account of their own ultimate negotiation aim,
- a conflict of the divergent (unknown) preferences between the parties,
- the possibility to improve the positions of both negotiation parties through a bargaining solution compared to a non-agreement, and
- an interaction process in which each party seeks to enforce its respective interests.

The majority of negotiation settings consists of two parties, called dyads, which can consist of as few as one person, or of multiple persons (Barry and Friedman 1998; Johnson and Cooper 2009; Naquin and Kurtzberg 2009; Zerres and Hüffmeier 2011). In professional negotiations in which parties consist of multiple individuals, the individuals should be selected based on their eligibility and their value for the negotiation (Backhaus and Voeth 2014; Hüffmeier et al. 2019; Voeth and Herbst 2015b). The parties negotiate either on their own or with the support of an independent third party (here and in the following Voeth and Herbst 2015b). Such third parties can either be called upon by one of the parties to assist this party, or by both parties for the purpose of conveying or determining a negotiation solution which is adequate for all participants. A common approach taken for the involvement of third parties is, for example, the inclusion of a so-called mediator, who is responsible for facilitating an agreement between the negotiation parties, in which process the latter retain their decision-making authority.

However, there are a number of settings, in which more than two parties are involved, called multiparty negotiations (Bazerman et al. 2000). In multiparty negotiations, preferences held by one bargaining party may contradict preferences held by some of the group members, but may be shared with preferences held by other parties (here and in the following Beersma and De Dreu 2002). Between parties displaying a mutual compatibility of interest, the formation of a coalition is an option. A coalition is a part of the entire group of negotiating parties that pools its resources and behaves as a single entity in order to enforce a decision for the whole group, in which case the interests of the coalition are in the foreground (Holsapple, Lai, and Whinston 1998; Murnighan and Brass 1991). Consequently, in those cases in which majority rule holds, a coalition formation goes along with an increase of the welfare of the majority at the expense of the minority (e.g., coalition negotiations in politics) (Thompson, Mannix, and Bazerman 1988).

Negotiators often do not negotiate for themselves, but rather on behalf of other people or organizations, called their "constituents" (Pruitt 1981; Pruitt and Carnevale 1993). As members of their respective groups, these so-called representatives, are intent on pleasing their constituents by striving to implement their positions (Enzle, Harvey, and Wright 1992). In matters of negotiations between representatives of two parties, communication takes place twice, since the representatives exchange views both with their constituents and with the representative of the opposing party (here and in the following Pruitt 1981). Since representatives of an organization usually negotiate with each other on a regular basis and share similar organizational positions, they often develop connections with their opponents. Representatives are present in such contexts as business negotiation settings because each company is represented by several employees with the task of conducting negotiations on its behalf.

In business settings, sales negotiations are an especially substantial instrument, as relevant parameters such as price or order volumes are determined based on them (Srivastava, Chakravarti, and Rapoport 2000; Voeth and Mayer 2018). Nevertheless, for business transactions in which products and services are standardized and have fixed prices, the relevance of negotiations is negligible (e.g. the purchase of a toothpaste) (Yang, Singhal, and Xu 2013). The importance of negotiations rises significantly when no market price exists, when they involve individualized rather than standardized products or services, or when further transaction features such as purchase quantities or delivery conditions need to be set (here and in the following Backhaus and Voeth 2014; Herbst, Voeth, and Meister 2011; Voeth and Herbst 2015a). For this reason, the relevance of negotiations is greater for industrial or business-to-business markets respectively in which conditions have to be scheduled, than for business-to-consumer markets with a considerable degree of standardization and set market prices. The importance of sales negotiations in industrial contexts is further attributable to high volumes and investments made by companies (Backhaus and Voeth 2014).

However, industrial markets also differ with regard to their negotiation potential. According to Backhaus's and Voeth's (2014) business type classification, industrial markets display diverging levels of complexity. While for example product businesses (e.g., sales of screws) resemble business-to-consumer transactions most closely due to their low level of specifity, project businesses are much more sophisticated with regard to negotiations (here and in the following Backhaus and Voeth 2014). Project businesses, such as the disposal of a production machine, have huge potential for negotiations, as the sales process usually precedes the production phase. In this manner, multiple negotiation objects are present, such as the price and the composition of the machine. These negotiations are usually long-term and are conducted by multiple persons on both sides of the bargaining table because the high degree of complexity involved in the purchase and sales decisions of companies calls for the expertise of a number of persons.

2.2 "Every Future Needs a Past" [4]

After the above presentation of the basics of negotiation, this subchapter aims at providing a brief overview of the historical development of negotiation research, for the purpose of enabling the reader to identify its different epochs. Research on negotiations has its origin in the 1940s, a time when it was shaped mainly by research pertaining to game-theory (Backhaus, Hummelsberger, and Wilken 2005; Caputo 2013a; Voeth and Herbst 2015b), especially by the game-theoretic approach of von Neumann and Morgenstern (1944), who assumed that agents in interdependent decision making have an individual utility function that they seek to maximize (Hausken 1997; Voeth and Herbst 2015b).

From there onwards, the 1960s and 1970s were particularly dominated by a plethora of social-psychological studies. The large amount of research conducted during this phase is reflected by the work of psychologists Rubin and Brown (1975), who documented more than 1,000 articles and books dealing with the social-psychological perspective of negotiation since the beginning of 1960. Social-psychological studies had their main focus on two sub-domains: *structural factors* and *personal factors* of negotiators (Bazerman et al. 2000) (see chapter 2.3.1). Among the most frequently examined structural variables were

[4] The title is taken from the English translation of Backhaus (2015) of the German book title "Zukunft braucht Herkunft" by Marquard (2003).

the negotiation medium, the negotiation location, and the number of bargaining objects (Rubin and Brown 1975). With regard to the personal factors, demographic characteristics and personality traits were given special priority as research objects (Backhaus, Hummelsberger, and Wilken 2005; Voeth and Herbst 2015b). Although these early social-psychological studies improved the general comprehension of negotiations, they did not have a lasting impact on negotiation research, as they did not build on a well-founded theoretical framework and led to contradictory results (Backhaus, Hummelsberger, and Wilken 2005; Herbst 2007; Voeth and Herbst 2015b).

A rethinking of behavioural negotiation research took place in the 1980s and 1990s with the development of the behavioural decision research (BDR) (Bazerman et al. 2000; Voeth and Herbst 2015b). BDR considers negotiations as interactive decision-making situations under uncertainty, and attributes the deviation of negotiators from economic-analytical models to individual irrational behaviour on their part (Bazerman and Neale 1991). Moreover, BDR formulates guidelines for the neutralization of these biases (here and in the following Voeth and Herbst 2015b). For this reason, BDR places its main focus on the cognitive processes of individuals while they are absorbing, processing and storing information. In particular, cognitive heuristics, also known as simplifying strategies, are the research objectives of BDR (Bazerman et al. 2000; Voeth and Herbst 2015b). These cognitive heuristics represent useful shortcuts, but they also lead to systematic and predictable biases (Bazerman et al. 2000; Tversky and Kahneman 1974).

However, because BDR ignored numerous factors of obvious importance in negotiations, such as social variables, it received criticism from a number of authors (Greenhalgh and Chapman 1995). This led to the re-inclusion of social-psychological variables, which this time were consistent with the BDR perspective (here and in the following Bazerman et al. 2000). Consequently, a rebirth of social-psychological research took place in the middle of the 1990s. This rebirth led to studies examining social relationships among negotiators, egocentrism and emotions in negotiations.

The 1980s and 1990s were furthermore characterized by more application-oriented research (Backhaus, Hummelsberger, and Wilken 2005). One main source on which practical negotiation research draws for the concrete examples it uses in order to deduce advice for real life negotiators, are scholarly paradigms (Backhaus, Hummelsberger, and

Wilken 2005; Voeth and Herbst 2015b). The strengths of practical negotiation research lie in its simplicity and in the concrete suggestions it makes for practitioners (here and in the following Voeth and Herbst 2015b). However, many of these approaches lack a theoretical foundation. By and large, they depict practical experience of negotiation experts. Such how-to manuals are not in keeping with the requirements of scientific work.

Lastly, from the 2000s onwards, topics such as ethics and emotions, and social and cultural aspects have played a major role in negotiation research (Brett and Thompson 2016; Tsay and Bazerman 2009).

2.3 The Negotiation Process in a World of Multiple Frameworks

The present subchapter seeks through a more precise description of the process of negotiation to shed light on these parts and factors so crucial for virtually all negotiations. However, much as all negotiations have these aspects in common, any negotiation process proceeds with a degree of specifity so considerable, that to make a generalizations pertaining to the negotiation process is a difficult matter indeed (Backhaus, Hummelsberger, and Wilken 2005). Be that as it may: unique unto itself as every negotiation is, its basic elements such as (multiple) negotiating parties or the interaction process remain stably universal (Fisher, Ury, and Patton 2011). Accordingly, multiple scholars have developed general frameworks in order to structure the negotiation process and to depict its defining aspects (e.g., Backhaus, Hummelsberger, and Wilken 2005; Backhaus and Pesch 2018; Dupont 1999; Neale and Northcraft 1991; Voeth and Herbst 2015b). Each negotiation framework has its own classification criteria and emphasis, and its own level of depth. For example, Voeth and Herbst (2015b) consider the negotiation process from a management perspective and structure it in five phases: a) analysis, b) organisation, c) preparation, d) implementation and e) controlling. In contrast, other negotiation frameworks distinguish between the negotiation process itself and the factors which influence its different stages (e.g., Backhaus and Pesch 2018; Dupont 1999; Hausken 1997). These fundamental factors – which in the following are referred to as influence factors – are characteristics of the involved negotiation parties, and structural variables such as the number of objects being negotiated (Dupont 1999).

Although multiple approaches to negotiation frameworks exist in the literature that differ from one another with regard to their concrete structuring, the main stages and influence factors of a negotiation reoccur across most of them (e.g., Backhaus and Pesch 2018; Dupont 1999; Voeth and Herbst 2015b). The thesis at hand endeavours to build on a framework which depicts the main recurring aspects throughout existing negotiation frameworks; moreover, it strives for a comprehensive classification and structuring of the negotiation process. As depicted in Figure 2-1, the main conception of this framework is derived from Backhaus and Pesch (2018), and also includes a number of other classification approaches (Agndal 2007; Hausken 1997; Levi 2001; Neale and Northcraft 1991; Voeth and Herbst 2015b). In the following, it will be used in order to describe the negotiation process and its influencing factors.

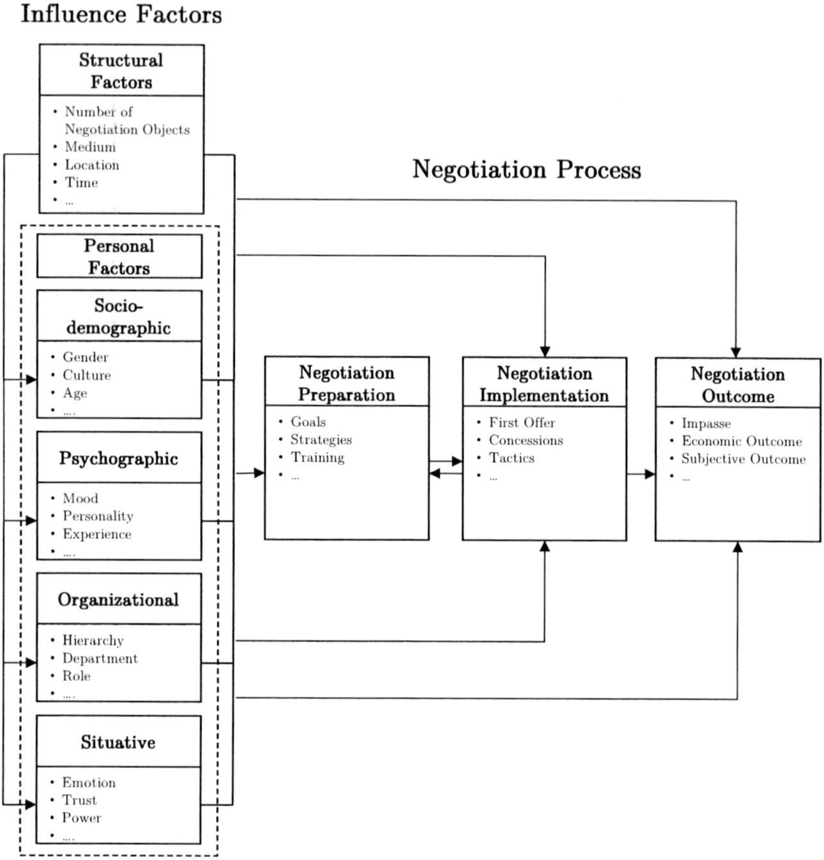

Figure 2-1: Framework of Negotiations
Source: Author's own illustration adapted from Backhaus and Pesch (2018), Voeth and Herbst (2015b), Agndal (2007), Levi (2001), Hausken (1997) and Neale and Northcraft (1991).

As it becomes obvious in Figure 2-1, a multitude of factors exist – for which a plethora of negotiation studies can be found in literature (e.g., Adair et al. 2004; Bowles, Babcock, and McGinn 2005; Brown and Baer 2011; Kong, Dirks, and Ferrin 2014) – that exert an influence on the negotiation process. The peculiarity of the influence factors is that they influence not only the negotiation process, but one another as well (see Figure 2-1). The influence factors in turn can be divided into structural and personal factors, whereby the latter can be further split up into socio-demographic, psychographic, organizational and situative factors (Backhaus and Pesch 2018; Dupont 1999; Levi 2001; Voeth and Herbst

2015b). The structural factors constitute the framework conditions of a negotiation, which usually remain constant throughout the negotiation (here and in the following Backhaus, Hummelsberger, and Wilken 2005; Neale and Northcraft 1991).[5] These factors are, for example, the number of negotiating parties, time restrictions, the negotiation medium, the negotiation location or the number of negotiation objects.

Various *personal factors* exist that have an impact on negotiations. Their strong influence has been worked out by many researchers (e.g., Barry and Friedman 1998; Elfenbein 2015; Sharma, Bottom, and Elfenbein 2013; Thompson, Wang, and Gunia 2010). The first category, socio-demographic factors, comprises all those social categories used to classify people (Levi 2001; McGrath, Berdahl, and Arrow 1995), and includes such things as the gender, age, culture, education or religion of negotiators (Levi 2001; Voeth and Herbst 2015b). Psychographic factors refer to the differences in individuals' values, cognitions, attitudes and behaviour (Levi 2001). This includes, among other things, the personality of individuals, their knowledge, experience, and moods (here and in the following Voeth and Herbst 2015b). Organizational factors consider the individual from the perspective of an institution and enable categorization of the person with regard to his or her standing within the organization. Factors such as organizational hierarchies and role obligations can be assigned to this category. Lastly, situative factors emerge from the interaction of the negotiating parties and arise as a result of situational conditions. The similarity of the bargainers, their power differential or the emotions triggered by the bargaining situation are examples of situative factors.

Furthermore, the negotiation process itself can be partitioned into multiple steps. The first step consists of a negotiation preparation phase (see Figure 2-1). Prior to each negotiation, a preparation is absolutely indispensable in order to determine the schedule for the subsequent bargaining (here and in the following Voeth and Herbst 2015b). This step includes, for example, the determination of the negotiation goals for oneself as well as for the opposing bargaining party, and the stipulating of negotiation strategies to be pursued during bargaining. Moreover, the advancement of negotiation skills with the

[5] In several cases, the structural factors can change during negotiation (Hausken 1997). For instance, the number of negotiation objects may be adapted or the number of negotiation parties may vary (Sebenius 1983).

help of training sessions can be assigned to this phase (Zerres et al. 2013). Subsequently, the negotiation implementation ensues. In this phase, the offer and concession making of the bargaining parties takes place (here and in the following Voeth and Herbst 2015b). In this stage, the behaviour of the negotiation parties is particularly important, since the chosen negotiation tactics, for instance, can influence the course of the bargaining significantly.

The negotiation process ends with a negotiation outcome where one can distinguish between an objective economic value and a subjective perceived value (Backhaus and Voeth 2014; Curhan, Elfenbein, and Kilduff 2009; Thompson 1990).[6] *Economic outcome (EO)* is the typical and substantial result of a negotiation as examined in most negotiation studies (e.g., Adair 2003; Backhaus, Hummelsberger, and Wilken 2005; Mannix 1993; Weingart, Bennett, and Brett 1993). EOs are those quantifiable key figures which can be derived directly from the negotiated agreement (Backhaus and Voeth 2014). They usually include the achieved profit for the negotiators (Reynolds, Simintiras, and Vlachou 2003; Thompson 1990; Wilken, Jacob, and Prime 2013). Nevertheless, EOs can also encompass delivery dates, purchase quantities or any other quantifiable figures (e.g., Rothman and Northcraft 2015; Yang, Singhal, and Xu 2013). EOs can be further divided into individual outcomes for one party and joint outcomes for all negotiating parties (Backhaus, Hummelsberger, and Wilken 2005; Mazei et al. 2015; Zerres and Hüffmeier 2011). Thereby, the determination of the joint EOs can take place through simple summation of the individual outcomes or ordinal scaling (Backhaus, Hummelsberger, and Wilken 2005; Pruitt and Lewis 1977; Thompson 1990).

However, the negotiation does not necessarily end in an agreement; it is also possible that the negotiation parties reach an impasse (O'Connor and Arnold 2001; Tripp and Sondak 1992). Although negotiation parties usually try to reach an agreement, several possibilities are conceivable in which an agreement becomes unlikely: Firstly, if the negotiating parties have perfect information and behave rationally, they would only come to an agreement if the agreement presented an improvement of the current situation of both parties (Nash 1950; Raiffa 1982; Tripp and Sondak 1992). In case of fully rational

[6] The subjective value is also designated as the social-psychological or socio-emotional outcome in literature (e.g., Thompson 1990; Zerres and Hüffmeier 2011).

negotiators and imperfect information, however, an impasse becomes possible even despite the fact that an agreement would improve both positions (Myerson and Satterthwaite 1983; Tripp and Sondak 1992). This also holds for irrational negotiators, who are more likely to occur in real life negotiation settings and who may leave the negotiation table even though a deal closing would satisfy their interests (White et al. 2004).

Due to the focus on EOs, subjective values were neglected for a long time (Bazerman, Curhan, and Moore 2001; Zerres and Hüffmeier 2011). However, several studies increased their research focus towards subjective value as a negotiation outcome (e.g., Curhan, Elfenbein, and Xu 2006; Oliver, Balakrishnan, and Barry 1994; Zerres and Hüffmeier 2011). *Subjective outcomes (SOs)* can be divided into multiple dimensions. A first dimension can consist in the feelings about the EOs, including whether it is desirable and balanced (here and in the following Curhan, Elfenbein, and Kilduff 2009). Or it can consist in the feelings toward oneself, for example when the negotiator has lost face during bargaining. A subjective value can also be the evaluation of the process itself, including the perception of whether one was treated fairly. A final dimension of SOs consists of feelings concerning the relationship between the negotiators with respect to trust, or to whether or not a solid basis is given for further collaboration. Unlike EOs, SOs can refer both to the result of the negotiation and to the process itself, so that they can also be achieved in cases of an impasse (Backhaus, Hummelsberger, and Wilken 2005; Thompson 1990).

In the following, specific factors that influence the negotiation process will be considered in more detail, along with as will be concrete features of the negotiation process itself. In the process, those components will be given particular attention which are pertinent to the course of the dissertation at hand.

2.3.1 Influence Factors

2.3.1.1 Structural Factors

A multitude of structural factors affect the negotiation process. For instance, time restrictions may arise due to deadlines or a maximum amount of time available to bargain, which often leads to time pressure (Mosterd and Rutte 2000). Additionally, the location of the negotiation needs to be set, which is frequently a contentious issue, as one's own home arena is widely seen as advantageous for the course and outcome of the negotiation (Brown and Baer 2011; Salacuse and Rubin 1990). The *number of negotiation objects* and the *negotiation medium* are two further factors that play an important role in negotiations (e.g., Bazerman, Magliozzi, and Neale 1985; Beersma and De Dreu 2002; Geiger 2014; Lewis and Fry 1977; Mor and Suppes 2011; Purdy and Nye 2000) and that will be discussed more extensively further below.

Number of Negotiation Objects: Distributive and Integrative Bargaining

One can differentiate negotiations on the basis of the number of negotiation objects, thus the literature makes the quite common distinction of negotiations into *distributive* and *integrative* ones (e.g., Lügger et al. 2015; Pruitt and Lewis 1975; Thompson, Peterson, and Brodt 1996). In distributive negotiations, the bargaining parties compete with regard to the allocation of a fixed pool of value (Luecke 2003), whereby an increasing utility for one party always goes along with a proportionate decrease in utility for the other party (Backhaus, Hummelsberger, and Wilken 2005; Zerres et al. 2013). Hence, distributive negotiations are often called win-lose negotiations, or zero-sum games (Brown and Baer 2011). A typical example of distributive bargaining is a price negotiation over a single product or service, where the seller wants to get the highest, while the buyer seeks the lowest price possible (Barry and Friedman 1998).

However, in many cases the win-set of a negotiation is variable rather than fixed (Voeth and Herbst 2015b). In these cases, the respective preferences of the bargaining parties are not diametrically opposed, and thus the negotiation possesses integrative potential (Backhaus and Voeth 2014). There are two ways for the integrativity of a negotiation to be realised. On the one hand, an integrative potential emerges if the negotiation object in question can be augmented (Voeth and Herbst 2015b). This may be the case if two parties have a single bargaining object, but have different preferences as to what parts they need from this object. For example, of two parties negotiating about a mobile phone, the one party wants only its display and the other party only the battery. Consequently, both parties could increase their utility if they were to extend the negotiation object "mobile phone" to the negotiation objects "battery" and "display". On the other hand, an integrative potential can be reached by negotiating directly about more than one object (Voeth and Herbst 2015b). This may be the case in a business negotiation whose aim is to determine multiple contract conditions. For one party the delivery date might be substantial, whereas what is important to the other party is the length of the contract. In contrast to a distributive negotiation, integrative bargaining allows an increase in the total utility of two parties without reducing the utility of the one negotiation party or the other (Lewicki, Barry, and Saunders 2015). For these reasons, integrative negotiations are so-called win-win-negotiations (Butler 1999; Carnevale and Pruitt 1992).

Considering real negotiation situations, bargainers are commonly not aware of the integrative potential of a negotiation so that possible gains are left on the table (Thompson 2015). Moreover, an optimal integration of preferences is rarely possible (Backhaus, Hummelsberger, and Wilken 2005). With regard to the two examples given above, the mobile phone example represents an optimal solution whereby in the delivery date and contract length negotiation both parties will still have diverging preferences for both negotiation objects. This circumstance is quite compatible with the fact that negotiators begin by applying integrative approaches, in order to increase the negotiation space so that they can conclude by bargaining distributively about the respective negotiation objects (Thompson 2015).

Negotiation Medium

Another structural factor is the negotiation medium. There are a number of different communication media through which the negotiation can take place. They can be classified into vocal, visual and/or text-mediated channels (Swaab et al. 2012). Furthermore, the media can be differentiated with regard to the physical proximity of the negotiators as well as the synchronicity of the media (Purdy and Nye 2000; Swaab et al. 2012). Synchronicity represents the ability to provide feedback to the other party in real time (Citera, Beauregard, and Mitsuya 2005; Johnson and Cooper 2009; Mor and Suppes 2011; Swaab et al. 2012). The most traditional negotiation medium is *face-to-face (FTF)* (Baltes et al. 2002; Daft and Lengel 1986; Galin, Gross, and Gosalker 2007). By negotiating FTF, the negotiation parties have both visual and vocal contact and the communication takes place immediately and at a single location common to all involved parties (Chen and Tseng 2016). Since FTF negotiations provide a high potential for trust and thus a potential desire for future connections, they are especially essential in contexts in which (business) relationships are yet to be established (Adler, Brahm, and Graham 1992; Backhaus and Voeth 2014; Naquin and Paulson 2003).

The prevalence of information technology (IT), offers a spectrum of *electronic (e-)negotiation media* apart from non-automatic FTF negotiations (Bazerman et al. 2000; Sokolova, Shah, and Szpakowicz 2006). Electronic negotiation channels comprise videoconferences, telephone calls, chats and e-mails among others, most of are computer-mediated channels (e.g., Johnson and Cooper 2009; Purdy and Nye 2000; Stuhlmacher and Citera 2005; Swaab et al. 2012; Yuan, Head, and Du 2003). The electronic channels vary with regard to their classification characteristics. They range from videoconferences, which are quite similar to FTF negotiations, the only difference being the lack of physical proximity (right upper corner of Figure 2-2), to e-mails, whose function is strictly text-based and asynchronous (bottom left corner of Figure 2-2) (Swaab et al. 2012).

Negotiation media exhibit different degrees of media richness which represents the amount of information that can be transmitted by a communication medium (Daft and Lengel 1986). Thereby it holds, that the higher the ability of a communication medium is to process multiple informational cues simultaneously and to provide direct feedback, the richer the underlying medium is (Daft and Lengel 1986; Li 2007). Cues can be of visual, vocal and/or written nature and encompass for example facial expressions, sound of voice or body language (here and in the following Daft and Lengel 1986). The richest communication medium that offers the most cues and direct feedback is FTF. From this richest medium downwards, the amount of available cues decreases starting with a missing FTF interaction (e.g., interactions via webcam), visual cues are absent (e.g., an interaction via phone call) and lastly only textual cues are available (e.g., an interaction via e-mail) (see Figure 2-2).

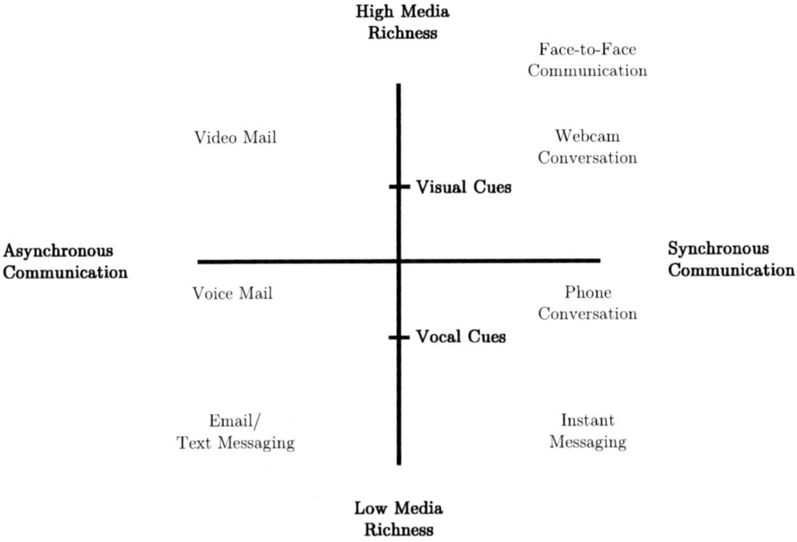

Figure 2-2: Classification of Negotiation Communication Channels
Source: Author's own illustration adapted from Swaab et al. (2012).

The digital era not only provides various computer-mediated communication channels, but also electronic assistance, such as negotiation support systems (NSS) (Sokolova, Shah, and Szpakowicz 2006). NSSs are designed to support and simplify the different steps of the negotiation process in order for the negotiators to be able to make more informed decisions (Bichler, Kersten, and Strecker 2003; Kersten and Lo 2001). With the help of NSSs, negotiations are partly automated, but the control over the process still lies with the bargainers (Schoop, Jertila, and List 2003). They follow in part a negotiation protocol which calls for recording the negotiation framework, including the specification of the negotiated objects and the bargainers' decision-making and communication activities (Bichler, Kersten, and Strecker 2003). In addition to NSSs, digitization extends the field of negotiations even further by the usage of software agents, which negotiate on behalf of human bargainers and may even replace them (Backhaus and König 2019; Bichler, Kersten, and Strecker 2003; Schoop, Jertila, and List 2003). Software agents are proactive software components that behave autonomously by following the programmed negotiation protocol (Vahidov, Kersten, and Saade 2014). They are constructed on the basis of various artificial intelligence techniques such as genetic or Bayesian models (e.g.,

Lau et al. 2008; Oliver 1997; Zeng and Sycara 1998).[7] Agent involvement can take place either in the context of pure agent-agent negotiations or in mixed agent-human bargaining settings, whereby the former case represents a rather unlikely setup in today's business settings or in those of the near future (Backhaus and König 2019; Yang et al. 2014). Agent-human bargaining situations are becoming more and more likely, due especially to the prevalence of such e-marketplaces as e-commerce platforms (e.g., Adam, Teubner, and Gimpel 2018; Cao et al. 2015; Yang, Singhal, and Xu 2013).

2.3.1.2 Personal Factors

In addition to the structural factors, a plethora of personal factors exist that influence the negotiation process. For instance, distinct personality traits and personal values are part and parcel of different behaviours of the negotiation parties that lead to an influence on the negotiation process (Neale and Northcraft 1986; Voeth and Herbst 2015b). Furthermore, the power position of the negotiators represents an important and widely examined personal factor (e.g., van Kleef et al. 2006; Magee, Galinsky, and Gruenfeld 2007; Overbeck and Kim 2013). Power relates to a negotiator's ability to check for and change one's own and other bargainers' conditions and outcomes by providing or withholding financial numbers expressing such as money, or by issuing or withdrawing punishments, such as job terminations (Keltner, Gruenfeld, and Anderson 2003; Magee, Galinsky, and Gruenfeld 2007). This implies that the power of one negotiating party is the other party's dependency, which leads to a stronger power position for the negotiating party that is less dependent on the other one (Bacharach and Lawler 1981; Wolfe and McGinn 2005).

Likewise, the factor of *emotions* strongly influences the negotiation behaviour and outcome (e.g., Pietroni et al. 2008; Shapiro 2004; Steinel, van Kleef, and Harinck 2008). As this constitutes an important part for the thesis at hand, it will be presented in more detail in the following.

Emotions – or so-called *emotional episodes* – are complex and structured processes (Frijda and Mesquita 1994; Russell and Barrett 1999). Due to the high heterogeneity of

[7] For more details about the design and artificial intelligence techniques used for software agents, please refer for instance to Kraus, Sycara and Evenchik (1998) or Adam, Teubner and Gimpel (2018).

the construct which is emotion, diverging opinions on the part of experts exist in literature which lead to differing definitions of the term (Keltner, Oatley, and Jenkins 2013; van Kleef, De Dreu, and Manstead 2004a; Russell and Barrett 1999).[8] Based on specific characteristics which recur in the definitions, the thesis at hand takes the following definition as a basis: An emotion is a psychophysiological phenomenon which is triggered by a conscious or unconscious perception of an object that is relevant to the concerns of an individual (Fischer, Shaver, and Carnochan 1990; Frijda and Mesquita 1994; Hatfield, Cacioppo, and Rapson 1992; Lazarus 1991). Such an "object" can be a person, a condition or an event, whereby it does not matter whether the object is real or imagined, in the past, present or in the future (Russell and Barrett 1999; Solomon 1993). Thus emotions are always object-related and the subjective experience of a given person (Frijda 2008; Lazarus 1991; Russell and Barrett 1999).

Emotional episodes can be caused by multiple reasons and can trigger the tendency towards a certain behavioural response (Solomon 1993). For instance, delaying a business negotiation triggers anger within the waiting negotiation party, which in turn can have a destructive influence on the course of the negotiation, since the angry party will be less willing to make concessions. Moreover, emotional episodes have a definite length with a beginning and an end point (Russell and Barrett 1999) and exhibit a relatively short duration (Keltner, Oatley, and Jenkins 2013). Emotional episodes are discrete, thus they can be categorized into distinct classes (Russell and Barrett 1999). For example, an unpleasant and wilful event which triggers opposition or retaliation is best categorized as anger (here and in the following Frijda and Mesquita 1994). Emotion labelling, however, is not always clear, so that one and the same emotion process can be categorized in different ways.

Emotional episodes need to be distinguished from *emotional dispositions* (e.g., Pesic 2016; Rhoades, Arnold, and Jay 2001; Voeth and Herbst 2015b). Emotional episodes represent situatively felt emotions which are triggered in a person through his or her interaction with the environment, and thus are context-dependent (Rhoades, Arnold, and Jay 2001; Ulich and Mayring 2003; Voeth and Herbst 2015a). In contrast, emotional dispositions

[8] A list of multiple definitions of the term *emotion* is provided by Solomon (1993) or Frijda (2008).

represent the general emotional tendency of an individual embedded in his or her personality (here and in the following Voeth and Herbst 2015b). They represent a constant and predictable factor. Thus, while dispositions are static, situatively triggered emotions are dynamic, as they vary from situation to situation (Rhoades, Arnold, and Jay 2001). In this way, for instance, fear is a currently felt situative emotion, whereas anxiety represents a general emotional tendency (Ekman 2004; Pesic 2016).

In addition, it is important to understand the distinction between the concept of emotions and the concepts of *mood* and *affect*, as the three are closely related to each other (Barry and Oliver 1996; Russell and Barrett 1999). Affect refers to a broad range of feelings that can be experienced during an emotion or a mood (Russell and Barrett 1999). Consequently, affect relates both to emotions and to moods, so that it is usually seen as an umbrella term for the two concepts (Barry and Oliver 1996; Barsade and Gibson 2007; Forgas 1992). In contrast, moods can be clearly differentiated from emotions, as they diverge in multiple aspects. Moods are long-lasting, less intensive than an emotion, and do not exhibit the object/event-relatedness necessary for the existence of an emotion (an example for a mood would be melancholy) (Ekman 1984; Forgas 1992; Lazarus 1991; Thompson 2015). As opposed to emotions which are specific, like "anger" or "fear", moods are more diffuse, such that one often simply distinguishes between "being in a good mood" or "being in a bad mood" (Thompson 2015).

A popular approach to classifying emotions is the circumplex model (Barrett and Russell 1998; Russell 1980; Russell and Barrett 1999). In its initial version, this model categorizes various affective feelings according to two dimensions: valence (ranging from unpleasant to pleasant) and arousal (ranging from activation to deactivation) (Rubin and Talarico 2009; Russell and Barrett 1999). Valence and arousal can be arranged in a two-dimensional circular model with a centerpoint representing neutrality on both dimensions (Rubin and Talarico 2009; Russell 1980). The inner circle of Figure 2-3 represents the classification of the affective feelings within the two dimensions (Russell and Barrett 1999). The peculiarity of this arrangement is that each affect faces its opposite in the circle (e.g., alert versus (vs.) fatigued or nervous vs. relaxed) (Russell 1980). Based on this classification, emotions can be assigned to those feelings which are predominantly present during the respective emotion.

As an example, the two basic emotions *anger* (as a negative emotion) and *happiness* (as a positive emotion) are assigned to the affective feelings in the outer circle of Figure 2-3 which deviate especially on the dimension of valence (Ortony and Turner 1990; Russell and Barrett 1999). The emotion anger is one of the most powerful of emotions, when one considers its intensive impact on the individual undergoing the emotion, along with its considerable influence on social relations (here and in the following Lazarus 1991). It can be described best as a belief that oneself or a person for which one cares has been unfairly treated leading to painful feelings and desire for vengeance. Anger has a high level of unpleasantness and a certain degree of arousal, and is accompanied by such affects as stress and nervousness (here and in the following Russell and Barrett 1999). On the contrary, happiness is a highly pleasant emotion that is typically accompanied by a lower level of arousal than is anger. Thus, happiness can be seen as a state of well-being with agreeable affective feelings ranging in their value from contentment to felicity (Hartmann 1934).

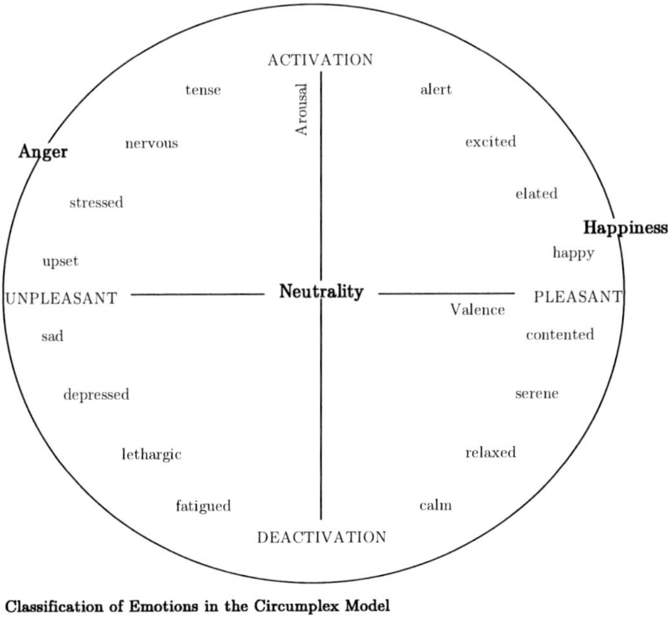

Figure 2-3: Classification of Emotions in the Circumplex Model

Source: Author's own illustration adapted from Barrett and Russell (1998), Russell and Barrett (1999) and Russell (1980).

In order to conceptualize the role of emotions in negotiations, scholars distinguish *interpersonal* and *intrapersonal effects* of emotions (e.g., van Kleef, De Dreu, and Manstead 2004a; van Kleef and Sinaceur 2013; Morris and Keltner 2000). Intrapersonal effects relate to the influence that individuals' own emotions have on their cognition, inner experiences and behaviour (here and in the following van Kleef and Sinaceur 2013). In contrast, interpersonal effects refer to the influence that the emotional expressions of an individual have on other individuals in a social context. As negotiations represent a form of social interaction, the interpersonal effects of emotions are highly relevant (Olekalns and Druckman 2014). In this context, it is important to emphasize that the expression of emotions which are perceived by the negotiation counterpart may reflect the truly felt emotions of an individual, but can also be strategically planned by the expresser in order to manipulate the counterpart (van Kleef and Sinaceur 2013).

2.3.2 Negotiation Preparation Phase

Having discussed the factors that influence negotiations, this subchapter now turns to the first phase of negotiating, namely the preparation phase. The main components of the negotiation preparation phase are the determination of the negotiation goals by the bargaining parties and the determination of the strategies to be followed during the negotiation implementation (here and in the following Voeth and Herbst 2015b). Strategies and goals are related, inasmuch as strategies deliver the direction of thrust for the bargainers in order to reach the goal sought for. Each negotiation party should not only determine its own goal but should also try to reveal or guess the one of the opposing party as goals influence the course of the negotiation. The negotiation goals depend thereby on the alternatives of the bargainers as they build the framing of the negotiation.

Strategies are defined as superordinate guidelines for the negotiation behaviour of the bargainers (Backhaus and Voeth 2014; Voeth and Herbst 2015b). They determine the broad objectives to be reached, as well as a general approach in order to achieve them (Pruitt and Carnevale 1993). Strategies need to be distinguished from tactics. Tactics represent concrete activities to be taken in order to implement the underlying strategy (Backhaus and Voeth 2014). Consequently, a hierarchical relationship exists between strategy and tactic. While the former sets the agenda, the latter fills it with more specific actions (Pruitt and Carnevale 1993). Thus, strategies can be assigned to the negotiation preparation phase, whereas tactics belong to the negotiation implementation stage.

Five basic *negotiation strategies* exist between which bargainers can chose (Lewicki, Hiam, and Olander 1998; Voeth and Herbst 2015b). These strategies can be categorized according to the degree to which the negotiators take into account their own interests and those of the opposing party (Carnevale and Pruitt 1992; Lewicki, Hiam, and Olander 1998). A visualization of this categorization can be found in Figure 2-4.

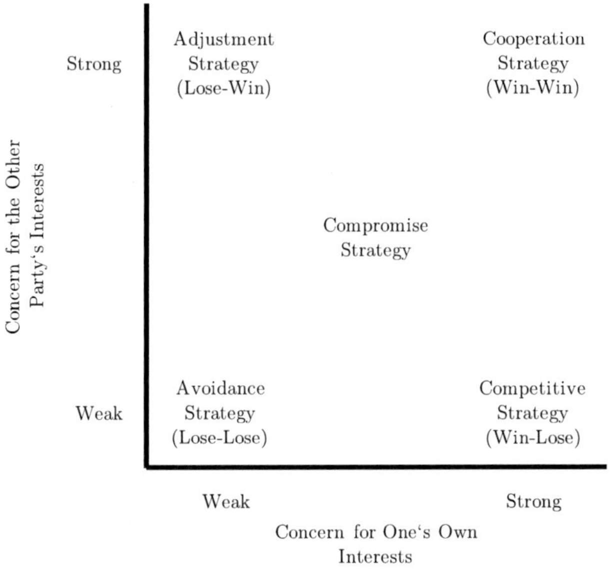

Figure 2-4: Basic Negotiation Strategies
Source: Author's own illustration adapted from Backhaus and Voeth (2014), Carnevale and Pruitt (1992) and Lewicki, Hiam, and Olander (1998).

One speaks of a competitive strategy if the concern for one's own interests is high as opposed to the low concern for the other parties' interests (Carnevale and Pruitt 1992). This strategy is especially used in distributive negotiations, in which each of the bargaining parties strives to maximize its own win-set without concern for any proportional diminution in the other party's win-set (Backhaus and Voeth 2014). The exact opposite strategy is an adjustment strategy that puts one's own interests in the background and focuses on the interests of the other party (Voeth and Herbst 2015b). This strategy becomes especially suited if follow-up negotiations are planned and if one negotiating party is interested in assuring a good position for the following negotiation (Backhaus and Voeth 2014; Voeth and Herbst 2015b). A cooperation strategy is appropriate if it is a negotiator's intention that both parties reach satisfying outcomes, which usually requires a negotiation for multiple bargaining objects (Backhaus and Voeth 2014; Pruitt and Carnevale 1993). Thus it can be found in integrative negotiations in which the win-set is enlarged in order to gratify all negotiating parties (see description in chapter 2.3.1.1). An avoidance strategy makes sense if a negotiator is not interested in closing a deal that is unimportant for him or her (Voeth and Herbst 2015b). In an avoidance strategy, neither party's interests are promoted, which brings about an impasse in the end. This may be the case if a supplier has vigorously utilized capacities or if a company has a principal-agent conflict leading to a refusal by the agents to negotiate seriously (Backhaus and Voeth 2014; Voeth and Herbst 2015b). Lastly, a compromise strategy represents a mixture of the other strategies (Voeth and Herbst 2015b).

The strategy chosen by the bargainers is closely linked to their respective *social value orientations* or *social motives*. An individual's social value orientation indicates his or her preferences towards particular outcome distributions between him or herself and others (De Dreu and van Lange 1995). Usually, literature differentiates between three types of social motives: cooperators/prosocials, individualists and competitors (Deutsch 1960a; De Dreu and van Lange 1995). In keeping with the previously described strategies, cooperators strongly care for each party's interest and want to maximize the joint outcome, whereas competitors are focused on the maximization of their own outcome at the expense of the opposing negotiation party (here and in the following Beggan, Messick, and Allison 1988; Kuhlman and Wimberley 1976; McClintock et al. 1973). Individualists are similar to competitors in their striving to maximize their own outcome; however, they differ from competitors, in that they – individualists – merely do not take into account the outcome of the other, while the competitor actively strives to get a relative advantage over the opponent at his or her expense.

2.3.3 Negotiation Implementation Phase

The implementation phase of negotiations is shaped mainly by the bargaining parties' offer-making activity (here and in the following Voeth and Herbst 2015b). The type of offer made is dependent on the stage at which it is made. Hence one may distinguish between the first (initial/opening) offer and subsequent offers. For each negotiating party it is attractive to submit the *first offer* (here and in the following Backhaus and Voeth 2014). It forces the opposing party to face the opening offer and to justify any deviating offers it might make. In order to submit an appropriate first offer, one's own goals should be determined and those of the opposing party estimated (at the best already in the negotiation preparation phase), which should be considered while negotiating (Voeth and Herbst 2015b).

The subsequent offers predominantly represent so-called *concessions* (Voeth and Herbst 2015b). A concession is defined as a change of offer made in order to support the interests of the opposing party, which at the same time means a reduction of one's own benefit (Pruitt 1981). As a rule, the negotiating parties arrive at a negotiation outcome acceptable to both sides only if they rely on a sequence of accommodating offers (Voeth and Herbst 2015b). Thus one's own concessions should encourage the opposing party to reciprocate; that is, any concessions one makes ought to enable the opposing part to respond with concessions of its own (Wall 1981). Accordingly, negotiating parties should not make concessions multiple times in a row without receiving concessions from the other party in between (Backhaus and Voeth 2014), whereby the application of reciprocal concessions entails the submission of concessions of comparable frequency, interval and seize to those of the opposing party (Esser and Komorita 1975; Voeth and Herbst 2015b). However, not every change in offer is a concession (Pruitt 1981). For instance, it is conceivable that bargainers backtrack from a previously made offer and ask for more than before.

The negotiation implementation phase is additionally characterized by the usage of various *tactics* which can be employed by the negotiators such as threats or deception (e.g., Boles, Croson, and Murnighan 2000; Pruitt and Carnevale 1993). As stated previously, the chosen tactics build on the predetermined strategies and seek to support them (here and in the following Pruitt and Carnevale 1993). In keeping with this, negotiation tactics can be assigned to the different strategies. For example, multiple tactics exist for the competitive and the cooperation strategies, in particular.

2.3.4 A Vast Set of Details

From the previous chapters, it has become obvious that negotiation research is characterized by a long history and a huge amount of influencing research areas and examined research topics. With respect to the research areas, one can find such a high breadth, since multiple fields such as private, business or political negotiations are examined by negotiation scholars (Voeth and Herbst 2015b). These research areas mainly encompass law, computer science/information systems, economic and management science, psychology, political science and sociology (Backhaus and König 2019; Bichler, Kersten, and Strecker 2003; Voeth and Herbst 2015b). The fields of computer science and information systems are predominant contributors to the construction of decision and negotiation support systems as well as of artificial negotiation software agents (e.g., Backhaus and König 2019; Holsapple, Lai, and Whinston 1998; Jennings et al. 2001; Kersten 1997). Law, political science, psychology and sociology are concerned with decision heuristics and studies focusing on aspects of the negotiation process (Bichler, Kersten, and Strecker 2003; Pruitt 1981). In contrast, economic and management science draw especially on the development of optimal strategies, the deduction of formal models, and the prediction of negotiation outcomes (Bichler, Kersten, and Strecker 2003).

These research disciplines give rise to a multitude of publications dealing with diverse research topics (see chapter 2.3). Nevertheless, the large majority of these publications present single scientific results, each with regard to a specific topic. To make this explicit, let us illustrate the fact by means of two examples. Firstly, van Kleef et al. (2006) investigate the influence of power on the emotions of anger and happiness, which in turn impact concession making on the part of bargainers. As a second example, Backhaus, van Doorn and Wilken (2008) examine the role played by team characteristics in the process and outcome of price negotiations. Such a high level of specificity is attributable to the publications' practice of picking out several particular topics and studying them in great detail. All these individual results flow into the widespread and generic field of research on negotiation, whose publications are more or less discrete from one another (Herbst 2007). Accordingly, the research field is defined by a high level of granularity.

For these reasons, it becomes quite difficult to gain an overview of relevant publications, research topics and findings within the field. As the publications stand more or less next to each other separately, attempts are needed to structure and aggregate the single results so that more profound and reliable insights become possible. Thus, approaches are necessary that depart from the granularity approach to negotiation research and seek to attain a superordinate view on the research field and its topics. Approaches of this kind are illuminated in the next chapter.

3 How to Deal With the Details?

3.1 Strengthening the Shoulders: What Do We Know about the Negotiation Giants' Research?

3.1.1 Overview of Prevailing Approaches

As shown in chapter 2, negotiation research has a long and comprehensive history with a broad range of topics and insights. In addition, negotiation research continues to grow, as in recent decades the number of its publications has increased considerably (Buelens et al. 2008; Thompson 2006) (see chapter 4.3.1). Thus, the research field exhibits a substantial size, accounting as it does for such a massive amount of publications and making thus the shoulders of the negotiation research giants quite large (e.g., Thompson, Wang, and Gunia 2010; Voeth and Herbst 2015b; Wanyama and Far 2007). Nevertheless, as criticized in chapter 2.3.4, the huge amount of studies presents a vast set of detailed but segregated knowledge. Thus, due to the mass of publications and a lacking organisation of their topics and findings, negotiation research is, in its present state, badly structured. Accordingly, it becomes rather difficult to gain an overview of the topics and characteristics of the research field for the purpose of deducing insights from previous scholars (giants) in a reliable fashion, and detecting well-studied concepts and blind spots (Agndal, Åge, and Eklinder-Frick 2017). It is consequently necessary to structure the insights, in order to gain a clear picture of the composition of negotiation research as concerns its content. Doing so will make it possible to reveal valuable insights as to the state of knowledge as well as to the intellectual development of negotiation research, and to strengthen the shoulders of the negotiation giants and increase their resilience.

A number of articles and books already have set out to structure the broad field of negotiation research. Nevertheless, these structuring approaches are characterized more or less all by three weak points, which can be assigned to three categories: 1) the applied structuring method of the studies, 2) their integration of negotiation topics, journals and research areas and 3) their consideration of multiple time periods. These three categories are introduced in more detail below. In the following, published structuring approaches will be evaluated according to each of the categories in order to highlight their potential weak points. Thereby, it is important to underline that this literature review does not aim for an all-encompassing overview of existing structuring approaches. Instead, it is aimed for a basic understanding about what kind of overviews have been published so far.[9] Thereby, the literature review is limited to peer-reviewed articles, as they have gained scientific approval by fellow scholars, so that their content can be seen as "certified knowledge" (Fernandez-Alles and Ramos-Rodriguez 2009). Furthermore, since a structuring study which is not recent does not reflect the status quo and lacks current topics and findings, only structuring approaches published within the time frame from 2009 to 2019 are included in the review.

1) The first category refers to the *structuring method* used by the studies in order to present the insights from negotiation research. Structuring approaches can be divided into qualitative and quantitative methods (Zupic and Čater 2015). Qualitative approaches or so-called "narrative reviews" aim at developing a literature overview by identifying and summarizing the publications relevant for the topic of investigation (Dalton and Dalton 2005). The entire process of narrative reviews takes place by the researcher personally and by and large unsystematically, as the researcher selects and analyses the publications (Fitzgerald and Rumrill 2003; Montori, Swiontkowski, and Cook 2003). For this reason, narrative reviews are prone to subjective biases by the scholar and often lacking in rigor (Tranfield, Denyer, and Smart 2003). Moreover, one of the major disadvantages of qualitative approaches lies in their limited capability to process and analyse publications, as the human capacity involved is restricted.

[9] Nevertheless, it was verified by the researcher of this thesis to her best knowledge and belief that no such approach, as it is conducted in Study I, has been published so far.

By contrast, quantitative approaches proceed in an almost completely objective manner and are highly systematic and reproducible, thus improving the quality of reviews (Zupic and Čater 2015). These so-called bibliometric methods, or *bibliometrics*, build on quantitative bibliographic data from relevant publications, all of which are included in the analysis with no prior assessment by the researcher (Verbeek et al. 2002). Moreover, bibliometric methods make use of multiple statistical methods to analyse the data, such as exploratory factor analysis, cluster analysis, multidimensional scaling or network analysis; hence no personal appraisals influence the analysis either (Samiee and Chabowski 2012; Zupic and Čater 2015).[10] Application of these methods makes bibliometrics well-suited to detecting interrelations between scientific publications and thus a reliable source for structuring research fields (Aparicio, Iturralde, and Maseda 2019; Boyack, Klavans, and Börner 2005; Cobo et al. 2015; Vogel and Güttel 2013; Zupic and Čater 2015). Lastly, bibliometrics allow for a much broader coverage of publications within the analysis, as the processing of the data takes place computerized, so that no human manpower is necessary (Samiee and Chabowski 2012).

2) The second category targets the manner in which and the extent to which the structuring approaches *integrate negotiation topics, journals and research areas*, whereby either a focused or a broad proceeding is possible. It is a matter of focus on a specific negotiation topic, journal or research area if only a specific topic, journal or research area is considered (e.g., topic: gender; journal: Journal of Marketing; research area: psychology), whereas a broad consideration does not limit itself to such specifications. In the following, a broad structuring approach is assumed whenever the scientific study accounts largely for topics, journals and research areas.

3) The third category represents the consideration of *multiple time periods* within the structuring approach. Possible study designs can be either a multi-period consideration or an investigation of an overall time frame that does not differentiate between periods (e.g., Jang, Elfenbein, and Bottom 2018; Laengle et al. 2018). The consideration of multiple periods provides the advantage of revealing the evolution of a research field, whereas

[10] Nevertheless, parts in which subjective influences of the researchers impact bibliometric analyses are the choice of the publications from which the references are drawn and the labelling of the topic clusters revealed for instance by a cluster analysis.

one overall time frame only delivers a static picture (Ding, Chowdhury, and Foo 1999). Consequently, multiple periods allow for a content-wise differentiation of multiple time frames and for the revelation of a dynamic development process within the research field. Therefore, the examination of distinct periods allows for more valuable insights and a greater level of detail.

Subsequently, 15 literatue reviews are assessed on the basis of the three categories. Table 3-1 summarizes the results.

Study	Structuring Method of Literature		Integration of Topics, Journals and Research Areas		Consideration of Multiple Periods		Size of Database	Overall Period Length*
	Qualitative	Quantitative	Focused	Broad**	Considered	Not Considered		
Gaspar, Methasani and Schweitzer (2019)	•		•			•	ns.	ns.
Jang, Elfenbein and Bottom (2018)	•		•			•	191 A	15
Laengle et al. (2018)		•	•		•		787 A	25
Agndal, Age and Eklinder-Frick (2017)	•			•		•	490 A	20
Karagözoglu and Urhan (2017)	•		•			•	46 A	ns.
Brett and Thompson (2016)	•		•			•	ns.	ns.
Gunia, Brett and Gelfand (2016)	•		•			•	ns.	ns.
Kennedy and Kray (2015)	•		•			•	ns.	20
Olekalns and Druckman (2014)	•		•			•	ns.	ns.
Caputo (2013a)	•		•			•	21 A	22
Caputo (2013b)	•		•			•	25 A	27
Herbst, Voeth and Meister (2011)	•	•	•			•	78 A	45
Tsay, Shu and Bazerman (2011)	•		•			•	ns.	ns.
Thompson, Wang and Gunia (2010)	•		•			•	ns.	ns.
Tsay and Bazerman (2009)	•		•			•	ns.	25
This Thesis		•		•	•		**66,294 R**	**58**

Legend: • = applies; A = articles; R = references; ns. = not stated; * = in years; ** no focus on specific topics, journals or research areas

Table 3-1: Structuring Approaches of Negotiation Research

Source: Author's own illustration.

1) Structuring Method of Literature

The first shortcoming represents the fact that nearly all structuring attempts are of qualitative nature, as the researchers personally review existing literature, so that the studies are prone to biases due to the judgments made by the scholars. In addition, the analysis is limited to a manageable amount of publications processible for humans (see Table 3-1). The only studies that account for a quantitative analysis of the articles are those by Herbst, Voeth and Meister (2011) and Laengle et al. (2018).

But Herbst, Voeth and Meister (2011), however, also start their examination qualitatively by conducting a qualitative content analysis, in order to determine the most relevant topics for business buyer-seller negotiations. Therefore, they use an adapted interaction model as their categorization scheme and distinguish the categories parties, atmosphere, environment, preparation, process and outcome. According to these predefined categories, the authors conclude that for the *negotiation parties* category, negotiator characteristics such as personality, intrinsic motivation, experience, managerial style and negotiation skills form the core of investigation. Research in the category of *atmosphere* focuses predominantly on inter-organizational power structures, the degree of conflict or cooperation between the parties, and the social distance between them as determined based on their cultural differences. The category of *environment* focuses on research such as market structures or technical developments, while the *negotiation preparation* encompasses publications that deal mainly with strategies, negotiation behaviour types, tactics, and electronic support systems. Finally, the category of *process* focuses on ethics or morality and fairness, and the category of *outcome* works out the most examined outcomes, which are a game-theoretical (distributive or integrative) outcome and a behavioural-scientific one (e.g., satisfaction). The scholars point out that such specific categories as the environment and the process are covered insufficiently in the research, whereas atmosphere and preparation have received greater attention. From the results of their qualitative analysis, the scholars deduce that interdependencies among the themes are hardly recognizable. Nevertheless, since content-analysis does not reveal in a reliable way the lack of interconnectedness the researchers refer to, the researchers complement their qualitative analysis by means of bibliometric methods, namely multiple quantitative co-citation analyses.[11] All these analyses reveal and confirm that the negotiation research findings are minimally interconnected and that they are collocated in an isolated manner. Nevertheless, for their definition of a co-citation analysis Herbst, Voeth and Meister (2011) depart from the prevailing definition in scientific research.[12]

Laengle et al. (2018) is the only article to conduct a thorough quantitative bibliometric approach. In their analysis, the authors proceed according to multiple bibliometric methods, such as bibliographic coupling, citation, co-citation and co-occurrence analysis, in order to detect the impact of the investigated journal *Group Decision and Negotiation*, along with its leading authors, publications, institutions, countries and main keywords, as well as their interconnectedness. The article works out aspects of exclusively structural nature without, however, determining topics that have received the greatest deal of research. Notwithstanding the fact that Laengle et al. (2018) only consider one journal in their investigation, the quantitative approach they take clearly stands out from the other studies by virtue of the size of its database, which comprehends 787 examined articles.[13]

2) Integration of Topics, Journals and Research Areas

The second shortcoming targets at the fact that many structuring publications focus either on one specific topic to be reviewed (e.g., the role of gender in negotiations), on a specific journal from which articles are considered or on a selected research area, so that they still rest on a rather granular level and do not take the important superordinate perspective claimed by this thesis.

[11] The different bibliometric methods will be explained in more detail in chapter 4.2.1.

[12] In particular, the criteria the scholars use to determine coherency in publication citing is not whether publications are cited together in the bibliography of another article, but rather whether publication pairs cite each other.

[13] As bibliometric methods rely on the bibliographic data of articles, the overall investigated database consists of the references of the 787 articles, hence it is much larger. However, Laengle et al. (2018) do not provide this number.

2a) Focus on Topics

In this way, a majority of the studies focus on a specific topic. Kennedy and Kray (2015) review the topic of *gender*. The scholars underscore that women achieve worse negotiation outcomes than do men because of three kind of barriers. One of these barriers are cognitive barriers, which encompass the negative stereotypes held about women's negotiation skills. The second kind of barriers are motivational barriers such as the active prevention of women from acting in a masculine field. The last kind of barriers are paradigmatic barriers, which target the way how negotiation studies are conceptualized. In this manner, the women's negotiation outcomes suffer from study designs that are usually short-term and that focus on economic outcomes. In order to overcome this paradigm, Kennedy and Kray (2015) emphasize that it should be focused more on women's strengths, which encompass their great concern for others and their cooperativeness.

Moreover, the topic of *deception* is reviewed (Gaspar, Methasani, and Schweitzer 2019). The scholars conceptualize deception as a multidimensional construct that is characterized by three dimensions: intentionality (self-interested or prosocial), content (emotional or informational) and activity (omission, commission or paltering). Moreover, the authors introduce the deception consequence model that depicts the consequences of deception on negotiation parties and negotiation outcomes. As the consequences of deception are dependent on multiple factors such as its being detected, or on the culture of the opposing party, it is underscored that it can lead both to positive and to negative outcomes.

Olekalns and Druckman (2014) focus on the topic of *emotion* in their literature review. They structure the literature around the four broad themes of the behavioural consequences of emotions; information processing; social interaction; and contextual influences on emotion. The scholars underline that the mostly examined emotions in negotiation research are anger and happiness, and that such factors as power or gender are among the more frequently examined influence factors of emotions.

Gunia, Brett and Gelfand (2016) review the negotiation literature about the role of *culture*. The authors develop a two-dimensional framework that distinguishes between intracultural and intercultural influences as well as between input, processes and outcomes.

They underscore that some input factors enhance win-win situations in intercultural negotiations, while others may hinder them (e.g., relationship building versus cultural distance). Besides, in intercultural negotiations the strategies used tend to be the ones normative for one's own culture. With regard to the outcomes, cultures differ as to whether they achieve distributive or integrative results.

Tsay, Shu and Bazerman (2011) consider the topic *naivety/cynicism* of the negotiators in their review. These scholars summarize decision errors that occur if a negotiator is too naive or too cynical. Errors on the side of cynicism are particularly harmful to negotiations, because they stimulate negative factors such as mutual distrust. The authors provide several aids for the purpose of overcoming these decision errors, such as perspective taking or communication.

Karagözoglu and Urhan (2017) review the literature on the role of the topic *stake size* in negotiations by focusing on distribution games and experiments. The authors conclude that study results are a long way from a clear idea about the effect of stake size on individual behaviour. The scholars argue that such uncontrolled individual characteristics as risk aversion and fairness issues, but also that overly small sample sizes could be responsible for this result.

Caputo (2013a) focuses on the topic of *cognitive biases* in negotiations. The author works out the most frequently examined biases in negotiation research, some of which are anchors, overconfidence, framing, status quo and self-serving bias. Lastly, Caputo (2013b) reviews the literature on *third party* inclusion in negotiations and focuses on integrative agreements. The author shows that third parties are mostly examined from the perspective of their supporting and conveying roles, rather than as main actors. Besides, most studies investigate third party inclusion from an individual perspective rather than from an organisational point of view.

2b) Focus on Research Areas

A number of structuring approaches consider topics broadly, but limit themselves with regard to the research areas they include. This holds for example for the review of Thompson, Wang and Gunia (2010) which focuses on descriptive research influenced by social psychology and organizational behaviour. The scholars systematize negotiation research along five categories: intrapersonal, interpersonal, group, organizational and virtual. Intrapersonal factors encompass power, gender and affect among others, whereas interpersonal factors include emotions, improvisation and trust. Factors such as group identity and culture are assigned to the group level of negotiation research, and choice of negotiation partner and reputation belong to the organizational factors. Lastly, the virtual level encompasses such elements as the included negotiation medium.

Brett and Thompson (2016) focus their review on two-party cooperative negotiations. The authors use Brett's (2000) model of culture and negotiation to structure negotiation research along the categories of interests, strategies, interactions and outcomes. Brett and Thompson (2016) first distinguish a distributive from an integrative negotiation and then go on to review the psychological and sociological factors that influence the four categories. These factors are cognitive and motivational biases, personality, social motives, emotions and moods, trust, reputation, gender, power and culture.

Lastly, the study of Tsay and Bazerman (2009) is limited to a perspective of BDR and social psychology. The authors consider negotiations from a historical perspective and review the development of Raiffa's (1982) decision-analytic perspective into the BDR-perspective of focusing on cognitive biases (see chapter 2.2). From there, the authors deduce the ensuing high importance of social factors such as social relationships and egocentrism.

2c) Focus on Journals

Three structuring studies focus on specific journals. In this manner, Jang, Elfenbein and Bottom (2018) focus on top ranked management and psychology journals. The scholars deduce a general negotiation model consisting of three main negotiation phases: a planning phase, the negotiation phase itself, and an implementation phase, which targets the realization of the agreement. The scholars examine which phases of the negotiation model are mainly investigated in empirical negotiation research and find that the bargaining phase is quite saturated with topics such as first offers, concession making, tactics and outcome value. However, the scholars also examine prescriptive textbooks and find that these books place greater focus on the preparation and implementation phases. Consequently, the researchers conclude that the planning and implementation phase are in need of further empirical research. In addition, they find that multi-party negotiations are rather neglected in the literature, as most negotiation studies focus on two parties.

Regarding the two quantitative approaches of Herbst, Voeth and Meister (2011) and Laengle et al. (2018), it is remarkable that they are restricted to specific journals as well. Herbst, Voeth and Meister (2011) focus solely on the ten highest ranking marketing journals and the article of Laengle et al. (2018) is even more concentrated and restricted, as its analysis is based on one single journal.

The literature review by Agndal, Age and Eklinder-Frick (2017) constitutes a special case, as their article is the only one that considers negotiation research from a broad perspective without limiting on specific journals, topics or research areas. They structure negotiation research according to the four categories of parties, context, process and outcomes, and work out for each category the predominant themes in the literature. With regard to the negotiation *parties*, the authors emphasize the high importance of individual variables such as personality, demographics (especially gender), negotiating style, training, experience, teams and attitudes (nearly half studies they examine focus on these topics). Moreover, they conclude that organizational variables such as organizational demographics or organizational climate influence the negotiation parties. Lastly, the researchers work out the importance of the relationship between the parties and show prior experiences, knowledge, feelings and power relations to be among those topics investigated frequently. With respect to the negotiation *context*, the authors point out the great importance of culture (almost one-third of the studies investigate this topic), the negotiation medium and negotiation support systems chosen, the location, the number of negotiation objects, and time. Regarding the category *process*, the researchers work out the main topics of negotiation preparation, the making of offers, and tactics. Lastly, the researchers take a closer look on negotiation outcomes and differentiate between an agreement and impasse, EOs versus SOs, as well as the degree of efficiency of the negotiation.

In summary, one can notice that nearly all structuring studies focus either on topics, journals or research areas. Besides, it becomes apparent from the present review that most of the structuring approaches either focus exclusively on the role of specific personal factors in negotiations, or work out the importance of these factors based on their investigation of the whole negotiation process. Based on these conclusions, it is possible to deduce a considerable degree of coverage of personal factors by negotiation research.

3) Consideration of Multiple Periods

The third shortcoming lies in the fact that nearly all approaches lack an indication of a temporal development of the research field, so that no sophisticated differentiation can be made between multiple epochs.

The article by Laengle et al. (2018) is the only one of the structuring approaches that splits the overall time frame into two periods for a more differentiated and deeper perspective. In this way, the authors divide the entire time period into two decades ranging from 1997 to 2006 and from 2007 to 2016. As the overall time length amounts to 25 years and starting in 1992, the authors neglect the time frame from 1992-1997 in their deeper analysis.

With respect to the overall time period, the study conducted by Herbst, Voeth, and Meister (2011) stands out from the other articles in the literature review, as it exhibits the longest entire time frame examined, covering a period of 45 years.[14]

3.1.2 Research Gaps

By having evaluated the structuring approaches according to the categories, the literature review has highlighted three shortcomings of existent studies. These shortcomings represent three central research gaps, which can be filled in large part by further analysis, so that they will be addressed by the thesis at hand. These gaps are (1) the lack of a quantitative approach, (2) the lack of a comprehensive approach with regard to topics,

[14] As different time frames are used for different journals, the longest time period is chosen in order to deduce the overall time frame.

journals and research areas and (3) the neglection of the dynamic structure within the research field. Each research gap will be considered in greater detail in the following.

(1) Lack of Quantitative Approach

As shown in Table 3-1, nearly all structuring studies choose a qualitative approach. Next to the huge amount of narrative reviews, only two studies quantitatively examine the research field with the help of bibliometric methods. This finding is similar to other research disciplines that criticize the fact that purely qualitative approaches are deployed in order to investigate a research field while quantitative techniques are neglected (e.g., Fernandez-Alles and Ramos-Rodriguez 2009; Samiee and Chabowski 2012). The two (partly) quantitative procedures are presented by Laengle et al. (2018) and Herbst, Voeth, and Meister (2011).

Herbst's, Voeth's, and Meister's (2011) use of multiple quantitative co-citation analyses, only takes place in a second step, for the purpose of verifying their findings from a qualitative content analysis. Besides, the scholars do not rely on the definition of co-citation analysis present in the literature; rather, they alter it into a quite simple interpretation for analysing the interconnections of publications. In their limitations Herbst, Voeth, and Meister (2011) themselves highlight the fact that a more sophisticated bibliometric method – such as a real co-citation analysis relying on bibliographic data – would be better suited to the purpose of examining the research field, as it delivers more profound results. In addition, the scholars do not follow a broad approach, focusing as they do exclusively on the marketing perspective of negotiations by concentrating on the top ten ranked marketing journals.

The only purely quantitative approach taken by Laengle et al. (2018) is even more restricted, as they build their analysis on a single journal, namely *Group Decision and Negotiation*. In addition, the authors apply nearly all existent bibliometric methods without justifying the validity of the methods used in each case (e.g., they apply co-citation analysis for journals and authors and bibliographic coupling for institutions and countries). However, the suitability of different bibliometric methods depends on the purpose of the specific investigation (Zupic and Čater 2015). As a consequence, their application should match the underlying data and the aim of the study, so that it is called for a sophisticated usage of bibliometric methods. Moreover and most importantly, rather than topical and structural developments within the research field, the authors deduce solely such features as main research institutions or journals.

Hence the literature lacks a quantitative deduction, by means of sophisticated bibliometric methods, of the structure of the negotiation research field. Such an examination would reduce the impact that subjective biases on the part of the researchers have on the outcome; would be highly transparent and reproducible; would allow for the detection of interconnections in the negotiation research field; and would enable the inclusion of a far larger amount of articles for processing and analysing purposes than is possible in qualitative reviews. The vast number of publications in the field of negotiation research is a clear demonstration of how urgently it calls for a quantitative bibliometric approach.

(2) Lack of Comprehensive Approach with regard to Topics, Journals and Research Areas

Nearly all studies are somehow focused on either negotiation topics, journals and/or research areas. As a rule, the studies focus on a particular research topic and thus narrow their scope of the research field. Further studies limit the research field to specific research areas either by including only descriptive research, BDR, or social psychology. Lastly, there are studies that restrict themselves with regard to the journals they include.

The only broader approach within the literature review is the one by Agndal, Age and Eklinder-Frick (2017), who do not focus on specific topics and journals and who include the identified business-relevant research areas. Nevertheless, these scholars' approach is marked by a number of disadvantages. Agndal, Age and Eklinder-Frick (2017) apply a qualitative approach that makes their investigation prone to subjective biases and makes it less transparent and reproducible. In addition, the detection of thematic interconnections is hardly possible and the number of studies it is possible to review is restricted due to the finite human processing capacity. On top of that, the researchers do not take dynamic development of the research field into account, which is attributable to the fact that they do not regard multiple time frames. Lastly, the overall examined time frame only dates back to 1995. As negotiation research has its roots in the 1940s, an overall time frame of 20 years is much too short when one bears in mind that the aim is to gain a comprehensive picture of negotiation research. Thus a more comprehensive structuring approach remains a desideratum.

(3) Neglection of Dynamic Structure within the Research Field

In respect of the consideration of multiple periods, it becomes obvious in Table 3-1 that only Laengle et al. (2018) split their investigation into different periods. Laengle et al. (2018) divide the overall time frame into two periods, each with a length of 10 years. However, as mentioned before, they only consider one journal and investigate an overall time frame of 25 years. Furthermore, the authors explain neither their subdivision into these two time frames, nor their exclusion of the time span from 1992-1997.

Consequently, the negotiation research literature lacks an approach that by means of an examination of the relevant topics, characteristics and structures of multiple periods would deal with the dynamic structure of the research field. A consideration of multiple periods would enable not only the revelation of the field's evolution, but also a substantive contentual differentiation of the time frames. Such an approach would require not only an adequate and justified segmentation into periods, but an overall length of time that is set with the fact in mind that negotiation research dates back to the 1940s and continues to grow to this day. Thus, an overly narrow time span is incapable of portraying the wide range of negotiation research.

Summing up, the need continues to exist for a both quantitative and broadly oriented structuring study on negotiation research that would reveal the main publications, topics and their interconnectedness by accounting for the dynamic development of the research field throughout all its long history. This dissertation aims to close these research gaps by addressing them in Study I, in order to strengthen the shoulders of negotiation giants (see chapter 4). To do so, it implements a quantitative bibliometric approach that consists of citation and co-citation analysis, based on which it determines the most-cited negotiation publications in a first step. From there it goes on to reveal the structure of the negotiation research field by determining its defining topics based on the time frames involved, as well as on these topics' respective developments and interrelations. In so doing, Study I does not focus on specific negotiation topics, journals or research areas; rather, it considers them in broad terms. In particular, this thesis stands out from the overview studies with respect to the size of its database as well as of the overall time frame examined: The underlying dataset exceeds the size of the review studies considerably by examining an amount of 66,294 references, over against a mere 21 to 787 articles examined in the studies reviewed (see here and in the following Table 3-1); and the overall time frame of 58 years surpasses the frames of the other studies, which range between 15 and 45 years.

3.2 Growing the Dwarf: Anything Goes? The Effects of Contradictory Results of Personal Factors on Negotiation Outcomes

3.2.1 Overview of Existing Approaches

Chapter 3.1 identified weaknesses in the resilience of the negotiation giants' shoulders which will be strengthened in the chapter 4 (Study I). This thesis has to this point forgone treatment of the dwarf-perspective, which will be contemplated further below, as the thesis aims to consider both perspectives (see chapter 1.3). The upcoming second study seeks to enrich the field of negotiation by conducting secondary research, in the process of which the dwarf benefits by being able to stand on the "trained" shoulders established in Study I. As underlined in chapter 2.3.4, the research field would benefit from an aggregation of the various single study results of specific topics, in order to get resilient outcomes. For the detection of the adequate topical focus for this purpose, there follows a more detailed look into the findings pertaining to personal influence factors in negotiations. The focus on this umbrella topic arises from the fact that it is characterized by multiple contradictory empirical research findings, delivers a particularly high amount of research studies, and represents a quite important subarea in negotiation research (Agndal 2007; Voeth and Herbst 2015b) (see chapter 1.1 and chapter 2.2). These inconsistencies currently give us cause to utter, if incredulously, the words: "Anything goes?". Thus the studies' findings need to be aggregated, in order to answer this question with a clear "no", and to deduce an overall result on which further scholars can rely.

Personal factors play a major role in negotiations (Agndal 2007; Voeth and Herbst 2015b). This is hardly surprising in light of the fact that the participation of individuals each influenced by their respective multiple personal factors – be it their personality, demographics, organizational roles or emotions – is fundamental to negotiations in the first place (Lax and Sebenius 1986; Sharma, Bottom, and Elfenbein 2013; Thompson, Wang, and Gunia 2010; Voeth and Herbst 2015b). Personal characteristics and values vary from individual to individual, leading to substantial variation in behaviour, which in turn exerts a heavy influence on the course of negotiation (Ames 2008; Lax and Sebenius 1986; Walton and McKersie 1965). Because of this high impact, organizations should plan carefully which representatives they want to include in their negotiation team, as the very success of negotiation depends on it (Voeth and Herbst 2015b).

Empirical research on the influence of personal variables in negotiations started in the early time of negotiation research, whose social-psychological era ranged from the 1960s to the 1970s (see chapter 2.2). In this time, a plethora of studies examined the effect of sociodemographic as well as psychographic individual characteristics on the negotiation process and outcomes (Herbst 2007). Nevertheless, after this phase the scholars Rubin and Brown (1975) reviewed around 1,000 socio-psychological articles in a narrative review and found no systematic relationship between individual variables and negotiation behaviour. Many subsequent studies relied on this and did not investigate these factors more intensively (Neale and Northcraft 1991; Pruitt and Carnevale 1993). However, a large number of researchers continued to see the importance of individual variables and claimed that scholars may have closed the book on the influence of individual variables on negotiations too soon (Barry and Friedman 1998; Elfenbein 2015; Lewicki, Barry, and Saunders 2015; Sharma, Bottom, and Elfenbein 2013). In this way, from the 1990s onwards, negotiation research was amplified in the field once more (e.g., Barry and Friedman 1998; Dimotakis, Conlon, and Ilies 2012; Elfenbein 2015; Fulmer and Barry 2004; Thompson, Wang, and Gunia 2010) (see chapter 2.2).

Especially from the early phases of research on individual variables, many findings exist that contradict each other, which were in particular the reason for the claim, made by a number of researchers, that individual variables exert no systematic influence on negotiation outcomes (Bowles, Babcock, and McGinn 2005; Pruitt and Carnevale 1993;

Thompson 1990). Moreover, such competitive findings have increased over time due to the rising number of research areas dealing with negotiations, to an increasing quantity of scholars who enter the research field, and due to pressure to publish research findings (Stuhlmacher and Gillespie 2006). These inconsistent findings exist for multiple individual variables, such as trust, organizational roles, gender, and culture (e.g., Holt and DeVore 2005; Kong, Dirks, and Ferrin 2014; Kray and Thompson 2005; Rahim 1986).

The main critical review conducted by Rubin and Brown (1975) is a narrative review that cannot combine study results and is only able to describe them verbally (here and in the following Sharma, Bottom, and Elfenbein 2013). However, if an overall reliable effect could be deduced for the respective personal characteristics, the belief of several scholars that no systematic relationship between individual variables and negotiation behaviour exists could be refuted. Accordingly, a considerable number of negotiation scholars have demanded the aggregation of the diverse study findings in this field, for the purpose of making progress in understanding the variables, and in order to reach a definitive conclusion concerning their impact (e.g., Kong, Dirks, and Ferrin 2014; Sharma, Bottom, and Elfenbein 2013; Stuhlmacher and Gillespie 2006; Walters, Stuhlmacher, and Meyer 1998). In order to synthesize dispersed findings, one method has emerged in recent decades, and has gained considerable attention: so-called *meta-analysis* (Eisend 2014; Rosenthal and DiMatteo 2001; Schewe, Hülsheger, and Maier 2014).

Meta-analyses are especially well-suited for aggregating the (partly) contradictory findings of primary empirical research studies in order to determine one overall effect size (Bijmolt and Pieters 2001; Chow 1987). They focus on the investigation of one specific topic and allow for the inclusion of a large set of empirical studies which makes their application attractive for well-studied themes (Borenstein et al. 2009; Farley, Lehmann, and Mann 1998; Geyskens et al. 2009). Since meta-analyses rely on a set of statistical procedures, which makes them more objective, reliable and rigorous than qualitative literature reviews, they can be classified as quantitative literature reviews (Aguinis et al. 2011; Chow 1987) (see chapter 3.1.1). Meta-analyses allow for the testing of hypotheses on a much larger scale, operating as they do with a large sample yielded by the overall aggregation of the single studies (Eden 2002; Ellis 2010). Besides, meta-analyses allow for the inclusion of moderators that show those cases in which the effects are present and those cases in which they are not (Cortina 2003; Cortina and Folger 1998).

Several research studies have already captured the need to bring clarity to the field of individual variables in negotiation research by conducting a meta-analysis. The present literature review has identified 13 relevant meta-analyses, which will be presented more in detail in the following (see Table 3-2). [15] From this on, blind spots are revealed for which meta-analyses are missing but which would profit from their conduction as well.

[15] For the same reasons as those presented in chapter 3.1.1, this literature review considers only meta-analyses published in peer reviewed journals. Furthermore, meta-analyses are included in the review only if they are explicitly conducted in the field of negotiations (e.g., Stuhlmacher and Walters 1999), or if they refer to related topics such as decision-making or conflict resolution and integrate an amount of negotiation studies in their determination of the overall effect size (e.g., Greer et al. 2018).

Study	IV		DV	Moderators	Quantities		
	Topic	Category			Studies	Sample Size*	Effect Sizes
Kugler et al. (2018)	Gender	D	B	Appropriateness to Negotiate, Inconsistency between Roles, SC	55	17,504	55
Greer et al. (2018)	Hierarchy	O	EO/SO	Team Task, Structure and Hierarchy	54	13,914	86
Mazei et al. (2015)	Gender	D	EO	Advocacy, Structural Ambiguity, Experience, Integrativeness, Self-initiated Negotiation	51	10,888	123
Kong et al. (2014)	Trust	S	B/EO/SO	Integrativeness, Level and Measure of Trust	38	9,154	68
Sharma et al. (2013)	Personality, Cognitive Ability, EI	P	B/EO/SO	Integrativeness, SC	ns.	ns.	ns.
Stuhlmacher and Morrissett (2008)	Gender of Mediator	D/O	SO	Relationship, Type of Perception, Power, Gender, SC	5	ns.	11
Orr and Guthrie (2006)	Anchoring	P	EO	Information, Expertise	18	1,259	18
Lyubomirsky et al. (2005)	Mood	P	B	-	8	574	17
Holt and DeVore (2005)	Culture, Gender, Organizational Role	D/O	B	Gender	36	ns.	ns.
Oosterbeek et al. (2004)	Culture	D	B	-	37	ns.	75
De Dreu et al. (2000)	Social Motives	P	B/EO	Resistance to Yielding, Gender, Integrativeness, SC	28	ns.	68
Stuhlmacher and Walters (1999)	Gender	D	EO	Opponent's Sex, Power, Integrativeness, Communication, SC	21	3,496	53
Walters et al. (1998)	Gender	D	B	Communication, Strategies, Opponent's Sex, SC	62	ns.	79
This Thesis	**Emotions**	**S**	EO/SO	Power, Competitiveness, Appropriateness of Emotion, Time Pressure, Medium, SC	94	16,795	267

Legend: IV = Independent Variable, DV = Dependent Variable, EI= Emotional Intelligence, B = Behaviour, EO = Economic Outcome, SO = Subjective Outcome, D = Demographic, P = Psychographic, O = Organizational, S = Situative, SC = Study Characteristics, ns. = not stated, * either individuals or teams

Table 3-2: Meta-Analyses of Personal Factors on Negotiation Outcomes

Source: Author's own illustration.

The meta-analyses will be described with respect to the following criteria: The examined main topics or independent variables (IVs) respectively will be presented, assigned to those individual difference category to which they refer. Therefore those categories presented in chapter 2.3 will be applied which rely on the classification by Voeth and Herbst (2015b), namely socio-demographic, psychographic, organizational and situative personal factors. Moreover, the examined dependent variables (DVs) will be distilled from the meta-analyses, as are moderators. For these variables, the significant findings will be outlined. Lastly, the meta-analyses will be characterized by a number of quantitative figures: the number of included studies, the overall number of investigated individuals, and the number of examined effect sizes that can be deduced from the studies.

Demographic Factors

As becomes apparent from Table 3-2, a large part of the IVs investigated by the meta-analyses can be assigned to individual demographic difference factors. Most of these studies meta-analytically review the effect of gender in negotiations. By taking a closer look at the meta-analyses on gender, each meta-analysis differs from the others with regard to its characteristics. In this way, Kugler et al. (2018) and Walters, Stuhlmacher and Meyer (1998) examine the influence of gender on negotiation behaviour, whereby the former target the initiation of negotiation and the latter the competitiveness of the negotiators. Besides, the investigated moderators of these reviews represent a completely different set. While the dataset exhibited by Kugler et al. (2018) contains the largest number of individuals of all the reviewed meta-analyses (17,504 individuals), Walters, Stuhlmacher and Meyer (1998) include the highest number of studies (62) in their analysis. Kugler et al. (2018) find an overall effect that women are slightly less prone to initiating a negotiation than men, however, the gender gap is smaller when there is little situational ambiguity with respect to the appropriateness of negotiating, when the situational factors are more consistent with the female instead of the male gender role (e.g., a cooperative negotiation), and when the publication is recent. Walters, Stuhlmacher and Meyer (1998) conclude that women are slightly more cooperative than men, whereby restrictions on negotiations such as limited communication lead to smaller gender differences. In cases in which the opponent uses a tit-for-tat negotiation strategy, women even negotiate more competitively than men.

In contrast to these two meta-analyses, Stuhlmacher and Walters (1999) and Mazei et al. (2015) examine the effect of gender on EOs. Mazei et al. (2015) build on the meta-analysis of Stuhlmacher and Walters (1999) and extend it by relying on a much larger sample. Stuhlmacher and Walters (1999) find that men negotiate slightly better EOs than women. Mazei et al. (2015) confirm this result and add a number of significant moderators to it. In this way, the superiority of men is reduced when bargainers have experience, receive information about the negotiation range and negotiate for others rather than for themselves. Their meta-analysis integrates the highest number of effect sizes, namely 123.

Lastly, the meta-analysis implemented by Oosterbeek, Sloof and van de Kuilen (2004) refers to the influence of culture on rejection rates in ultimatum games. The researchers find significant differences in rejection rates, such that the Western United States have lower rejection rates than the Eastern United States and that Asia has far higher rejection rates than the United States in general.

Psychographic Factors

Dealing as they do with multiple topics, meta-analyses in the category of psychographic IVs, are less focused than those in the demographic category. Thus Orr and Guthrie (2006) meta-analytically review the effect of anchoring on EOs and find a positive influencing effect, which is even amplified when information and experience on the part of the negotiators are low.

De Dreu, Weingart and Kwon (2000) investigate the influence of social motives on contentious and on problem-solving behaviour, as well as on joint EOs. Prosocial individuals are found to be less contentious, to concentrate more on problem-solving and to realize higher joint outcomes than those following an egoistic motive. These findings are positively moderated by a high degree of resistance to yielding, by inducements offered to the participants in the experiment, and by publication of the research.

Sharma, Bottom and Elfenbein (2013) consider the following multiple IVs in their meta-analysis: personality traits, cognitive intelligence, emotional intelligence (EI) and attitudes, and they examine the effect of these IVs on cooperative behaviour, as well as on

EOs and on SOs. They divide these variables into subcategories. Referring to personality traits, the scholars find a positive influence of extraversion, agreeableness, as well as self-efficacy on SOs, on cooperative behaviour for the two former and on individual EOs for the latter. Moreover, trust has a positive effect on cooperative behaviour, whereas Machiavellianism has a negative influence. Cognitive intelligence impacts cooperative behaviour and all outcomes positively, whereas EI has a negative effect on individual EOs and a positive one on SOs. Lastly, with respect to attitudes, intolerance toward ambiguity and self-esteem have a positive effect on cooperative behaviour, while the willingness to take risks has a negative influence. Nevertheless, the study of Sharma, Bottom and Elfenbein (2013) focuses predominantly on broad coverage of psychographic factors at the expense of the meaningfulness of the overall effect for several variables, as results are merely deduced from datasets with less than 5 effect sizes (and base in some cases on just one effect size).

Lastly, Lyubomirsky, King and Diener (2005) meta-analytically review the effect of mood on competitive negotiation behaviour. However, this meta-analysis encompasses a mere eight studies. Based on this paucity, the researchers find that a positive mood increases the likelihood of a cooperative behaviour.

Organizational Factors

Regarding the organizational variables, Greer et al. (2018) and Holt and DeVore (2005) both consider the role of hierarchies in their meta-analyses. Greer et al. (2018) focus on the effect of hierarchies on team performance and viability (EOs and SOs), whereas Holt and DeVore (2005) investigate the effect on the chosen conflict resolution style.[16] Holt and DeVore (2005) furthermore examine two independent demographic variables, namely culture and gender. Greer et al. (2018) show that hierarchy negatively influences team performance and viability. The effect on team performance is amplified by the moderating effects of greater membership instability, skill differentiation, and hierarchy mutability. However, the moderating role of a greater task ambiguity leads to a positive effect of hierarchies on team performance. Holt and DeVore (2005) find that competing is used

[16] The conflict resolution styles correspond to the negotiation strategies presented in chapter 2.3.2.

less with peers; that adjusting and avoiding are used more towards superiors; and that both compromising and cooperation are used more with peers and subordinates than with superiors. With respect to moderation according to gender, the scholars conclude that men are more prone to using a forcing style towards their superiors than are women.

Stuhlmacher and Morrissett (2008) examine the effect of a mediator's gender on the disputants' perception of this mediator. Male mediators are perceived more positively than female mediators whereas when a power differential between the disputing parties exist, the impact of the mediator's gender is weaker. The meta-analysis by Stuhlmacher and Morrissett (2008) examines a mere five studies and thus, represents the smallest meta-analysis within the reviewed set.

Situative Factors

From Table 3-2 it becomes apparent that factors from the category of situative variables are those that are meta-analysed the least. Only the meta-analysis by Kong, Dirks and Ferrin (2014) can be assigned to the category, which investigates the influence of trust on distributive or integrative behaviour, as well as on EOs and on SOs. These scholars find that the effect of trust on distributive behaviour is negative, whereas its influence on integrative behaviour, individual and joint EOs, and on SOs is positive. The impact of trust on joint EOs and integrative behaviour is positively moderated by integrative potential, and its effect is stronger for dyadic than for individual trust.

As can be deduced from the literature review, meta-analyses that can be assigned to the category of situative IVs have been largely neglected in negotiation research. Situative factors arise as a result of situational conditions in a negotiation setting, whether they concern the similarity of the negotiators, the power differential of the bargaining parties or the renewed delegation of negotiators if they successfully negotiated with a negotiation partner in the past (Voeth and Herbst 2015b) (see chapter 2.3). Thus, they emerge from the interrelation of the characteristics and the (past) interaction of the negotiating parties. The presence of negotiating parties and their interaction are main features of negotiations (see chapter 2.1), thus underscoring the importance of situative factors. Hence this thesis seeks to build on this insight by focusing on a situative factor for which it seems especially important to conduct a meta-analysis: the topic of *emotions*, the relevancy of which is particularly highlighted by the results of the co-citation analysis, as it is worked out as the most frequently investigated topic of the detected research themes within the last decades (see chapter 4.3.2).

However, this increasing interest on emotions in negotiation research is not only emphasized by the results of Study I, but also underscored by several negotiation scholars (e.g., Barry 2008; Barry, Fulmer, and van Kleef 2004). Deviating from the assumption of early negotiation scholars that bargainers behave rationally by maximizing their negotiation outcomes without being influenced by emotions (see chapter 2.2), latest researchers have identified that bargainers can hardly blank out or ignore their emotions or the emotional expressions of their counterparts; thus negotiations can be seen as a highly emotional setting (Bazerman and Chugh 2006; Elfenbein et al. 2007; Voeth and Herbst 2015b). Consequently, large quantities of (empirical) research exist, that examine the effect of emotions in negotiations (Barry 2008; Pesic 2016). Emotions influence negotiations diversely, as they have an impact on negotiation behaviour such as the offer making or the choice of a bargaining tactic, and on negotiation outcomes, whether EOs or SOs (e.g., Barry and Oliver 1996; Lawler and Yoon 1995; Olekalns and Druckman 2014). In addition, emotional expressions provide information about the bargaining partner that can be used to make negotiations more predictable and easier to handle (Elfenbein et al. 2007).

A multitude of different emotions are investigated by negotiation scholars. One can notice that, while the sole object of a number of studies is positive and negative emotions in general (e.g., Hine et al. 2009; Kopelman, Rosette, and Thompson 2006; Lawler and Yoon 1995; Shapiro 2004), which means that these studies distinguish at the level of valence only rather than at the level of arousal as shown by the circumplex model presented in chapter 2.3.1.2., it is nevertheless the case that the majority of negotiation scholars do indeed further differentiate between concrete emotions. Thus emotions such as fear, disappointment, anger, happiness or guilt are repeatedly an object of negotiation researchers' investigation (e.g., van Kleef, De Dreu, and Manstead 2006; Olekalns et al. 2005; Pesic 2016; Sinaceur et al. 2013). However, the frequencies by which these emotions are examined differ strongly. As underlined by the results of Study I and by a variety of negotiation scholars, the emotion that is by far the most frequently examined is anger (Adam and Shirako 2013; van Kleef, Anastasopoulou, and Nijstad 2010; Schroth 2008; Sinaceur et al. 2011). This fact is quite obvious, as negotiations depict competitive social interactions in which parties have divergent interests and negative interdependences, so that anger is likely to arise (van Kleef, Anastasopoulou, and Nijstad 2010). Thereupon, happiness is highlighted to represent the second most frequently investigated emotion (e.g., Druckman and Olekalns 2008; van Kleef 2007; Olekalns and Druckman 2014; Tng and Au 2014). As anger is predominantly compared to happiness or emotional neutrality, the frequent consideration of happiness can be explained by this fact, among others (e.g., van Kleef, De Dreu, and Manstead 2004b; De Melo, Carnevale, and Gratch 2011; Rothman and Northcraft 2015). With regard to the level of its valence, anger is quite contrary to happiness or emotional neutrality: anger is an unpleasant emotion, while happiness is a quite pleasant one and emotional neutrality lies exactly inbetween (see chapter 2.3.1.2), clarifying why anger is frequently set into relation to the two.

The high importance of emotions in negotiations is mirrored by a plethora of research studies conducted; however, a number of these empirical studies is characterized by inconsistent findings (Olekalns and Druckman 2014; Overbeck, Neale, and Govan 2010) (see here and in the following chapter 4.3.2). These contradictions hold especially true for the effect that expressions of emotion have on the negotiating counterpart over against the perception of one's own emotions (Kopelman, Rosette, and Thompson 2006; Sinaceur and Tiedens 2006). For example, with regard to the emotions of anger and happiness, studies find on the one hand that anger leads the opposing party to indulge, so that higher concessions can be achieved compared to a display of happiness (van Kleef et al. 2006; van Kleef, De Dreu, and Manstead 2004a; Sinaceur and Tiedens 2006). On the other hand, it is shown that the display of positive rather than negative emotions leads to a higher extraction of concessions from the opponent and a better chance for agreement (Hine et al. 2009; Kopelman, Rosette, and Thompson 2006).

In addition, the lack of consistent findings most likely arises from the various different moderators which are considered by the studies (Barry 2008; Hine et al. 2009; Olekalns and Druckman 2014). For example, if anger targets the negotiator rather than the negotiation object, the probability that the opponent will make concessions decreases and an impasse becomes more likely (here and in the following, van Dijk et al. 2008; Lelieveld et al. 2011; Pesic 2016; Steinel, van Kleef, and Harinck 2008). The opposite holds for happiness, as it yields higher concessions when it is directed at the person rather than at the offer. Furthermore, the power differential between the negotiating parties influences the effect of emotions. In this way, low power individuals concede more to negotiators who express anger and claim less value, than when the latter expresses happiness (van Kleef et al. 2006; van Kleef and Côté 2007; Sinaceur and Tiedens 2006). Finally, a further factor that impacts the effect of emotions is the chosen negotiation medium; thus e-negotiations are more successful if bargainers express positive emotions rather than negative emotions (Brett et al. 2007; Friedman et al. 2004; Olekalns and Druckman 2014).

Multiple negotiation scholars offer literature reviews that qualitatively summarize the role of emotions and their moderators in negotiations in order to deliver a clearer picture of their impact (e.g., Barry 2008; Druckman and Olekalns 2008; Hunsaker 2017; van Kleef and Côté 2018; Li and Roloff 2006; Olekalns and Druckman 2014; Thompson et al. 2001). However, as already criticized in chapter 3.1.1, qualitative literature reviews are based on subjective assessments as the selection and evaluation of the reviewed articles take place by the researchers. Particularly, literature reviews, lacking as they do the possibility to calculate an overall reliable effect of the reviewed studies, have no possibility to depict an integrated view on the topic on which further scholars can rely. They can no more than to verbally list the diverse results, and are incapable of any quantitative synthesis of the empirical findings.

Taking together all these aspects, a meta-analysis on the diverse study results of emotions on negotiation outcomes seems to be meaningful and overdue, in order to determine an overall effect and to identify influencing moderators.

3.2.2 Research Gaps

This literature review on meta-analyses focusing on personal influence factors in negotiations has revealed a shortcoming with regard to the examination of situative factors. Much as situative factors play a major role in negotiations, there is only one single meta-analysis in the literature that deals with the topic of trust. In the previous chapter, emotions were identified as a situative factor for which an aggregation of study results appears to be particularly important. From this circumstance there arise two central research gaps which would strongly benefit from further analysis. These gaps are (1) the lack of an overall effect of emotions on negotiation outcomes and (2) the lack of an approach of pooling moderating variables. Each research gap will be considered in more detail in the following.

(1) Lack of an Overall Effect of Emotions on Negotiation Outcomes

As shown in the previous chapter, a plethora of studies exists that examines the role of emotions – in particular anger and happiness – on negotiation outcomes. These studies are particularly lacking in consistent findings, arriving as they do at diverging impacts that both emotions exert on the negotiated outcomes. Owing to the fact that qualitative review studies fail to quantitatively aggregate empirical study results, such an approach – which is a meta-analysis – remains a desideratum in the literature. Two related constructs of emotions have been reviewed meta-analytically by negotiation scholars, however, both need to be clearly delimited from emotions. The one construct investigated by the meta-analysis of Sharma, Bottom and Elfenbein (2013), is EI in negotiations (see Table 3-2). EI is a psychographic factor that targets the accurate recognition and processing of emotions by an individual (Hume 2012). Consequently, the construct of emotional intelligence does not distinguish between different emotions, nor does it provide any consideration of their diverse influences on negotiation outcomes. The other construct is considered by Lyubomirsky, King and Diener (2005), who synthesize the study results of the related concept of mood (see Table 3-2). But mood – as a psychographic factor which is not triggered by a specific object, and as something that is less intensive than emotions and that cannot be subdivided into discrete conditions in the way possible emotions can – is clearly not the same thing as emotions (see chapter 2.3.1.2).

(2) Lack of Jointly Investigating Moderating Variables

As discussed above, a considerable amount of moderating variables exist that influence the relationship between emotions and negotiation outcomes. The main effect is influenced differently, depending on which moderator one is dealing with. Nonetheless, the contextual dependence of the effect of emotions in negotiations has only been empirically taken up by primary studies until now. It is characteristic of these studies that both their sample size is small (compared to meta-analysis' sample size) and that they are restricted to the consideration of only one or a few moderators. Thus, an approach is lacking that would pool important moderating variables that may exert an influence on the main effect. This approach can be delivered by a meta-regression analysis. As a component of meta-analysis, meta-regression allows for the collocation of a variety of moderators. Due to the aggregation of the sample sizes of the primary studies, the effect of the moderators can be deduced from a much larger amount of individuals, which in turn leads to greater reliability. This permits the investigation of the overall main effect to be complemented by the valuable information of a plethora of moderating variables.

To sum up: for the purpose both of revealing a reliable overall effect and of examining important moderating factors that have an influence on this effect, the need exists for a quantitative synthesis of empirical research findings capable of examining the influence of emotions in negotiations. This dissertation aims to close the research gaps identified by addressing them in Study II (see chapter 5). Study II focuses on the most frequently examined emotion anger – compared to happiness as second most investigated emotion and emotional neutrality – as well as on the interpersonal effect of emotions. As a well-established method, a meta-analysis is implemented which aggregates the effect sizes of the investigated studies. Seeing as emotions shape both EOs and SOs, Study II addresses each of them as separate DVs (see here and in the following Table 3-2). In addition, a meta-regression analysis is conducted that accounts for the influence of moderating variables (e.g., power, integrativeness and study characteristics).

The meta-analysis of this thesis joins the higher ranks of the other meta-analyses conducted for purposes of investigating individual variables with regard to the examined quantities. With its consideration of 94 studies encompassing 267 effect sizes and 16,795 individuals within the sample, Study II is among the most comprehensive meta-analyses implemented in this field. Based on such a large dataset, Study II enables the dwarf to grow, as it empowers the author of this thesis to look further and to generate additional knowledge by standing on the giants' structured shoulders. Nevertheless, without the giants' prior work, this would not be possible.

3.3 The Basic Decision: Strengthening the Shoulders or Growing the Dwarf?

The previous chapters depicted on the one hand the need for an adequate structuring of the various single findings yielded by the detailed research that has accumulated in the field of negotiations. Negotiation research stands to benefit enormously from such an approach, as it would strengthen the shoulders of the negotiation research giants', thus making them more resilient. On the other hand, it was underscored that the derivation of a clear overall effect of *emotions* on negotiation outcomes is disturbed by the contradictory findings on this personal influence factor. This would justify growing the dwarf, who could be carried upward by the additional insights gained through an aggregation of the inconsistencies. So the basic decision as to whether to strengthen the shoulders of the giants or to grow the dwarf can easily be made: Let us do the one without neglecting the other – as the field of negotiation research would profit equally from both.

In the following, two empirical analyses will be conducted in order to close the gaps presented in chapter 3 and consider both perspectives – the giants' and the dwarf's. The approximate course of both studies and the way they interrelate is visualized in Figure 3-1. The first study seeks to reveal the structure and evolution of negotiation research (chapter 4). To this end, it implements the two bibliometric methods of citation analysis and co-citation analysis. The citation analysis forms the basis for the co-citation analysis by extracting the most-cited publications, a portion of which are used by the co-citation analysis, which in turn identifies the main topics of the negotiation research field.

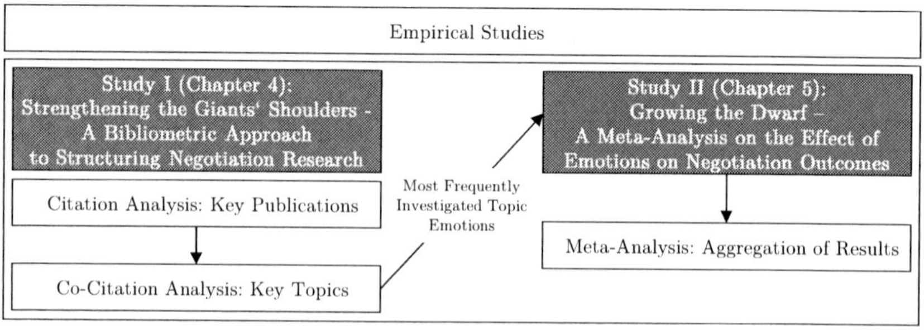

Figure 3-1: Basic Structure of Empirical Studies
Source: Author's own illustration.

The two empirical studies complement each other by Study II building on the results of the co-citation analysis from Study I, which identifies the topic of *emotions* as the most frequently investigated negotiation theme over the last two decades (see chapter 4.3.2). As this topic consists of multiple empirical studies with diverging results, the aim of Study II is an aggregation of empirical study findings, in order to determine clear insights pertaining to the influence of emotions on negotiation outcomes (see chapter 5). As a methodology that quantitatively synthesizes (partly) contradictory findings of independent primary empirical research studies for the purpose of determining a single overall effect size, meta-analysis is the methodology most adequate to this approach (Bijmolt and Pieters 2001; Chow 1987). As the most frequently examined emotion in negotiation studies, anger will be the object of this meta-analysis and it will be compared to the second most frequently investigated emotion happiness and emotional neutrality (see chapter 3.2.1). Moreover, multiple moderator variables, which influence the main effect (e.g., the power level of the emotion receiver or the competitiveness of the situation), will be integrated into the model.

The combination of co-citation and meta-analysis is somewhat new in the literature; it is uncommon for the two approaches to be implemented together. However, there are a number of studies that work out how meaningful the combination of both approaches is (Belter 2016; Janssens et al. 2019; Janssens and Gwinn 2015; Zupic and Čater 2015). In this regard, Janssens and Gwinn (2015) investigate in how far a co-citation analysis is able to generate the set of relevant studies to be integrated in a meta-analysis. To do so, the authors compare the results of their co-citation analysis for a specific topic with a self-determined dataset of studies which are considered relevant for meta-analysis. The scholars thereby find that a considerable amount of 82% can be retrieved through co-citation analysis. As finding all eligible studies is one of the most time-consuming parts of meta-analyses, the prior application of a co-citation analysis is an efficient strategy for the reduction of effort (Mullen and Miller 1991).

The second advantage arising from a previously conducted co-citation analysis is that it provides a content basis for the meta-analysis. Since co-citation analysis deduces highly defining research topics of a field by incorporating large sets of scientific publications (Zupic and Čater 2015), these topics can be used to determine the theme of a meta-analysis. Moreover, with an increasing number of publications within the research topics, the probability of inconsistent findings among the studies increases, which leads to scientific debates. Since meta-analyses target research topics for which many different and inconsistent empirical findings exist (Cooper and Hedges 2009; Geyskens et al. 2009), this requirement can adequately be covered by a preceding co-citation analysis.

4 Study I: Strengthening the Giants' Shoulders – A Bibliometric Approach to Structuring Negotiation Research

4.1 Research Objectives and Study Overview

The first study conducted in this thesis aims at establishing the structure of negotiation research and its development over time in order to strengthen the shoulders of the negotiation research giants. Other structuring studies that have developed an overview of the research field display at least one of three deficiencies: They concentrate on specific topics; they do not consider a temporal development; they qualitatively review a limited quantity of negotiation literature (see chapter 3.1.1). This is why the thesis at hand wants to enrich negotiation research by accounting for all these aspects. By quantitatively evaluating the referencing behaviour of scholars and thereby considering evolutionary trends in most-cited negotiation references pertaining to content, this study is the first to deliver such an approach. Thereby, it bases on a large dataset of 66,294 references.

To do so, the most-cited publications and defining topics are determined, as is their development across four consecutive time frames. These time frames are deduced by reference to the classification of negotiation research presented in chapter 2.2. Bibliometric methods are typically well-suited to the above mentioned purposes (e.g., Backhaus, Lügger, and Koch 2011; Ding, Chowdhury, and Foo 2000; Morlacchi, Wilkinson, and Young 2005). Bibliometrics allow for the evaluation of quantitative bibliographic data on publications, and thereby enable the analysis of the most-cited references, detect changes and developments of the research discipline and identify key topics (Fernandez-Alles and Ramos-Rodriguez 2009; Verbeek et al. 2002).

Through the development of the Internet and the resulting access to huge electronic bibliographic databases, the scope and popularity of bibliometric methods have seen a strong increase (Cronin 2001), with many studies within the social sciences making use of bibliometrics in order to structure diverse disciplines (Schäffer, Binder, and Gmür 2006; Zupic and Čater 2015). The growing number of scientific publications over recent decades makes a bibliometric approach particularly attractive, as this approach enables easy structuring and aggregation of large quantities of articles (Jinha 2010; Larsen and von Ins 2010; Samiee and Chabowski 2012). Hence a bibliometric examination of the negotiation research field distinguishes itself as an adequate methodology in accounting for the large and still growing number of negotiation studies.

By doing so, scholars are able to get concise information about the most relevant and representative negotiation publications and topics, their interconnectedness and their evolution over time enabling future studies to draw on the results gained in this one. This will lead to affiliations of empirical studies capable of delving into smaller negotiation topics that on account of their size would otherwise remain under-investigated. But by the same token quite impactful negotiation topics with many (contradictory) findings could be used in further studies to aggregate the different findings to one overall effect. Indeed, the thesis at hand is an instance of the second case, as Study II conducts a meta-analysis that synthesizes the findings of the most frequently investigated topic as identified in Study I.

Study I is structured in the following way: Firstly, the methodological background and research design are presented in chapter 4.2, in which the reader receives an introduction to bibliometric methods in general (chapter 4.2.1), and to citation and co-citation analysis in particular (chapter 4.2.2). Subsequently, the sample and data collection proceeding are described (chapter 4.2.3). Chapter 4.3 presents the results of Study I by differentiating between the results of the citation analysis (chapter 4.3.1) and the results of the co-citation analysis (chapter 4.3.2). Finally, Study I closes with a discussion of the results in chapter 4.4.

4.2 Methodological Background and Research Design

4.2.1 Introduction to Bibliometric Methods

Bibliometrics were first introduced by Pritchard in 1969 as "the application of mathematical and statistical methods to books and other means of communication" (Pritchard 1969, p. 349). Since then they have been established as a frequently applied form of analysis for literature on which many research areas rely (e.g., Aparicio, Iturralde, and Maseda 2019; Backhaus, Lügger, and Koch 2011; White and McCain 1998). Bibliometrics (usually) encompass the collection, processing and analysis of quantitative bibliographic data from scientific publications (here and in the following Verbeek et al. 2002). Scientific publications can be any kind of written material either in physical or digital format that deals with scientific research activities (e.g., books, conference proceedings and articles in scientific journals). Since a research discipline is developed on the basis of a collection of relevant studies, the analysis of the most influential ones reveals the change and growth of that discipline as well as different research streams within it, and allows for the identification of key topics (Fernandez-Alles and Ramos-Rodriguez 2009). Thus the main aim of bibliometric methods is the examination of the structure of a specific scientific discipline (Roth and Gmür 2004; Schäffer, Binder, and Gmür 2006).

Bibliometric analyses can be implemented either as a snapshot of a specific time period, or they can be used to describe a research field over a specified time frame with multiple periods (Gmür 2003; Samiee and Chabowski 2012; Walter and Ribière 2013). With regard to the determination of the period lengths, it is necessary to choose an adequate time span. Since research contributions need time to establish themselves, examined periods should not be shorter than five years (van Raan 1996).

Bibliometric techniques are characterized by several key advantages. First, their quantitative approach enhances the results of qualitative literature studies (Marsilio, Cappellaro, and Cuccurullo 2011). As worked out in chapter 3.1, this strong point encompasses the reduction of subjectivity, in particular (Samiee and Chabowski 2012; Wang et al. 2016). Second, bibliometrics adequately mirror and validate the intuitive opinions of experts (Marsilio, Cappellaro, and Cuccurullo 2011; Nerur, Rasheed, and Natarajan 2008; White and McCain 1998). Lastly, since the processing and analysis of the data take place by statistical methods and thus make human manpower superfluous, such a quantitative approach allows for a much broader coverage of publications within the analysis (Samiee and Chabowski 2012).

In this way, bibliometrics are especially suited if the underlying dataset is large, which holds in particular for well-studied research fields (e.g., 100,000 references). Such an amount of bibliographic data is definitely too high to be processed manually. This fact is visualized by point P_1 in Figure 4-1, which considers the level of information capable of being revealed from bibliographic data in dependence on the degree of applied simplification or data reduction. Therefore, unless a data reduction takes place, the level of information stays on a rather low level.

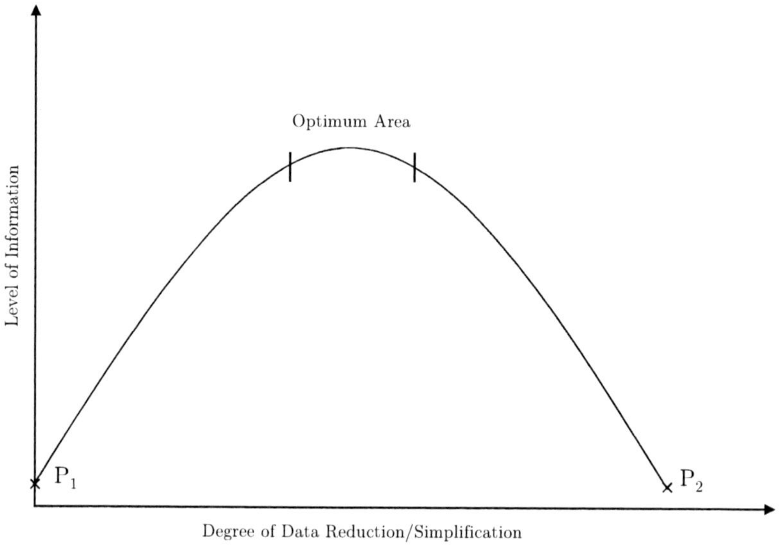

Figure 4-1: Relationship of Data Reduction and Information Level in Bibliometrics
Source: Author's own illustration.

Bibliometrics draw on this, and with their multiple multidimensional reduction methods offer the possibility to boil down the huge number of data to an evaluable size. The data reduction involved takes place in such way that the information from the various references is condensed. By doing so, bibliometrics allow for a strong increase in the level of information (see here and in the following Figure 4-1). At a certain level of simplification, they reach an optimum area in which the information level can be maximized, as an ideal amount of data reduction is achieved. However, if the data are over-simplified, it is at the expense of the level of information, so that in an extreme case one reaches point P_2 in Figure 4-1. Thus, the relationship between the level of information and the reduction of data is described best by an inverted U-shape curve. Accordingly, the major advantage of bibliometrics lies in their ability to process a huge bibliographic dataset in such a way that the data are simplified and condensed, while the level of information is optimized. It is, however, first necessary to determine the optimal data reduction area in which this state holds.

Manifold bibliometric analysis methods exist that differ from one another in the information base used for their investigation. The analysis methods commonly used are co-word analysis, co-authorship analysis and citation-based analysis (van Eck and Waltman 2014). Co-word analysis examines the number of articles in which two words co-occur either in the title, the abstract or the keywords, and determines in this way the interrelations of the co-occurring words (Leung, Sun, and Bai 2017; Whittaker 1989; Zupic and Čater 2015). Thus, the analysis is based on the actual content of the documents and delivers research concepts, but does not allow for the detection of the scientific structure of the field (e.g., interconnection of publications) (Zupic and Čater 2015). Co-authorship analysis connects authors with each other whenever they co-author, and is in fact limited to collaboration between authors (Raasch et al. 2013; Zupic and Čater 2015).

Citation-based analyses can further be divided into citation analysis, co-citation analysis, and bibliographic coupling (van Eck and Waltman 2014), whereby citation and co-citation analysis are the bibliometric methods used most frequently in bibliometric studies (here and in the following Zupic and Čater 2015). Citation analysis determines the influence of publications on the basis of their citation counts. Co-citation analysis and bibliographic coupling both seek to examine the intellectual structure of the research field by determining relationships and main themes; but they differ in their approach and focus (Backhaus, Lügger, and Koch 2011; Batistič, Černe, and Vogel 2017; Zupic and Čater 2015). While co-citation analysis measures the similarity between cited documents, bibliographic coupling determines the similarity between citing publications (Small 1973; Vogel and Güttel 2013). Thus if two citing publications point to the same reference, they are coupled (here and in the following Vogel and Güttel 2013). It follows that bibliographic coupling examines the relationships between more recently published documents, while co-citation analysis investigates older structures. Hence bibliographic coupling is especially suited for the detection of current trends, as opposed to co-citation analysis, which is particularly suitable for long-term analysis (Glänzel and Thijs 2012; Vogel and Güttel 2013).

As explained in chapter 3.1, the thesis at hand aims at revealing the intellectual structure and main topics of the negotiation research field from a retrospective, long-term perspective. Due to the previously described aspects, co-citation analysis is suited best for this purpose and will be implemented in the following. As a rule, citation and co-citation analysis are implemented jointly, and the citation analysis forms the baseline for the co-citation analysis (e.g., Leung, Sun, and Bai 2017; Marsilio, Cappellaro, and Cuccurullo 2011; Wang et al. 2016). In this dissertation the same approach will be taken. The typical procedure adopted in citation and co-citation analysis is visualized in Figure 4-2 and will be described in the following chapter.

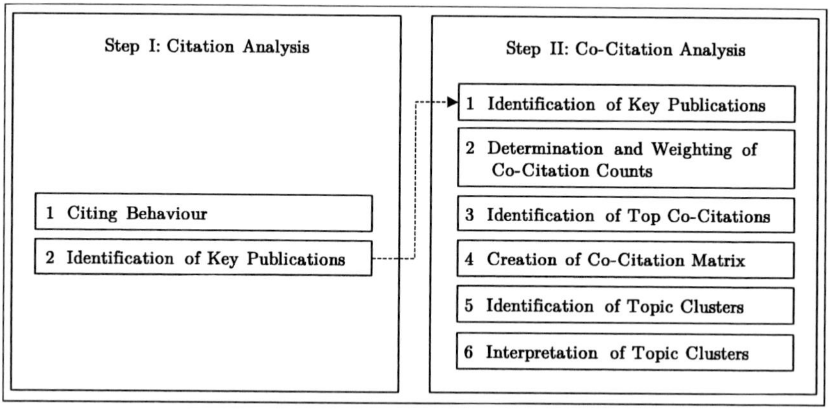

Figure 4-2: Typical Procedure of Citation and Co-Citation Analysis
Source: Author's own illustration.

4.2.2 Citation and Co-Citation Analysis: Two Tools of Key Importance to Explore Central Topics and Publications

4.2.2.1 Citation Analysis

There are various reasons for authors to make citations (e.g., the substantiation of statements or the criticism and correction of previous (own) work) (Garfield 1965). Citations express a relationship between the cited and the citing document, thus positing reliable and valid indicators for scientific interaction between researchers (Garfield 1979; Smith 1981). Citation analysis usually starts by providing key figures emerging from the overall citing behaviour displayed by the citing documents (here and in the following Backhaus, Lügger, and Koch 2011). These key figures deliver a first impression of the data basis and depict the relationship between the citing articles and their references. These key figures are 1) the determination of the average number of references per article, 2) the average age of references and 3) the self-citation ratio.

The main assumption of citation analysis is that the *frequency* by which a publication is cited represents its *relevance* and *influence* on the research field (Culnan 1986; Üsdiken and Pasadeos 1995). Consequently, highly cited research is seen as more influential for the research field. However, the frequency with which a publication is cited does not consider whether the citation is made because of approval or criticism by the citing paper (here and in the following Roth and Gmür 2004). Nevertheless, the cases of criticism are rather rare and carry hardly any weight in the big dataset of citations examined in citation analysis, so that these cases are negligible. It follows that citations are a measure of impact, and that the citation analysis' main objective is to identify a research field's most influential publications (Zupic and Čater 2015).

Usually, studies conducting a citation analysis provide top x-lists of the most-cited research (Zupic and Čater 2015). These top x-lists draw in the present citation analysis on the so-called *citation value (CV)*, which sets the individual citations of a publication in relation to the number of works examined (here and in the following Backhaus, Lügger, and Koch 2011). Thus, the CV increases with the number of citations a publication gets. Consequently, a higher CV mirrors higher importance of the unit of analysis under consideration. In addition, the CV is normalized to values between 0 and 1. In cases of citation analysis over multiple periods, this normalization accounts for the varying number of works examined per period and thereby allows for a comparison between the different time frames.

Typically, citation counts are weighted equally in most citation studies (Verbeek et al. 2002). The only citations that can possibly form an exception are self-citations (Smith 1981; Verbeek et al. 2002). Self-citations can have diverse motives. On the one hand, self-citations can be used to rely on previous relevant work representing real influential citations and contributions (Porter 1977). On the other hand, they may also follow self-serving aspects with the intent of increasing one's own popularity and accumulating citation counts, which would lead to an inflation of citation scores (Porter 1977; Verbeek et al. 2002). The literature displays different procedural modes with regard to self-citations. While self-references receive equal treatment like other citations from a number of studies (e.g., Pilkington and Meredith 2009), there are other studies that give them less weight (e.g., Backhaus, Lügger, and Koch 2011), or even exclude them from analysis altogether (e.g., Üsdiken and Pasadeos 1995).

Nevertheless, a citation analysis lacks the possibility to identify networks of interconnections. This is the purpose of a co-citation analysis, which analyses the similarity between publications on the basis of their co-occurrences in bibliographies in order to detect their interconnections (Üsdiken and Pasadeos 1995).

4.2.2.2 Co-Citation Analysis

Characteristics of Co-Citation Analysis

A co-citation analysis reveals relationships between scientific publications by examining their referencing behaviour (here and in the following Schäffer, Binder, and Gmür 2006; Vogel and Güttel 2013). The main assumption of the analysis is that the number of co-citations represents the degree of content-based proximity between two publications. A co-citation exists if two publications are cited together in a document, that is, if both are listed in the bibliography of the citing document (Small 1973) (see Figure 4-3). Consequently, the higher the number of articles that cite two publications jointly, the higher the co-citation count becomes for these two publications.

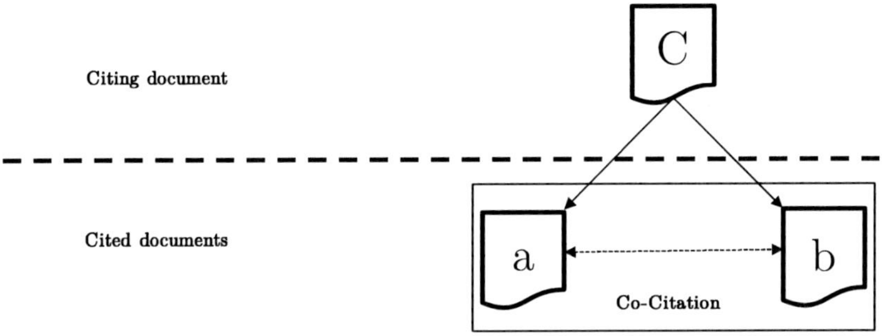

C = citing document; a = document a within reference list of C;
b = document b within reference list of C

Figure 4-3: Explanation of a Co-Citation
Source: Author's own illustration adapted from Vogel and Güttel (2013) and Zupic and Cater (2015).

The main purpose of a co-citation analysis is to determine the intellectual structure of a research field (McCain 1990; Nerur, Rasheed, and Natarajan 2008; Small 1980). The technique allows for the identification of groups of scientific publications which are representative of the main themes of the research domain (Backhaus, Lügger, and Koch 2011). This was verified by Mullins et al. (1977) and McCain (1986), who examined the performance of co-citation analysis by asking researchers whether or not the identified co-citation networks represent their perception of the research field. The big advantage of such an approach is that co-citation networks are research networks that have been created unconsciously by the citations of publications, rather than by formal organizational links (Crane 1972; Vogel 2012).

The co-citation counts are determined for a fraction of the most-cited publications (here and in the following Backhaus, Lügger, and Koch 2011).[17] At this point, the co-citation analysis resorts to the results of the citation analysis, which has produced a ranking of the CVs of the publications. For this purpose, a cut-off value needs to be determined based on the publications' respective CVs and marking the minimum value required for

[17] In the present thesis, the bibliometric software Bibexcel is used in order to extract the co-citation counts (Persson, Danell, and Wiborg Schneider 2009).

a publication to be integrated into the following steps.[18] Thus, only highly relevant documents are included in the analysis. This ensures that the optimum area presented in chapter 4.2.1 is reached, in which the data reduction is implemented such that the level of information can be maximized.

The usage of absolute co-citation counts for the purpose of further analysis is not recommended in the literature, since the raw counts do not generate accurate and distinct cluster results (Boyack, Klavans, and Börner 2005; Gmür 2003; Klavans and Boyack 2006; Perrey, Schäffer, and Becker 2012). Thus, a normalization of the raw co-citations is required (Zupic and Čater 2015). The normalization metric most commonly used is the Pearson correlation coefficient (r) (e.g., Chang, Huang, and Lin 2015; Kärki 1996; Vogel 2012; White and McCain 1998).[19] It increases with each 1-1 combination (both publications are cited by the same article) and each 0-0 combination (neither of the publications is cited by the article), but decreases for 0-1 and 1-0 combinations (Gmür 2003). Nevertheless, use of the Pearson correlation coefficient has been the subject of controversial debates. While some researchers outline that Pearson's r is a valid and robust measure of similarity (White 2003), others reveal that the correlation coefficient is not the right choice if a study wants to find clusters of similar objects based on co-citation patterns – which is the main aim of co-citation analysis (Ahlgren, Jarneving, and Rousseau 2003). In addition, Pearson's r equal treatment of 1-1 and 0-0 combinations is quite misleading, as the information that a co-citation exists is much more valuable than the information that both articles were not cited together.

In order to circumvent the problems stated above, another standardization approach, the so-called *Co-Citation (CoCit)-Score*, is highly suitable and is receiving increasing attention from researchers (e.g., Backhaus, Lügger, and Koch 2011; García-Lillo et al. 2017; Gmür 2003; Meyer, Lorscheid, and Troitzsch 2009; Perrey, Schäffer, and Becker 2012; Raasch et al. 2013). The CoCit-Score is calculated as the ratio of raw co-citation counts

[18] The optimum amount of publications to be considered for the present co-citation analysis is determined in chapter 4.3.2.

[19] Several other standardization measures exist, such as Salton's cosine or the Jaccard index, which are not considered in this study. For more information on these measures please refer to e.g., Leydesdorff (2008) or Schildt, Zahra, and Sillanpää (2006).

and the minimum and mean values of the individual citations (Roth and Gmür 2004) (see Formula 4.1), whereby it is normalized to values between 0 and 1 (Gmür 2003):

$$CoCit_{ab} = \frac{(Co - Citation_{ab})^2}{Minimum\ (Citation_a : Citation_b)\ x\ Mean\ (Citation_a : Citation_b)} \quad (4.1)$$

$CoCit$ = Co-Citation-Score
a = document a
b = document b

The CoCit-Score exhibits several advantages: It considers the possible unequal weighting of co-citation counts by accounting for the direct citation effects. Thus, the CoCit-Score allows for a much greater robustness of results (Gmür 2003). Besides, it only measures actual co-citations and does not include 0-0 combinations in the calculation. Moreover, it is suited best for the main aim of co-citation analysis, which is the identification of communities and research areas within the discipline (here and in the following Gmür 2003). Lastly, it is also convenient with regard to the identification of those documents that are most representative of the identified communities. It is for these reasons that the thesis at hand uses the CoCit-Score as its normalization method of choice. Thereupon, the top co-citations, that is, those co-citations that exhibit the highest CoCit-Scores, are identified.[20] This step ensures for a second time that the optimum area presented in chapter 4.2.1 is reached, and that only relevant publications will appear in the co-citation networks.

Analysis and Visualization

For the co-citation relationships with the highest CoCit-Score, the normalized co-citation counts are arranged in a matrix structure (here and in the following Zupic and Čater 2015). This matrix represents a similarity matrix that serves as the dataset for the following analysis. In order to identify the subgroups of the research field, various dimensionality reduction methods exist for processing the matrix. The most commonly used analysis methods are exploratory factor analysis (EFA), multidimensional scaling (MDS), hierarchical cluster analysis (HCA) and social network analysis (SNA) (Cobo et al. 2011; Zupic and Čater 2015). However, by applying EFA, a clear distinction between topic

[20] Also the threshold for the top co-citations of the present analysis is determined in chapter 4.3.2.

groups becomes difficult, as objects can belong to multiple groups, which constitutes a challenge to the interpretation (Zupic and Čater 2015). In cases of high structural complexity in particular, EFA does not generate a true representation of the subgroups within the research field (Gmür 2003). Since the dataset of the present co-citation analysis is complex due to its size, this drawback is circumvented by relying on MDS, HCA and SNA.

In a conceptual two-dimensional space, MDS optimizes distances between publications by solving the minimization problem posed by distances between pairs of articles having to match the given similarities as closely as possible (Börner, Chen, and Boyack 2005; Zupic and Čater 2015). One especially frequently employed means of such optimization is the so-called *spring-embedded algorithm*, which represents an advancement of standard MDS (e.g., Calero Medina and van Leeuwen 2012; Pilkington and Meredith 2009; Vogel 2012). By means of spring-embedded algorithms, the distances are approximated based on the path length between any pair of publications, which in turn is determined by the number of other publications between the pair (Vogel 2012; Vogel and Güttel 2013). The algorithms cause a force of attraction, which brings connected publications together and thus positioning those publications closer together that are highly co-cited and resemble one another (Backhaus, Lügger, and Koch 2011; Calero Medina and van Leeuwen 2012). Consequently, the spring-embedded algorithm is especially helpful in visualizing network structures (here and in the following Zupic and Čater 2015). These algorithms, however, do not automatically assign publications to a specific group. For this purpose, HCA is best suited.

HCA offers the possibility of placing similar publications into a group or cluster, respectively (Zupic and Čater 2015). The underlying cluster algorithm provides no single partitioning, but rather a hierarchy of cluster solutions, from which the researcher has to determine the optimal one; for example, by examining the merging process graphically in a dendrogram, or by applying stopping rules (Backhaus et al. 2016; McCain 1990; Zupic and Čater 2015). Hierarchical cluster analysis can be further divided into agglomerative and divisive algorithms (here and in the following Backhaus et al. 2016). While agglomerative algorithms assume the finest partitioning of objects prior to clustering, (id est (i.e.), each object represents its own cluster), divisive methods spring from a rough partitioning of objects (i.e., all objects are in one cluster). Agglomerative algorithms are by far more studied than divisive methods; however, they exhibit some weaknesses with regard to the detection of communities, and thus they tend to neglect peripheral objects even when it is obvious that they belong to a community (Newman and Girvan 2004).

A more recent approach applied to co-citation analysis is *social network analysis* (e.g., Pilkington and Meredith 2009; Vogel and Güttel 2013; Wang et al. 2016). The aim of social network analysis is to reveal the interconnections between individuals in order to depict the structures of specific societies (Borgatti et al. 2009). This approach is highly similar to that of co-citation analysis, as the latter detects interconnections between publications or authors for the purpose of identifying the structure of the respective research field (Backhaus, Lügger, and Koch 2011; Batistič, Černe, and Vogel 2017). A network consists of so-called *nodes*, which represent the unit of analysis (such as publications, in the case of a co-citation analysis) and *edges*, which symbolize the (co-citation) relationships between the nodes (Otte and Rousseau 2002; Zupic and Čater 2015). An exemplary network is visualized in Figure 4-4, whereby the nodes are depicted by bubbles and the edges by the connecting lines. Network analysis differentiates between directed and undirected networks, depending on whether the direction of edges is important for the structure (Otte and Rousseau 2002). In the case of co-citation analysis only undirected relationships are meaningful, since by definition co-citations do not provide any directions. Therefore, edges between the nodes are simple lines without any arrows.

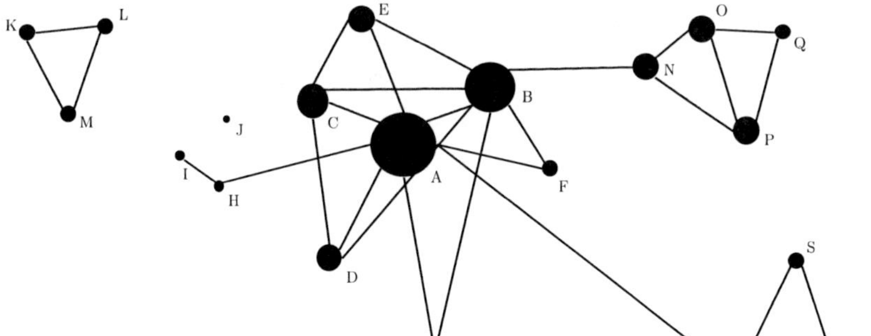

Figure 4-4: Exemplary Co-Citation Network
Source: Author's own illustration.

In addition, networks can consist of so-called *pendants* and *isolates*. Pendants are nodes linked by a single edge to another node (here and in the following Hanneman and Riddle 2005). They thus represent the end node of chain-like extensions and can be compared to an appendage. In Figure 4-4, this holds for node I. Isolates are nodes that are not linked to other nodes at all and hence are even less connected than pendants (in Figure 4-4 node J). As the emphasis of co-citation analysis is on the exposure of community structures, pendants and isolates are not defining for interconnected relationships; on the contrary: they even disturb the focus on the core network, as multiple pendants and isolates make it more difficult to indicate the more relevant cases that are linked to each other at least twice. For these reasons, pendants and isolates are commonly deleted in co-citation analyses (e.g., Backhaus, Lügger, and Koch 2011).

One of the most studied concepts in social network analysis is *centrality*, which determines the importance of a node in a network (Borgatti 2005; Valente et al. 2008). Many different kinds of centrality measures exist, of which *degree* and *betweenness centrality* are two common measures (Valente et al. 2008). Degree centrality is defined as the number of edges that a node exhibits, and is therefore an indicator of this node's impact within the network (Borgatti 2005). Degree centrality can be visualized in the network by the size of the node, whereby a larger node represents a higher number of edges and thus a higher degree centrality (Jeong, Song, and Ding 2014) (see Figure 4-4). Betweenness centrality is defined as the number of times one specific node needs to be passed in order to get from one node to another node via the shortest way (here and in the following Freeman 1979; Wang et al. 2016). In other words, betweenness centrality measures the frequency by which one node lies on the shortest paths between two other nodes. Nodes with a high degree of betweenness centrality are those which are of relevance for the knowledge flow of the network (Wang et al. 2016).[21]

On the basis of betweenness centrality, Girvan and Newman developed a divisive clustering algorithm that addresses the problem of the agglomerative clustering methods presented above (Girvan and Newman 2002; Newman and Girvan 2004). For their approach, Girvan and Newman adapt the standard definition of betweenness centrality to

[21] For an explanation of other centrality measures and more detailed information on social network analysis in general, see for example Borgatti, Everett and Johnson (2018).

edge betweenness, which counts the number of times an edge builds the shortest path between a pair of nodes (here and in the following Girvan and Newman 2002; Wallace, Gingras, and Duhon 2009). The Girvan-Newman algorithm divides the starting partition (all objects in one cluster) into subgroups by cutting the edges with the highest betweenness centrality. After each removal, the betweenness centrality of the edges affected by the removal is calculated anew, and again the edge with the highest betweenness centrality is removed. This process is repeated until no edge remains. In order to determine the best cluster solution, Newman and Girvan (2004) defined a measure of quality called modularity (Y), which assesses the meaningfulness of each given division into subgroups (Wallace, Gingras, and Duhon 2009).

$$Y = \sum_{j} (e_{jj} - q_j^2) \tag{4.2}$$

Y = modularity
j = subgroup j
e_{jj} = fraction of the edges in the network that link nodes in subgroup j to other nodes in subgroup j
q_j^2 = expected value of the fraction if the connections between the nodes are determined randomly, whereby the quantities of edges in the network and the subgroup division are held constant

The modularity measures the fraction of the edges in a network that connect nodes within the same subgroup, minus the expected value of the fraction when a random distribution of the same number of edges within the same community structure is assumed (here and in the following Newman and Girvan 2004). If Y= 0, the cluster solution delivers no better division than chance would, whereas values approaching 1, which is the maximum, indicate a strong cluster structure. The values usually found in the literature range between 0.3 and 0.7. Thus the higher the modularity, the more this underlying cluster solution should be preferred. Based on the Girvan-Newman clustering algorithm, the exemplary network presented above in Figure 4-4 would be divided into 4 clusters (A-G (after the removal of pendants and isolates); K-M; N-Q and R-T). The Girvan-Newman algorithm is especially strong in detecting communities in complex systems consisting of large numbers of objects, which makes it attractive for the present analysis (Klavans and Boyack 2017; Wallace, Gingras, and Duhon 2009).

The detected subgroups can be assessed on the basis of their *size* and *density*. The size is determined by the number of included nodes (Gmür 2003) (e.g., 7 nodes for the subgroup A-G in Figure 4-4), whereby a larger size goes along with a higher significance of the cluster (Backhaus, Lügger, and Koch 2011; Schäffer, Binder, and Gmür 2006). Density, on the other hand, quantifies the connectivity and proximity of nodes within a community and assigns a level of homogeneity of knowledge to it (Otte and Rousseau 2002). Density is calculated as the ratio of the actual co-citation relationships in a subgroup to the maximum possible number of co-citations (Gmür 2003). The more nodes are connected with each other, the denser the cluster will be (e.g., the cluster A-G in Figure 4-4 has a density of 61.91% which depicts a rather high density (Backhaus, Lügger, and Koch 2011)). Increasing network sizes are usually accompanied by decreasing density because of the commonly weaker connections (Schäffer, Binder, and Gmür 2006).

4.2.3 Sample and Data Collection

Generation and Description of Sample

In order to generate the database for the present analysis, the publication sample needs to be identified from which to draw the bibliographic data. The bibliographic databases Social Science Citation Index (SSCI) and Science Citation Index Expanded (SCIE) are used for this purpose. Besides the fact that these databases are the ones most commonly used for bibliographic studies in business sciences (e.g., Backhaus, Lügger, and Koch 2011; Chabowski, Samiee, and Hult 2013; Di Guardo and Harrigan 2012; Nerur, Rasheed, and Natarajan 2008; Zupic and Čater 2015), they are for a number of reasons more advantageous than other databases. First, they go back to 1945, which makes their usage quite appealing for the present analysis (here and in the following Clarivate Analytics 2019a).[22] Second, SSCI and SCIE are provided via the Web of Science (WOS) user interface, which makes the extraction of the bibliographic data very easy and user friendly.[23] Third, SSCI and SCIE are quite comprehensive, the former containing over

[22] For instance, the second most frequently used database Scopus lacks citation data of publications prior to 1970 (Elsevier 2019; Zupic and Čater 2015).

[23] In contrast, the Google Scholar database lacks an application programming interface that would enable easy extraction of bibliographic information (here and in the following Zupic and Čater 2015). Consequently, it cannot be directly used for bibliometric analysis, which would require an additional programming component for downloading the data.

3,200 journals across 55 social science disciplines, and the latter comprehending 8,850 journals across 150 disciplines (Clarivate Analytics 2019b).

The following information relevant for the subsequent analysis can be collected per article from WOS (here and in the following Clarivate Analytics 2019a): title, author name(s), publication year, keywords, abstract, classification categories, and a list of cited references.[24] With respect to the classification categories, WOS makes so-called research areas and WOS categories available for use. Both classification approaches assign one or more categories to a publication, which allows for a more accurate retrieval and analysis of articles, as this enables them to be filtered according to their categories. The difference between the two classification methods lies in their specificity. While research areas represent rather broad categories, WOS categories are frequently more specific and can be assigned to a research area (e.g., research area: business & economics; WOS categories: business, business (finance), economics) (Clarivate Analytics 2019c; d).

Moreover, the overall time frame to be examined in the analysis needs to be set. Since negotiation research represents such a long-established and comprehensive field, the aim of Study I is to conduct a long-term analysis and to stand out from other studies by using the broadest time frame (see chapter 3.1). 1960 is chosen as the starting point of the analysis, as it makes up the beginning of a specific epoch in negotiation research (see chapter 2.2).[25] Concerning the endpoint of analysis, late October 2018 is set, since this point in time represents the termination of the search. As this study's aim is to reveal the evolution of negotiation research, it further divides the overall time frame into multiple periods. This is especially meaningful, as the overall time period amounts to 58 years in which developments are quite likely. As worked out in chapter 2.2, special branches of negotiation research have already been classified by scholars into specific time periods (e.g., Backhaus, Hummelsberger, and Wilken 2005; Voeth and Herbst 2015b). Based on this division, this thesis classifies the publications into four periods.

[24] In some cases not all information is provided.

[25] It was worked out in chapter 2.2 that negotiation research dates back to the 1940s; but since only a handful of articles focusing on negotiations is provided by WOS prior to 1960, this amount is negligible for the analysis at hand (Clarivate Analytics 2019a).

The first time period to be examined extends from 1960-1979. Within these 20 years, structural factors (e.g., deadlines or the inclusion of third parties) and personal differences between negotiators in relation to the negotiation outcome were the particular focus of the analysis (Bazerman et al. 2000). The next division is from 1980-1999 (Backhaus, Hummelsberger, and Wilken 2005; Bazerman et al. 2000). In this period, behavioural decision research (BDR), which deals with deviations on the part of negotiators from optimal behaviour in an economic sense on account of irrationality, had a strong presence (Bazerman et al. 2000; Kahneman and Tversky 1979). Moreover, social-psychological topics were addressed, such as social relationships or emotions in negotiations (Backhaus, Hummelsberger, and Wilken 2005; Bazerman et al. 2000). The millennium was seen again as a turning point in which especially social and cultural context factors were investigated (Brett and Thompson 2016). However, with regard to the division of the remaining period from 2000-2018, there exists no literature review or anything similar that works out differing focuses and offers a recommendation for classification. One option could be to use this period of 18 years, in order – for the sake of comparison – to have the same 20-year time span as the other two periods. However, negotiation research provides so many more articles for recent times that the portion of articles belonging to this period would far exceed those of period 1 and 2. Consequently, for the sake of having reasonably balanced periods the time frame is further divided into period 3, which ranges from 2000-2009, and period 4, from 2010-2018.

In addition, both the object of analysis and the language of the publications need to be determined. With respect to the object of analysis, it is relied exclusively on journal articles. This decision is made because of the "certified knowledge" journals provide (here and in the following Fernandez-Alles and Ramos-Rodriguez 2009). The fact that the articles are reviewed by fellow researchers prior to their publication provides some kind of guarantee for this certified knowledge, as the articles have gained the approval of other researchers. In addition, journal articles are what WOS offers in large part. The paucity of books, conference proceedings et cetera (etc.) in WOS is negligible. With regard to the language, English is chosen, as it is the leading scientific language, and as it prevails in journals (Maeße, Pahl, and Sparsam 2017; Narvaez-Berthelemot and Russell 2001).

In order to generate the relevant articles, a trinomial approach is implemented, as illustrated in Figure 4-5 and described in the following.

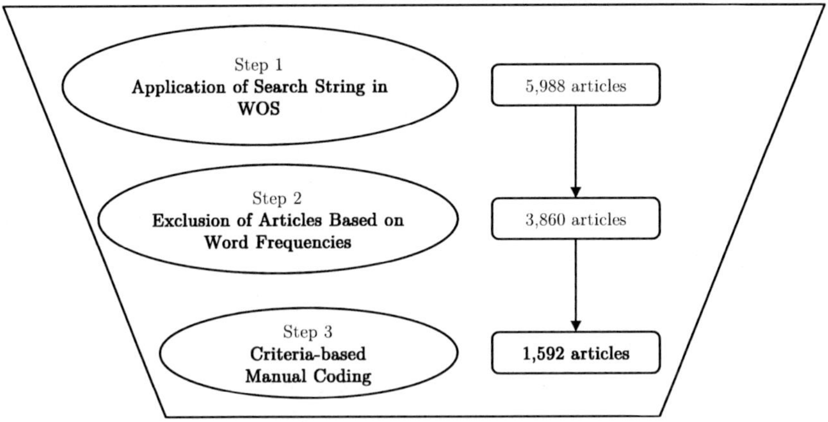

Figure 4-5: Trinomial Approach of Sample Generation
Source: Author's own illustration.

In a first step, a search string is developed for WOS that determines the relevant search words, research areas and WOS categories (see Appendix A-1). Similarly to other negotiation publications that conduct literature searches, the two most common English words for negotiations "negotiat*" and "bargain*" are used as search words in their truncated forms (characterized by the asterisk) (e.g., Agndal, Åge, and Eklinder-Frick 2017; Caputo 2013a).[26] The truncation enables all possible endings of the words to be considered, so that all inflected forms are included in the search (e.g. negotiator, negotiation, bargainer, bargaining etc.). Unlike other studies that search only in the title of publications (e.g., Agndal, Åge, and Eklinder-Frick 2017), the search words are deployed to title, abstract and keywords. This approach is meaningful, as articles dealing with negotiations do not necessarily have "negotiat*" or "bargain*" in the title (e.g. "Lose-Lose Agreements in Interdependent Decision Making" by Thompson and Hrebec (1996)).

Apart from that, the search string specifies the research areas pertinent to the analysis. At this point, it is important to point out again that this dissertation is located within

[26] Further search words related to negotiations are also applied, such as decision making, deal, agreement etc. Nevertheless, their application leads to the inclusion of too many irrelevant articles, and so the decision is made not to consider these terms.

the scope of business economics, and so it aims to gain insights relevant for business negotiations. However, not only business economics proper, but also those research areas which have a formative or shaping influence on it contain business-relevant negotiation topics. According to several scientists, these shaping research areas are psychology, sociology and recently, with the advent of digitization, computer science (Backhaus and König 2019; Herbst 2007; Voeth and Herbst 2015b).[27] Negotiation research in psychology and sociology focuses on the individual and relationship level, which can likewise play an important role for business people and their relationships (Neale and Bazerman 1991; Rubin and Brown 1975; Thompson, Wang, and Gunia 2010). In addition, computer science research on negotiations is suited to business contexts, as the digital era continues to change the business world in undeniable ways, so that technological support and software agents become ever more relevant for business negotiation settings (Backhaus and König 2019). Based on this appraisal of scholars, these shaping research areas are included in the analysis next to business economics itself.

Since these domains are quite broad, not all fields are involved in the analysis in their entirety; instead some are broken down to their subcategories (WOS categories), in order to determine which subcategories to consider. This procedure results in the exclusion of subcategories like psychoanalysis or macroeconomics, which do not represent areas of influence in business negotiations (Agndal, Åge, and Eklinder-Frick 2017). On top of that, the search string is completed by precluding articles that contain terms in their title, abstract or keywords that specifically belong to one of the excluded research areas (e.g. "politic*", "law*", "medic*"). The implementation of the search string identifies a total of 5,988 articles that fulfill the search criteria.

A huge amount of these articles, however, do indeed include the terms "negotiat*" or "bargain*" in the title, the keywords or the abstract, but do not deal with the topic at all. For example the publication "At the intersection of idealized youth and marginalized almost-adulthood: how girls negotiate young motherhood in London" (Hamilton et al. 2018) exhibits the search word "negotiat*" in the title but covers negotiations in no

[27] It cannot be precluded that further aspects relevant to business negotiations can be drawn from other research areas. Nevertheless, a delimitation is necessary, as otherwise the integrated articles would be too far away on average with respect to their content. Thus, it is focused on the research areas worked out by the literature as especially relevant for business negotiations.

sense. The second step of the procedure addresses this problem. In the first place, the frequencies of the terms "negotiat*" and "bargain*" are determined for each article in the respective title, abstract and keywords.[28] Subsequently, the frequencies are summed up per article. Manual sampling is used to examine whether a high number of negotiation words serves as an indicator for an article's relevance to the analysis. In fact, the investigation reveals that a strong positive relationship exists between the quantity of the words and the relevance of the publications. Thanks to the examination, it is especially conspicuous that the publications with "negotiat*" or "bargain*" only once in the title, the abstract and the keywords are the ones that deal by far least with negotiations. By taking this into account it is possible to reduce the total of articles to be examined by 2,128, from 5,988 to 3,860.[29]

However, although this proceeding reduces the amount of irrelevant publications, it still does not suffice for the purpose of generating the final dataset: an examination of the researcher still reveals the inappropriateness of many publications as pertains to content. For this reason, the third and last step entails manual examination of the complete body of remaining articles with regard to their suitability for the analysis. If provided by WOS, title, abstract and keywords of the articles are used as decision support the same as was done in steps one and two of the procedure. If no abstract or keywords are provided, either because they are not supplied by the article itself or because they are not recorded in the WOS database, the whole article is screened. Similarly, if it is not clear whether an article is relevant on the basis of title, abstract and keywords, the article itself is likewise screened. According to a set of inclusion and exclusion criteria predefined by the author of this dissertation, the articles' appropriateness is assessed (for the whole set of criteria see Appendix A-2). These criteria are formulated to give concrete instructions in order to reduce room for interpretation and subjectivity. Exemplarily, several exclusion criteria are presented subsequently:

- Research fields like "law" or "politics" have already been excluded in the processing step before; however, articles tagged with "business" or "psychology" may also deal with

---

[28] If no abstract or keywords are available for a publication in WOS, the publication is not considered for the proceeding of the second step.

[29] It cannot be ensured that there are some relevant articles within the 2,128 excluded publications. However, the manual sampling verifies that the amount is rather small and therefore negligible.

macro-political or labor negotiations, for instance. In this respect, it also holds for the manual coding that publications dealing with such topics as labor-union or political crisis negotiations are not considered for further analysis (e.g., Bacon and Blyton 2002; Bar-Zion 2004).

- Articles should be excluded that honour the life work of negotiation researchers, since these articles focus more on the researcher at hand than on the negotiation topics (e.g., Adair et al. 2016; Tomlinson et al. 2017).

- As the word "bargain" has a double meaning, all articles focusing on the notion of a bargain as a cheap offer should be omitted, too (e.g., Cox, Cox, and Anderson 2005; Kwon, Schumann, and Fairhurst 2010).

- All articles should be excluded from further analysis that do not show a clear focus on negotiations at all. This comprises, among other things, publications that merely deal with negotiations in a subordinated example by focusing, for instance, on mathematic or IT approaches in which negotiations are of low priority by developing algorithms or mathematic solutions to a problem (e.g., Adabi et al. 2013; Tseng and Shih 2018).

- Lastly, articles dealing with auctions are excluded. Much as auctions are seen as a sub-form of negotiations by multiple researchers (e.g., Cao et al. 2015; Citera, Beauregard, and Mitsuya 2005), the nature of auctions contradicts the definition of negotiation in several ways, for which reason their inclusion is not seen as meaningful for this study. These contradicting aspects are on the one hand the auction process (e.g., the time frame, the auction object(s) etc.), which is managed by the auctioneer and thus by only one side, whereas negotiations are formed by all parties included (here and in the following Kersten and Lo 2001). On the other hand, the perceptions and preferences of the included parties can change during negotiation through the communication and interaction of the parties, whereas auctions are clear from this process.

The application of the inclusion and exclusion criteria to the remaining 3,860 articles reduces the data set by a further 2,268 publications, to 1,592 relevant articles. It is noteworthy that more than two-thirds of the publications originate from period 3 and period 4 (see Table 4-1). In contrast, period 1 in particular is characterized by a small sample size of 100 publications. Since the aim of the study is to analyse the references of the identified publications, a total of 35 articles have to be excluded over the entire time period due to lack of references on their part. This results in a final number of 1,557 articles with a total of 67,679 references.

	Period 1 1960 - 1979	Period 2 1980 - 1999	Period 3 2000 - 2009	Period 4 2010 - 2018	Total
Relevant Publications	100	394	549	549	1,592
Unsuitable Publications*	3	17	6	9	35
Final Number of Publications	97	377	543	540	1,557

* Publications without References

Table 4-1: Description of Publication Sample
Source: Author's own illustration.

For each reference, WOS delivers the following information (provided the information is recorded in the database): the surname of the first author and the initials of his or her prename(s) (e.g., Adler NJ), an abbreviated form of the journal, book or conference title (e.g., J Bus Ethics), its publication year and its digital object identifier (DOI) (e.g., DOI 10.1007/S10551-008-9774-4). A DOI represents a distinct and permanent identifier for an object and has as its main application online sources (Chandrakar 2006; Paskin 2010). Thus the big advantage of a DOI is the easy and clear determination of a publication without necessarily having other information about the article at hand (i.e., if one knows the DOI, there is no need to know the title, author, publication year etc. of a publication since all this information is contained in the DOI itself). The structure of a DOI depends on the particular organization (e.g., a publisher) (Chandrakar 2006; Paskin 2010).

Data Cleaning of References

Since the 67,679 references make up the baseline for the citation and co-citation analysis, data cleaning and preprocessing are required prior to the analysis, in order to ensure the quality of the data. And since the citation and co-citation analysis draw on the frequencies of (co-)citations, the data provided per reference needs to be aligned in order to a faultless final frequency count. To this end, a semi-automatic correction procedure is conducted. In a first step, the semi-automatic procedure is composed of an automatic preprocessing step, which is implemented by means of the bibliometric software *Bibexcel* (Persson, Danell, and Wiborg Schneider 2009). Step two, a manual preprocessing of the references, then accounts for such adaptations that cannot be implemented via Bibexcel. The manual step is conducted by the author of this dissertation, along with an assistant.

In summary, correction steps are conducted automatically whenever possible in order to reduce manual effort and biases.

Bibexcel is used especially for the alignment of the authors' names. To allow for different ways of writing an author's name (for instance, in the event the name is sometimes written in all capital letters and other times with only the first initial as a capital letter), names are aligned by adapting them to a consistent format (e.g., ACKERMANN F to Ackermann F). Moreover, only the first name of an author is kept, and middle names are removed (e.g., Adair WL to Adair W). Doing so has the advantage that two references of one and the same person can be identified as identical.

Then follows the implementation of manual preprocessing. First, all references are eliminated that provide no information about the author(s) at all (either because nothing is stated or because the author is anonymous) or that only exhibit a senseless concatenation of letters rather than an author name (e.g., 4%ght). In the style of the (co-)citation analysis of Backhaus, Lügger and Koch (2011), any reference without a specific author is deleted as well, because no possibility exists to attribute the real writer to the publication. Such publications are articles published by institutions and organizations, for example. Secondly, misspellings occurring in author names are corrected. If it is obvious that a specific author is meant and just misspelled, the wrong letters are corrected (e.g., in the event that the author "Thompson" is meant but written as "Thomson" instead, the missing "p" is added). Checks are run for the purpose of avoiding wrong adaptations, i.e., to ensure a name being truly misspelt. Therefore, it is verified with the help of the DOI, whether two similarly written names refer to one and the same author. If no DOI is available, the publication year and journal/book/conference title serve as an identifier to prove a misspelling. If no comparable case exists, the author name is not adapted, because of insufficient knowledge to document a misspelling (staying with the example given above, it is possible that the name "Thomson" is written correctly, so that "Thompson" and "Thomson" are indeed two different authors).

Another adaptation step consists in correcting or completing publication years. Thus a further check is run by DOI – or, if DOI is not available, by author name and journal title – as to whether for identical publications the year is either missing and should be added, or wrong and hence in need of correct adaptation. Furthermore, unpublished references are eliminated, since their missing publication year and DOI would distort the results. In addition, publication years are aligned for different editions of monographs per period. Since multiple editions of monographs are frequently cited within a single period, they need to be adapted in order to be shown as related. For this reason, the editions are adapted to the most recent one in each period, in keeping with the approach of other bibliometric studies (e.g., the book "Essentials of Negotiation" by Lewicki, Barry and Saunders is adapted for the time period 2000-2009 for all references to the latest possible edition in 2009) (e.g., Backhaus, Lügger, and Koch 2011). Those publication years that are adapted on account of a more recent edition are not included for the determination of the average age of references for the citation analysis, as this would distort the results.

Since citation and co-citation analysis draw on the frequencies of the units of analysis, a different assignment of one and the same document would complicate further analysis. For this reason, DOIs are aligned and complemented whenever possible. For instance, DOIs are aligned if different writings occur for what is actually one and the same DOI. This includes such cases as those in which some DOIs are stated with a double occurrence of the term "DOI" in front of the remaining part, as opposed to its occurrence just once in other cases (e.g., DOI DOI 10.2307/1969529 vs. DOI 10.2307/1969529). In addition, some actually identical DOIs differ because of an additional slash exhibited by several DOIs (e.g., DOI 10.1037//0022-3514.40.1.31 vs. DOI 10.1037/0022-3514.40.1.31). Furthermore, if it is obvious that two publications are equal (by virtue, say, of author name, publication year, journal title identical), but only one of them has a DOI, the missing DOI is added accordingly.

The last adaptation step consists in an adjustment of the journal, book and conference title abbreviations. Since not all references provide the same abbreviation for the journal, book or conference titles, different abbreviation versions need to be aligned to one and the same version (e.g., Org Behav Hum Decis vs. Org Behav Human Deci). In addition, this step excludes theses from the dataset. Since WOS indicates no more than that a reference is indeed a thesis (sometimes with further details about the university), any theses included could be inadequately differentiated.

The preceding steps lead to an overall exclusion of 1,385 references, which reduces the overall amount of references from 67,679 to 66,294. Of this amount, 1,478 references belong to the first period, 12,452 to the second period, 21,587 to the third period, and 30,777 to the fourth period (see Table 4-2). These 66,294 references constitute the database for the citation and co-citation analysis.

	Period 1 1960 - 1979	**Period 2 1980 - 1999**	**Period 3 2000 - 2009**	**Period 4 2010 - 2018**	**Total**
Number of References	1,478	12,452	21,587	30,777	66,294
%	2.23%	18.78%	32.56%	46.43%	100%

Table 4-2: Description of Reference Database
Source: Author's own illustration.

In the following analyses, the DOI is used if at all possible, as the clarity with which it identifies a publication enables the most accurate characterization of a document. For those cases which do not provide a DOI, a combination of author surname, publication year and journal/book/conference name is applied.

4.3 Results

Prior to presenting the results, the manner needs to be determined in which self-citations are handled. As presented in chapter 4.2.2.1, multiple ways exist for dealing with self-citations (they can be excluded from analysis, they can be weighted or they can be treated the same as other citations). The thesis at hand will follow the approach of treating self-references in the same way as other citations, for the following reasons: First, the dataset of references provided by WOS only contains the information about the first author. Consequently, special treatment of self-references would bias the results, as only self-citations made by the first authors could be considered, while those made by co-authors would not be detectable. Secondly, a weighting of self-citations is tested with a value of 0.5 for the purpose of decreasing their influence. However, this procedure has barely any influence on the results of the analysis, as can be seen exemplarily in Appendix A-3. This goes along with a study by Glänzel and Thijs (2004), who find that even an exclusion of self-citations hardly changes the results.

4.3.1 Citation Analysis

In the following, the results of the citation analysis are presented, beginning with an overview of the citing behaviour across the time periods, and including the average number of references, the average age of references, and the self-reference ratio. Then follows the main part of the citation analysis, which focuses on the most-cited publications.

Citing Behaviour

First, the average number of references serves as an indicator for the characterization of the four examined time periods. Table 4-3 records a continuously growing average number of references from period to period. Starting with an average citation count of 15.24 per article for 1960-1979, the average number of references for 2010-2018 amounts to 57 per publication, which corresponds to a growth rate of 274%. This development indicates on the one hand an enlargement of the knowledge base in the field of negotiation research over the time periods, which is indicative of a dynamic research field that in turn stimulates further research. On the other hand, an explanation for this pattern is the improved procurement and dissemination of publications through electronic databases in the Internet over the years.

	Period 1 **1960 - 1979**	**Period 2** **1980 - 1999**	**Period 3** **2000 - 2009**	**Period 4** **2010 - 2018**
Average Number of References	15.24	33.03	39.76	57.00
Average Age of References (Years)	7.54	11.79	13.01	15.75
Self-Reference Ratio	7.92%	6.93%	7.07%	6.21%

Table 4-3: Citing Behaviour
Source: Author's own illustration adapted from Backhaus, Lügger and Koch (2011).

The average age of references increased over the time periods. While for the first period the average age of references is 7.54 years, it is 15.75 years for the fourth period (+109%). This may represent a rather stagnating field of negotiation research, which would contradict the dynamic and growing field presented above. However, such an increase in aging is quite common for research fields – including negotiation research (see the following results) – in which classic works play a major role (Backhaus, Lügger, and Koch 2011). Consequently, the increasing average age of citations can be ascribed best to the importance of classic works in bargaining research.

With regard to the self-citation ratio, a slight decrease becomes visible from period 1 with 7.92% to period 4 with 6.21%.[30] The overall reduction of 1.71 percentage points can also be attributed to the increasing number of publications, as well as to their improved accessibility and publishing possibilities for the more recent time periods. Due to the higher number of publications available, scholars have a broader range of authors to cite, so that they tend to become less dependent on the citation of their own prior work. Despite the slight decrease in the self-reference ratio, the values are still quite high compared to citation analyses in other research fields, which speaks for a lower level of maturation in negotiation research (e.g., Backhaus, Lügger, and Koch 2011).

Most-cited Publications

In the following, top 10 lists of the most-cited publications are provided for each period and are analysed with regard to their development. Moreover, it is important to keep in mind that the number of citations received reflects the impact of a publication (see chapter 4.2.2.1).

In order to identify the most-cited and therefore most influential publications, the publications are sorted in descending order for each time period according to their CV (relative count of citations), as described in chapter 4.2.2.1. Table 4-4 depicts the top 10 most-cited publications per period by providing the information about their rank, first author and publication year, absolute citation count, and CV. As the WOS provides only the name of the first author of a publication, the present thesis will, in order to avoid confusion, retain this notation throughout the citation and co-citation analysis such that it matches the output of WOS (the entire bibliographic data of the top 10 most-cited publications of all periods can be found in Appendix A-4). In cases in which a single period contains identical CV values, a single rank is occupied several times, as these publications are of equivalent importance. Since top 10 lists are provided, no further publications are supplemented where equal CVs occur. The colour design in Table 4-4 is arranged in the following way: Publications appearing for the first time in period 1 (2, 3, 4) are highlighted in red (blue, yellow, green) across all time periods. This colour scheme

[30] It has to be taken into account that the self reference ratio is only calculated on the basis of the first author. Thus, its interpretation should be taken with caution as it rather represents a general tendency.

simplifies the interpretation of the overall relevance of the publications, as well as their impact development over time. Prior to the analysis, it is important to emphasize that the identification of the most-cited publications takes a crucial role, constituting as it does the foundation of the co-citation analysis (see chapter 4.2.2.2).

Taking a closer look at the CVs, it becomes obvious that the range of CV values diminishes from time period to time period (e.g., from 12.37% to 41.24% in period 1 versus from 11.67% to 20.37% in period 4). It shows the way in which publications move closer together in their relevance and impact from period to period (see here and in the following Table 4-4).

Period 1: 1960 – 1979

Rank	Publication	Count	CV (%)
1	Siegel (1960)*	40	41.24%
2	Schelling (1963)*	25	25.77%
3	Walton (1965)*	19	19.59%
4	Deutsch (1960)	15	15.46%
4	Thibaut (1959)*	15	15.46%
6	Kelley (1967)	14	14.43%
6	Liebert (1968)	14	14.43%
6	Pruitt (1969)	14	14.43%
9	Pruitt (1970)	12	12.37%
9	Deutsch (1962)	12	12.37%
9	Homans (1974)*	12	12.37%

Period 2: 1980 - 1999

Rank	Publication	Count	CV (%)
1	Pruitt (1981)*	125	33.16%
2	Raiffa (1982)*	119	31.57%
3	Fisher (1981)*	111	29.44%
4	Rubin (1975)*	107	28.38%
5	Walton (1965)*	106	28.12%
6	Lax (1986)*	81	21.49%
7	Pruitt (1975)	62	16.45%
8	Lewicki (1997)*	58	15.39%
9	Rubin (1994)*	55	14.59%
10	Neale (1991)*	53	14.06%

Period 3: 2000 – 2009

Rank	Publication	Count	CV (%)
1	Fisher (1981)*	124	22.84%
2	Raiffa (1982)*	109	20.07%
3	Pruitt (1981)*	92	16.94%
4	Lax (1986)*	84	15.47%
5	Thompson (2008)*	82	15.10%
6	Lewicki (1997)*	76	14.00%
6	Pruitt (1993)*	76	14.00%
8	Walton (1965)*	66	12.15%
9	Baron (1986)	56	10.31%
10	Thompson (1990)	55	10.13%

Period 4: 2010 – 2018

Rank	Publication	Count	CV (%)
1	Fisher (1981)*	110	20.37%
2	Raiffa (1982)*	93	17.22%
3	Lewicki (1997)*	79	14.63%
4	Walton (1965)*	77	14.26%
5	Galinsky (2001)	75	13.89%
5	Pruitt (1993)*	75	13.89%
7	Pruitt (1981)*	74	13.70%
8	Van Kleef (2004)	71	13.15%
8	Thompson (2008)*	71	13.15%
10	De Dreu (2000)	63	11.67%

First Appearing in Period 1
First Appearing in Period 2
First Appearing in Period 3
First Appearing in Period 4

*Book
CV = Citation Value

Table 4-4: Most-Cited Publications
Source: Author's own illustration.

With regard to the evolution of publications in the top 10 lists, one phenomenon in particular catches the reader's eye: From period 1 only one publication reappears in the following three periods, namely Walton (1965). Thus an almost complete shift in the most-cited publications from period 1 to period 2 is noticeable. This phenomenon corresponds to statements made in the literature that numerous publications prior to 1980 did not establish themselves and hence have minor impact on negotiation research (e.g., Backhaus, Hummelsberger, and Wilken 2005; Herbst 2007; Voeth and Herbst 2015b).

Period 3 consists predominantly of a mixture of new publications and those from period 2. This is also the case for period 4, which combines new publications, a few from periods 1 and 3, and mainly the ones from period 2. Accordingly, the publications from period 2 seem to be quite defining for the overall time frame of 1980-2018. This is also demonstrated by the fact that the top three ranks in period 3 and 4 are derived from period 2. Because of the subordinate relevance of period 1, the focus for purposes of identifying the most influential publications lies on periods 2 through 4.

The most-cited publication across all time periods is Fisher's (1981) *Getting to YES: Negotiating Agreement without Giving In.*[31] It ranks third in period 2 and first in periods 3 and 4. Thus its impact increased subsequent to period 2, so that it constitutes the publication with the highest impact in the more recent periods. In this publication, a method called principled negotiation is introduced, which consists of four main principles that serve as a guideline for negotiators in the different phases of the negotiation process. The four principles are: 1. Separate the people from the problem, 2. Focus on interests, not positions, 3. Invent options for mutual gain, and 4. Insist on using objective criteria. It can therefore be concluded that principled negotiation aims at creating successful negotiations by illustrating correct negotiation behaviour, which in turn is characterized by an issue-related rather than an individual-related focus. This approach delivers a combination of both hard and soft bargaining: Bargain hard on objects and soft on people.

The second most frequently cited publication across the time frames is *The Art and Science of Negotiation* by Raiffa (1982). It continuously ranks second across the time periods 2, 3 and 4. The publication considers real decision-making perspectives of negotiators and demonstrates how optimal solutions can be reached within different settings, even if negotiators do not behave in a fully rational manner or if uncertainties and asymmetries of information are present. Raiffa's book deals in considerable detail with distributive and integrative negotiations, as well as with support from third parties in conflict situations. The third most influential publication is *Negotiation Behavior* by Pruitt (1981), which ranks first in period 2, third in period 3 and seventh in period 4. Thus a diminishing importance of this work can be deduced from the rankings. As if Raiffa

[31] In cases of books with multiple editions, the oldest edition is cited in the analysis, for the sake of a consistent notation.

(1982), *Negotiation Behavior* is structured according to distributive and integrative negotiations. The peculiarity of these three publications is that they are all books.

This case is not only for the highest ranking publications, but also for nearly all publications in the top 10 lists (as indicated by the asterisk in Table 4-4). Period 2 consists of nine books in the top 10 ranking, period 3 ranks eight books and period 4 seven books. Only period 1 ranks as many books as journal articles. The strong influence of books can be explained in a twofold manner. On the one hand, books usually deal with multiple topics, which makes it more likely that they will be cited by each scholar who investigates one of these topics, and leads to an overall high share of citations. On the other hand, books that aggregate the wide-ranging knowledge of negotiation scholars are quite popular on account of their comprehensive approach, which facilitates the knowledge generation in the field. As most such books reappear throughout the time periods, it can be deduced that these books have established themselves as standard works within negotiation research: They define the research domain, dominating as they do the most relevant publications (not including period 1). This high relevance of standard works was already indicated in the previous section *Citing Behaviour*, which shows the increase in the average age of cited references from period to period.

However, main topics that dominate negotiation research cannot be worked out by referring to the influential books. Such books combine a plethora of different themes, but do not sufficiently work out either the degree to which these themes have been explorated, or their importance for the research field.[32] But the identification of relevant negotiation topics per period does deliver valuable insights, because it enables the working-out both of peak phases of specific topics and of thematic developments over the time periods. This calls for an adequate methodology which allows for such identification. The deduction of these main themes and their interrelations is the main task of co-citation analysis (see chapter 4.2.2.2). Accordingly, this analysis will be conducted in the following subchapter.

[32] Neither does a closer look at the few journal articles ranking in the top 10 lists make any thematic focus apparent.

4.3.2 Co-Citation Analysis

Collocation and Size of Co-Citation Networks

As shown in chapter 4.2.1, it is necessary to determine the amount of data to include in the analysis in a way that enables optimization of the level of information. To this end, prior to presenting the results of the co-citation analysis this section will present several key values of the co-citation networks.[33] It touches both upon the number of publications included in the analysis and on the number of co-citation relationships further examined. The size of the analysed research networks is presented last.

In order to determine the number of publications per period to include in the co-citation analysis, attention is drawn to the ranking of the citation analysis publications within the respective time frame. Taking into account that publications with a high ranking and thus with a high CV are included in the co-citation analysis, ensures that only highly relevant publications are examined. However, no single concrete reference value exists in the literature, which would determine how many of the top ranked publications should be considered (e.g., Backhaus, Lügger and Koch (2011) choose the top 300 ranked publications per period, whereas Schäffer, Binder and Gmür (2006) set a threshold value of 1.6% for the CV). Consequently, it needs to be analysed which value is suited best for the present investigation, whereby the following three criteria should be optimized: 1) The periods should have a balanced publication size, in order to adequately interpret and compare each period's result, 2) Multiple meaningful topic clusters should be identified, and 3) As topics should be deduced preferably from larger clusters, since such clusters represent more impactful themes, the number of clusters that base on less than five publications should be small.

With these criteria at hand, one can underscore that threshold values are not able to fulfill criterion 1), as they result in strongly deviating numbers of publications among the periods. For example, by applying the threshold value of 1.6% to the given dataset, recommended by Schäffer, Binder and Gmür (2006), period 2 would account for approximately 200 publications, whereas period 4 would encompass around 400 publications.

[33] In the following, the whole output of a period is referred to as a network, while the topics within the networks are referred to as clusters.

Accordingly, threshold values are not further considered. By contrast, relying on the top x ranked publications allows for the fulfillment of criterion 1), as this approach enables the creation of balanced periods. For this purpose, Backhaus's, Lügger's and Koch's (2011) reference value of the top 300 ranked publications is examined with regard to criteria 2) and 3), as well as its surrounding values of 200 and 400. The top 300 publications best fulfill the criteria, as they result in more meaningful clusters than the top 200 and deliver less small clusters than the top 400 (see Appendix A-5). Thus the meaningfulness is seen strongest for the top 300 publications, so that this number is taken.

Taking the 300 top publications in each period leads to the distribution displayed in Table 4-5. The deviations of the values in Table 4-5 from 300 are due to the fact that many publications exhibit the same CV, thus excluding the possibility of a strict cut-off point at publication number 300. Instead, the closest possible number to 300 is used. The degree of density within a network (4), which is defined as the ratio of the actual number (3) to the potential number of co-cited pairs (2), rises significantly with progressing time (14.81% in period 1 to 55.62% in period 4). This mirrors that the interconnectedness and the thematic proximity of the publications are on a higher level in more recent periods.

		Period 1 **1960 - 1979**	**Period 2** **1980 - 1999**	**Period 3** **2000 - 2009**	**Period 4** **2010 - 2018**
(1)	Publications Included in Co-Citation Analysis [Top 300 Based on CV]	214	260	286	317
(2)	Potential Number of Co-Cited Pairs (1)x((1)-1)/2	22,791	33,670	40,755	50,086
(3)	Actual Number of Co-Citations	3,376	11,972	17,315	27,857
(4)	Density [(3)/(2)]	14.81%	35.56%	42.49%	55.62%
(5)	Top 1.25% Co-Citations Based on CoCit-Score with min. 3 Absolute Co-Citations (or Top 175 Co-Citations)	175	175	221	349
(6)	Publications in Network	76	129	147	159
(7)	Eliminated Isolates and Pendants	22	66	71	69
(8)	Visualized Publications [(6)-(7)]	54	63	76	90

Table 4-5: Description of Co-Citation Networks

Source: Author's own illustration adapted from Backhaus, Lügger and Koch (2011).

Before creating the co-citation matrix, the number of co-citations to be further investigated needs to be set. The co-citation pairs to receive further consideration are those with the highest CoCit-Score and thus with the greatest similarity. Likewise pertaining to the determination of publications to be included (1), no uniform concrete reference value exists in the literature. Therefore, Backhaus's, Lügger's and Koch's (2011) reference value of the top 1.25% co-citation pairs highest in CoCit-Score is tested and compared to a sensitive value of 2%, as their reference value of the top 300 publications turned out to be sensible for the present case as well and the underlying datasets are comparable in size (Backhaus, Lügger and Koch (2011): 55,283 references; this thesis: 66,294 references). Both cluster solutions lead to comparable results (see Appendix A-5). Nevertheless, the 2% cluster solution leads to a weaker resilience of the clusters, as it is characterized by more small clusters some of which originate predominantly or even exclusively from the same first author, or which are not interpretable at all (see for example the circled clusters in the network of period 4 in Appendix A-6). Thus the value of the top 1.25% co-citation relationships is seen as superior, as it does not exhibit these weak points, so that it is applied in the following. However, at least 175 co-citation relationships should enter the analysis in cases in which the top 1.25% lead to a lower amount of co-citation relations (as it holds for period 1 and 2), as the number of co-citation relationships per period should not be too low to generate sensible networks and clusters (here and in the following Backhaus, Lügger, and Koch 2011). In addition, at least three co-citations should be present between two publications in order to be included in the analysis, even when the CoCit-Score is high. This ensures the inclusion of co-citation pairs truly close in content. Bringing these aspects together, one receives the values (5) in Table 4-5.[34]

Lastly, the co-citation networks can be created. Passing through the threshold of the co-citation relationships, the number of the publications in the networks (6) decreases to 76 in period 1, to 129 in period 2, to 147 in period 3, and to 159 in period 4. The difference between the values (6) and (8) is due to the fact that the networks are adjusted for pendants (publications with only one relation/edge to another publication) and isolates

[34] The network of period 4 based on the top 2% co-citation relationships (Appendix A-6) shows exemplarily that the cluster solutions of the other values examined do not differ significantly from the ones chosen, so that the robustness of the subsequently presented results is confirmed.

(publications with no relation/edge to another publication). As both isolates and pendants hinder the exposure of community structures, they are not included in the co-citation analysis (see chapter 4.2.2.2). Consequently, the visualized publications in the networks amount to 54 in period 1, 63 in period 2, 76 in period 3 and 90 in period 4.

Interpretation of Co-Citation Clusters

The results of the co-citation analysis are visualized in a network structure with publications as nodes and co-citation relationships as edges.[35] Moreover, the topics of the clusters are interpreted with reference to their *content* as well as to the relevant key figures *size*, *density* and *centrality* of social network analysis, which were introduced in chapter 4.2.2.2. In this manner, the size of a cluster mirrors its relevance within the network. Moreover, the size of each node represents its degree centrality in the respective network. The greater the size of the node, the larger the number of co-citation relationships with other nodes. For this reason, it can be assumed that the content of publications with comparatively large nodes is defining for the thematic focus of the respective cluster (Backhaus, Lügger, and Koch 2011). In this manner, for each cluster those publications will subsequently be considered in more detail which overshadow other publications in the respective cluster with respect to their node size. Nevertheless, the other publications within the cluster are also checked for confirmation of this topical orientation. According to the orientation detected with respect to content, the clusters are labeled in the following.[36] Furthermore, large nodes tend to be centrally located, as opposed to less co-cited publications, which find themselves at the edges of the network. This is due to the spring-embedded algorithm that minimizes the distances between the publications. The shape of the nodes indicates the period in which the publication appears for the first time in the network (square for period 1, up triangle for period 2, circle for period 3 and down triangle for period 4).

---

[35] For the analysis and visualization of the networks, UCINET 6 and its visualization tool NetDraw are used, which represent the most frequently applied analysis and visualization softwares for co-citation analysis (Borgatti, Everett, and Fremman 2002; Zupic and Čater 2015).

[36] The labelling of the clusters is an interpretative and subjective process. Consequently, slight differences could arise through labelling done by a different person. However, the topical focus of the clusters is in nearly all cases so clear that this fact is negligible.

Co-Citation Network Period 1 (1960-1979)

The co-citation network of period 1 is visualized in Figure 4-6. It comprises 54 of the most-cited 214 publications in this period and thus represents the smallest network of the four periods. The cluster solution provided by the Girvan-Newman algorithm divides the network into three clearly selective clusters which are partly connected (Y=0.52).[37] The cluster sizes vary between 9 and 26 publications and for each cluster a clear thematic focus can be deduced from the publications (here and in the following see Table 4-6).

Period 1

Cluster	Size	Density	Avg. Degree Centrality	Avg. Year	Topic
I	9	0.39	3.11	1962	Threats
II	19	0.31	5.58	1968	Representatives
III	26	0.26	6.54	1967	Offer Making

Table 4-6: Results of Period 1
Source: Author's own illustration.

Cluster I, the smallest cluster, has the topical focus *threats* in negotiations, whereby threats are messages which signal that one will punish the opposing party if it fails to conform (Pruitt and Carnevale 1993). With an average (avg.) publication year of 1962, this topic is the most mature one in this period. The most defining publication of this cluster is Deutsch (1960), as its node size is largest, which means it has the largest amount of co-citation relationships. The publication examines the effect of threats on the achievement of an agreement and on profit gained within a bargaining game, and finds that in settings in which either one or both of the bargainers use threats it becomes more difficult to reach an agreement, and both individual and collective payoffs turn up negative. This holds especially for bilateral negotiations.

[37] The modularity of the best cluster solution is indicated for each period. Representing as it does the best cluster solution, the indicated modularity value is the highest value reached within the respective period.

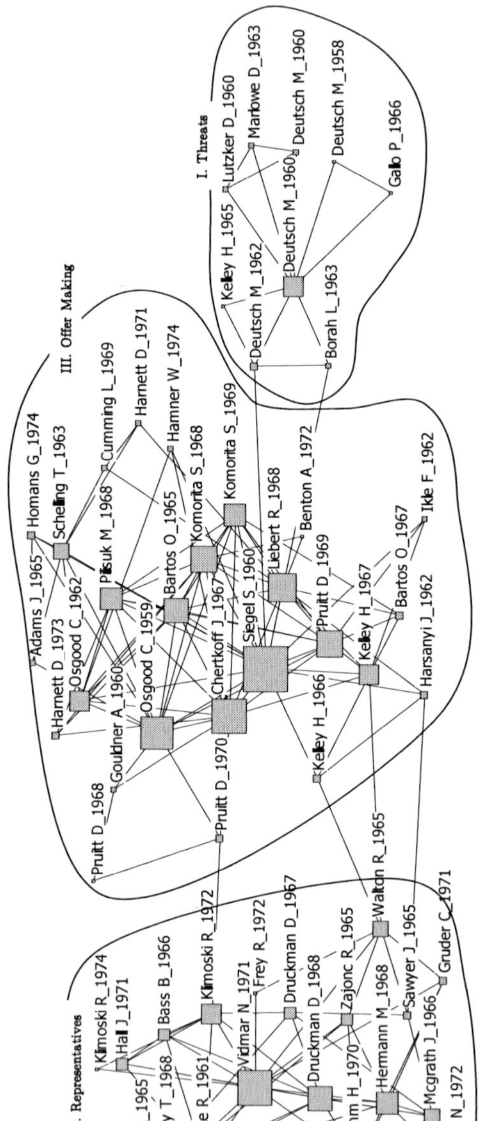

Figure 4-6: Network of Period 1

Source: Author's own illustration.

The second cluster deals with *group representatives* in negotiations. This cluster contains a total of 19 publications, hence it is twice as large as the first one. With an average publication year of 1968, this topic is the most recent one in period 1. The most defining publication is Vidmar (1971). Vidmar (1971) works out that the presence of a representative role obligation leads to a negotiation performance that is inferior to one that lacks a representative function. However, this negative effect can be improved by the employment of mediators. The other publications of that cluster deliver several explications for why representatives show worse performance than non-representatives. The most commonly provided explanation is that representatives are usually observed and evaluated by the group they represent, which is the one with which they are most likely to exhibit a high degree of cohesiveness (e.g., Klimoski 1972). Accordingly, they try to behave in a way in which they are least likely to lose face, which entails low risk-taking behaviour in their efforts to reach agreement for their constituents (e.g., Kogan 1972; Lamm 1970).

Lastly, the thematic focus of the third and largest cluster is on *opening offers and concessions*. With a size of 26 publications, this cluster is the second largest across all time periods. When one looks more closely at the defining publications of the topic, it becomes obvious that the cluster exhibits multiple publications with a larger node size and, accordingly, with a higher impact. As the most predominant publications, Siegel (1960), Osgood (1959), Chertkoff (1967), Liebert (1968), Komorita (1968) and Pruitt (1969) are presented in more detail in the following. Siegel (1960), who stands out most from the cluster, ranked first among the top 10 publications in the citation analysis. This book ascertains that a negotiator seeking to maximize his/her gain should make only small concessions, because this leads to larger concessions on the part of the opposing party. Komorita (1968) confirms this effect, and works out that concessions are inherently important in a bargaining situation, since concessions must take place if progress is to be made in reaching an agreement. By contrast, the second most defining publication, Osgood (1959), ascertains an opposite effect. This publication underlines that reciprocal behaviour is prerequisite reducing tension in an interaction. Accordingly, the opponent will reciprocate small concessions with small concessions of his or her own, and large ones with large concessions. Pruitt (1969) delivers yet another approach and finds that the concession rate of the opponent does not influence one's own concessions. Moreover, the publication underlines that concessions decrease when time elapses in negotiation. Hence in the literature there is disagreement concerning concession making behaviour. Chertkoff (1967) shows that a high initial offer on the part of a seller in a buyer-seller negotiation is superior for the seller, because it is accompanied by a greater possible distance of movement. Additionally, infrequent concessions are superior to frequent ones. Lastly, Liebert (1968) shows that uninformed negotiators with no knowledge of the other party's payoff structure use the opponent's initial offer in order to set their own negotiation goal. In cases in which a fair initial offer is made, they make less favorable following and final offers. This cluster's size and the presence of publications within it that belong to the top 10 most-cited publications (e.g., Kelley 1967; Liebert 1968; Pruitt 1969; Siegel 1960) are indicative of the relevance of this topic for period 1960-1979.

Co-Citation Network Period 2 (1980-1999)

Period 2 consists of 63 publications, whereby the cluster sizes vary between 3 and 15 publications. Its network is characterized by seven clusters (Y=0.77) (here and in the following see Figure 4-7). Accordingly, more than twice as many clusters or topics are identifiable and relevant for period 2 than for period 1. Four of the seven clusters are unconnected, while three of them are partly interconnected. However, the interconnections between these three clusters are quite negligible compared to those of period 1, as only one link exists between them. Consequently, the interrelatedness of topics in period 2 is hardly given.

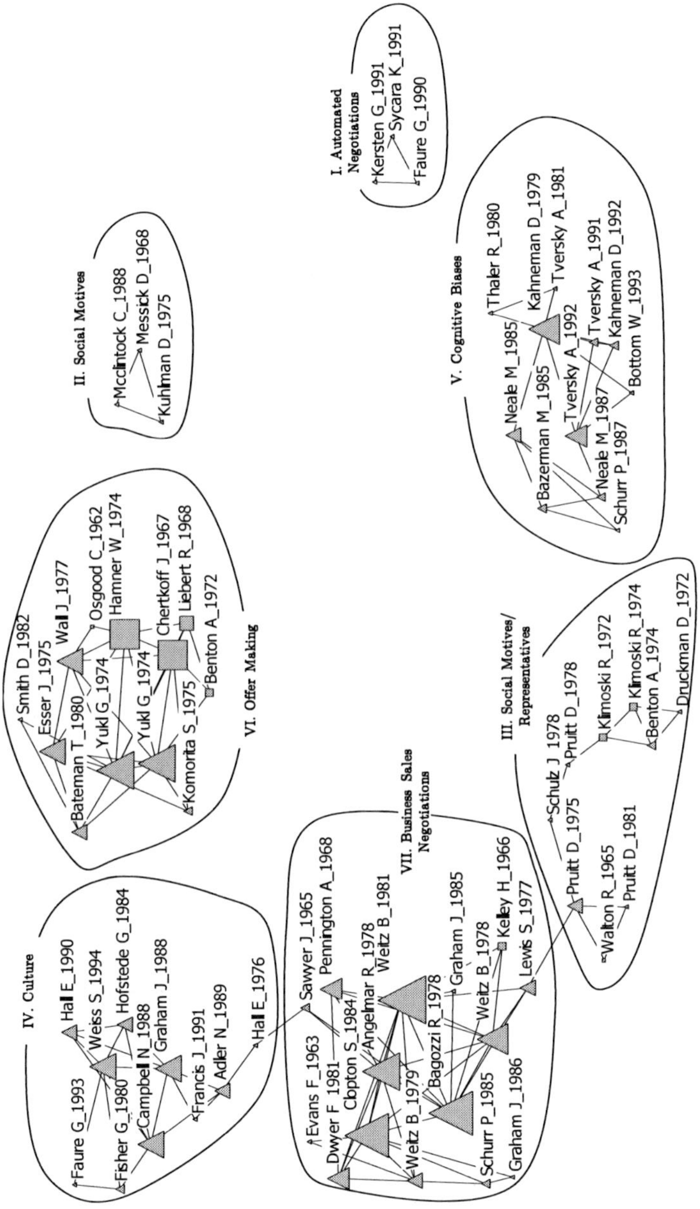

Figure 4-7: Network of Period 2

Source: Author's own illustration.

Two small clusters exist in period 2 that are based on three publications each. Cluster I deals with *automated negotiations*, while cluster II focuses on *social motives* (here and in the following see Table 4-7). The topic of *automated negotiations* represents the most recent theme in period 2, with an average publication year of 1991. This reflects the fact that this topic has just evolved in this period, while the foundation of other topics was laid earlier. The publications dealing with *automated negotiations* focus on automatic restructuring approaches of negotiations taken with the help of both artificial intelligence and the advancement of NSSs. The developed restructuring algorithms replace manual approaches and allow such things as the expansion of the space of feasible settlements by reacting flexible to changes and requirements (Kersten 1991; Sycara 1991). The topic of *social motives* examines both theoretically and empirically the role of *individualistic*, *competitive* and *prosocial* motives in decision making situations. Among the publications, the relationships between strategies and social motives is examined, which leads to a significant interaction between the two (e.g., a prosocial motivation and a cooperative strategy) (Kuhlman 1975).

Period 2

Cluster	Size	Density	Avg. Degree Centrality	Avg. Year	Topic
I	3	1.00	2.00	1991	Automated Negotiations
II	3	1.00	2.00	1977	Social Motives
III	9	0.31	2.44	1974	Social Motives/ Representatives
IV	10	0.42	3.80	1987	Culture
V	11	0.33	3.27	1987	Cognitive Biases
VI	12	0.46	5.00	1973	Offer Making
VII	15	0.35	4.93	1977	Business Sales Negotiations

Table 4-7: Results of Period 2
Source: Author's own illustration.

The second smallest cluster has a size of nine publications. The peculiarity of this cluster is that it is thematically two-tiered. While the upper part of the cluster deals anew with the topic of *social motives*, the lower part covers the topic of *group representatives*. The cluster resembles more a chain than a strongly interconnected cluster, as the two topics are linked solely by means of the publication by Pruitt (1978). The cluster's density is rather low (0.31, the smallest value in period 2), which underlines even more strongly the dichotomy of that cluster. The dominating publication on the topic of *social motives* is the one by Pruitt (1975), which confirms the results of the small *social motives* topic presented above. The topic of *group representatives* already appeared in period 1, but there it was larger (with 19 publications), and thus its impact was much higher than in the current, second period, meaning that its influence has diminished. Moreover, none of the publications stands out from the others, and the defining publications of period 1 are no longer present. The thematic focus, however, has not changed.

Cluster IV is similar in size to cluster III, as it encompasses 10 publications. However, its density is much higher at 0.42. Its thematic orientation is *culture*. This topic represents the second most recent theme in the time frame of 1980-1999, with an average publication year of 1987. As becomes visible in Figure 4-7, the cluster displays several dominant publications (nodes of greater size): Graham (1988), Campbell (1988) and Weiss (1994). Weiss (1994) lists options for how to conduct cross-cultural negotiations in a sensitive way. The publication underscores that the mode of negotiation one chooses is dependent on familiarity with the opponent's culture. In situations in which there is little familiarity, for instance, an advisor could be helpful. Graham (1988) and Campbell (1988) examine differences between Americans and Asian/European bargainers when it comes to negotiation behaviour and outcomes. The main finding is that while problem-solving strategies in American negotiations lead to significantly higher satisfaction and exert a positive influence on the outcome of the negotiation, this effect is not visible for European or Asian negotiations. Moreover, differences also occur between the examined Asian and European countries, which indicates that geographic proximity does not automatically induce similar negotiation behaviour (e.g., in China competitive negotiations lead to better results than in Korea and Japan).

The third largest cluster is cluster V, which focuses on *cognitive biases*, and especially on *framing* in negotiations. Framing refers to the way in which individuals display different perceptions and behaviour depending on how situations are presented to them (Caputo 2013a). Cognitive biases, with its average publication year of 1987, represents a quite recent topic in period 2. The publications by Kahneman (1979) and Tversky (1992) are highly defining for this cluster. The authors of these publications are themselves characteristic of the cluster, appearing as they do multiple times as authors of different publications within it (e.g., Kahneman 1992; Tversky 1981; Tversky 1991). Their frequent appearance can be ascribed to the fact that both authors developed the model on which the topic relies, which seeks to account for irrationality on the part of decision makers. The two defining publications mentioned here introduce and refine this model while dealing with decision making under risk. The main rationale of these articles is that decision makers are averse to risk-taking when it comes to decisions involving sure gains, and willing to take risks when it comes to choices involving sure losses. Accordingly, the framing of decision making situations as either positive (gains) or negative (losses) exerts an influence on decisions makers' behaviour. By means of this procedure, it can be confirmed, for instance, that negative framing significantly decreases negotiators' concession behaviour, as they are more risk-seeking and even willing to accept an impasse, whereas in positively framed negotiations more transactions are completed (Neale 1985; Bazerman 1985). The substantial importance of the topic of *cognitive biases* in period 2 reflects the strong influence of BDR at this time, as it defined the research agenda of the 1980s and 1990s and focused predominantly on cognitive heuristics (see chapter 2.2). Therefore, it makes sense for this topic to emerge in period 2.

Cluster VI is the second largest of time period 2 and takes up *offer making*, which is the largest topic of the preceding period. The average publication year of this topic is 1973, which is the earliest of all average publication years exhibited in period 2, and that underscores the fact that the main development of this topic took place prior to 1980-1999. The fact that the size of the cluster (12 publications) is much smaller than in period 1, indicates that its importance has decreased. However, this topic's cluster displays the greatest density for this period, with a value of 0.46, which in turn is indicative of the most strongly interconnected topic. Moreover, the topic is characterized by multiple larger nodes and exhibits the highest average degree centrality of 5. These defining publications are Yukl (1974a), Yukl (1974b), Chertkoff (1967), Hamner (1974), Esser (1975) and Wall (1977). The dominating publications of this period are different than in period 1: only Chertkoff (1967) remains in this list. However, some of the other (less dominating) publications reappear, such as Hamner (1974), whose influence has even increased; or Liebert (1968), which shows decreasing impact. Yukl (1974a) and Yukl (1974b) confirm the findings by Liebert (1968) presented in period 1. In addition, they show that a small concession strategy is superior in situations in which information on the opponent's payoff structure is incomplete, when negotiations take place under time pressure, or in the event that frequent concessions are made, as the effects of this strategy are similar to those of a hard initial offer (e.g., the creation of a lower negotiation goal on the part of the opponent). By contrast, Hamner (1974) finds that in high pressure contexts negotiators make larger concessions, as they are accompanied by higher payoffs. Lastly, Esser (1975) and Wall (1977) confirm the perspective of Osgood (1959) in period 1, according to which large concessions should be reciprocated (Esser 1975) or rewarded (Wall 1977).

The largest cluster in period 2 (15 publications) is cluster VII. It is the only cluster with a quite broad thematic focus, namely *business sales negotiations*. The most defining publications are those by Weitz (1981), Bagozzi (1978), Clopton (1984) and Angelmar (1978). While Weitz (1981) works out a framework for the purpose of evaluating the effectiveness of salespersons, Bagozzi (1978) emphasizes that a rethinking of marketing has taken place towards social activity. Both publications target the buyer-seller relationship and work out the fact that negotiations are an important part of this social interaction. Clopton (1984) examines the effect of the seller's negotiation strategy (competitive vs. cooperative), the type of information the seller provides (clear vs. ambiguous), and the influence of an internal monitoring system belonging to the buyer (process vs. outcome monitoring) on negotiation behaviour and outcome. The buyer's outcome is to a significant degree positively affected in cases in which a cooperative strategy is coupled with the practice of providing ambiguous information. In addition, the buyer's negotiation behaviour is significantly influenced by an interaction between information provided by the seller and a system for monitoring outcomes. On the other hand, Angelmar (1978) develops a content analytic system based on a theoretically category system for the purpose of analysing bargaining communication. The former system aims at examining negotiations by working out specific bargaining behaviours in order to enhance the learning possibilities for managers to improve their bargaining skills.

Since the cluster sizes in period 2 are quite close together starting from cluster III to cluster VII (ranging from 9 to 15 publications), a main topic of this time period cannot be identified as was done for period 1.

Co-Citation Network Period 3 (2000-2009)

Period 3 includes 76 publications in its network. The cluster solution provided by the Girvan-Newman algorithm contains eleven clusters, which makes it by far the largest cluster solution of all time periods (Y=0.77). When one takes a closer look at the network structure in Figure 4-8, it becomes obvious that the eleven clusters are not interconnected at all, as they much rather stand next to each other separately. The second characterizing feature of the time frame 2000-2009 is the large number of small clusters that exhibit an equal impact. Nine clusters do not exceed the number of nine publications within a topic, so that only clusters X and XI stand out, with 17 and 21 publications (see Table 4-8).

Period 3

Cluster	Size	Density	Avg. Degree Centrality	Avg. Year	Topic
I	3	1.00	2.00	2000	Automated Negotiations
II	3	1.00	2.00	2001	Negotiation Training
III	4	0.83	2.50	1996	Coalitions
IV	4	0.83	2.50	1976	Offer Making
V	4	0.83	2.50	2000	Power
VI	4	0.67	2.00	1998	Culture
VII	4	0.67	2.00	1992	Reciprocity
VIII	5	0.90	3.60	1991	Ultimatum Games
IX	7	0.71	4.29	1997	Gender
X	17	0.44	7.06	1999	Emotions (esp. Anger)
XI	21	0.23	4.67	1991	Social Motives

Table 4-8: Results of Period 3
Source: Author's own illustration.

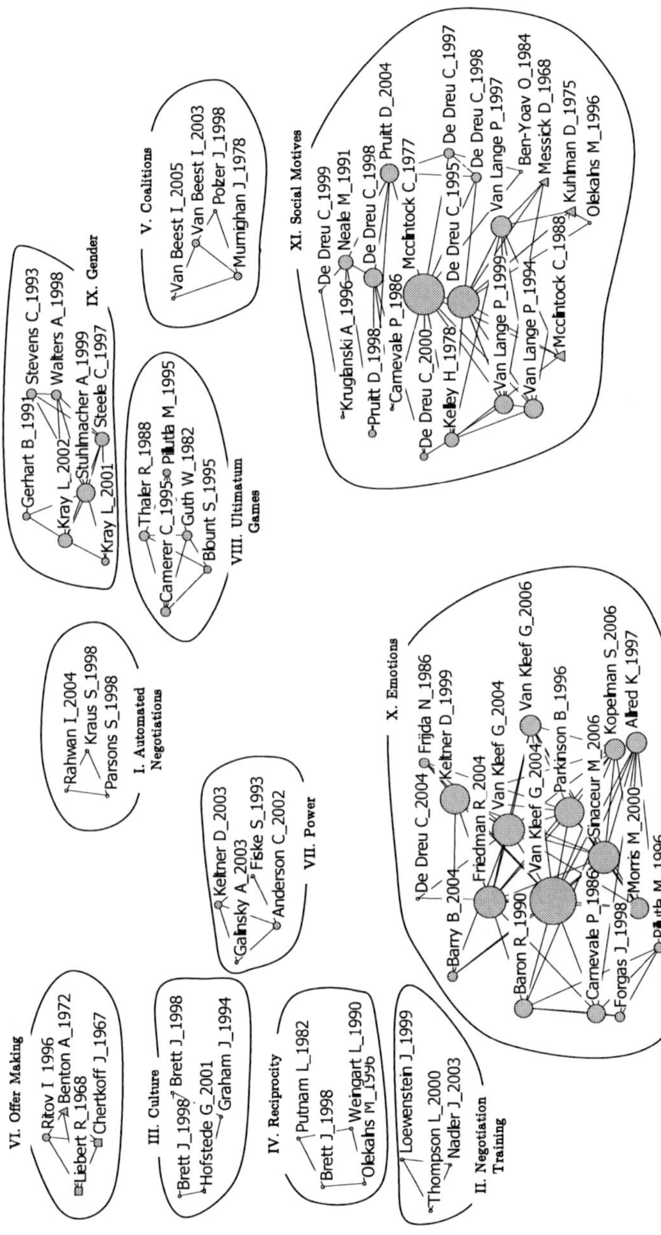

Figure 4-8: Network of Period 3

Source: Author's own illustration.

The two smallest clusters of period 3 are cluster I and II, with 3 publications each. While cluster II represents a new topic, *negotiation training*, cluster I deals once more with *automated negotiations*, which was also a small topic in period 2. It is remarkable that these two topics have the latest average publication years in period 3 (2000 and 2001, respectively). The publications on the topic of *negotiation training* work out the most effective ways of learning negotiation skills, finding analogical and observational learning to be more effective than learning by experience only (Nadler 2003). When learning analogically, negotiators compare training cases and derive their common solution (Thompson 2000). Based on the analogies worked out in this way, the learner can transfer knowledge he or she has gained to subsequent negotiation tasks (Loewenstein 1999). Unlike its analogical counterpart, observational learning relies on observation of the negotiation skills of others (Nadler 2003). A closer look at the publications within the *automated negotiations* cluster reveals that a complete shift in publications has taken place from period 2 to period 3. The topical focus likewise shifted, and now lies on the conception of automated negotiation agents that are based on an argumentation logic. As opposed to game-theoretic approaches in which agents only exchange proposals, agents based on argumentation are capable of exchanging such additional information as their mental states or beliefs (Rahwan 2004). Thus it is possible for agents to justify their own negotiation positions and to influence those of the others (Rahwan 2004).

A special characteristic of period 3 is its five clusters (cluster III – cluster VII) consisting of four publications each. Some of their topics appear for the first time in period 3; however, several topics have appeared in the previous periods as well. The new topics are *reciprocity*, *coalitions*, and *power*, whereby *power* is a quite recent theme with an average publication year of 2000. The publications on the topic of *reciprocity* examine the effect of reciprocal behaviour and communication between individuals in negotiations. It is found that negotiators reciprocate both distributive (e.g., single issue offers) and integrative tactics (e.g., multiple-issue offers) (Weingart 1990). For this reason, an impasse becomes more likely if distributive tactics are used (Putnam 1982). Moreover, reciprocity is worked out as a factor that is critical for moving from distributive to integrative negotiation (Putnam 1982). With respect to *coalitions*, the publications find that on the one hand parties with compatible interests are more likely both to form coalitions and to reach significantly higher individual gains for both sides (Polzer 1998). On the other hand, publications on coalitions find that when losses are being negotiated, larger coalitions are formed than when gains are the object of negotiation, the aim being to prevent others from losses, which holds especially for prosocials rather than for individualists (van Beest 2005; 2003). Lastly, the publications on the topic of *power* deal more superficially with their topic as the other ones within the network do with their respective topic. They work out that power leads to goal-directed behaviour and such actions as expression of one's attitudes (Anderson 2002; Galinsky 2003). In addition, power leads to the experience of more positive emotions and to an accurate perception of whether or not one is liked by one's negotiation partner (Anderson 2002).

The topic of *culture* has already appeared in period 2, while *offer making* has even emerged in both preceding networks in periods 1 and 2. With regard to the topic *offer making*, its influence already vanished from period 1 to period 2. This development continues in period 3, as it reduces its size from 12 to four publications. Consequently, the impact of this topic decreases even further, such that no additional findings can be deduced. The topic of *culture* also shows a decrease in importance from period 2 to period 3, as its size diminishes from a medium number of ten publications to a cluster of a mere four publications. Moreover, the key publications on the topic in period 2 are no longer among the publications in period 3. Nevertheless, this cluster partly revisits the topical emphasis of period 2 by conducting a cross-cultural examination of the influence of negotiators' problem-solving approaches on satisfaction and profit, and confirms that Asians reject this approach, as worked out in period 2, while European countries like Germany now show a significant positive effect of this same approach on the partner's satisfaction. In addition, it is shown that, due to such cultural differences as power or deviations along lines of self-interest, joint gains are lower in intercultural negotiations.

Two further new topics in period 3 are the ones in cluster VIII and cluster IX, which exhibit a size of five and seven publications, respectively. Cluster VIII deals with *fairness in ultimatum games*[38], while cluster IX examines the role of *gender* in negotiation settings.

[38] An ultimatum game can be described as follows: A proposer gets the task of splitting up an item (e.g. money) between him or herself and a responder (here and in the following Kubota et al. 2013). The responder can then accept the offered division or refuse it. In the case of acceptance, the players obtain the allocation determined by the proposer. However, if the responder refuses the offer, neither party receives anything.

While in cluster VIII the node sizes are quite regular, in the *gender* cluster three publications stand out slightly. These are the publications by Stuhlmacher (1999), Steele (1997) and Kray (2002). The cluster *fairness in ultimatum games* deals with deviations by individuals from rational behaviour in ultimatum games. One of the main criteria adduced for irrational behaviour is that of intended fair behaviour on the part of the individuals (e.g., Camerer 1995; Güth 1982). If an offer is made that is not perceived by the counterpart as fair, this counterpart is much more likely to reject the offer. However, it is also found that fair behaviour of this kind is a mere strategy that is applied for the purpose of avoiding the other party's rejection in order to increase one's own reward (Pillutla 1995). With respect to the cluster *gender*, Stuhlmacher (1999) conducts a meta-analysis and aggregates the effect sizes of various negotiation studies that examine the influence of gender on negotiation outcomes. The meta-analysis finds that men reach significantly better results than women, independent from multiple investigated moderators, such as the opponent's sex or the relative power of the bargainers. One possible explanation for why men outperform women is delivered by Steele (1997) and Kray (2002). The former compares in how far the performance of females and males is shaped by gender stereotypes. The publication finds that women who are told that it is assumed they will perform worse than men are negatively affected by that stereotype and actually do perform worse. In contrast, the performance of women who do not allow themselves to be influenced by this stereotype threat is equal to that of men. Kray (2002) takes up the examined stereotype threat and confirms it in a negotiation setting. Women who were told that their skills are essential for negotiation success performed significantly better than men. In contrast, if they were told that their skills were ineffective, they negotiated worse (the same holds for men).

The first cluster which clearly stands out from the others in period 3 in terms of size and importance is cluster X. With 17 publications, it represents the second largest cluster of the period. Its topical focus is *emotions*, with a special emphasis on anger. The second most examined emotion is that of happiness. This confirms the proposition of scholars that anger and happiness are the most frequently examined emotions in negotiation literature (van Kleef and Sinaceur 2013) (see chapter 3.2.1). With its average publication year of 1999, this topic is very recent. A plethora of negotiation studies within this cluster are concerned with empirical examination of the effect of emotions in negotiations (e.g., van Kleef 2004a; Sinaceur 2006). In addition, the cluster contains a number of conceptual/basic publications, such as the one by Frijda (1986). One can also find a few publications that deal with moods in negotiations (e.g., Carnevale 1986; Forgas 1998), along with studies dealing with positive or negative affect in bargaining settings which – in these cases, at any rate – refer to moods as well (e.g., Baron 1990). Their occurrence in the cluster can be ascribed to the conceptual relatedness of emotions, moods and affect which was worked out in chapter 2.3.1.2. Moreover, most of the publications investigate not the intrapersonal perspective of emotions in negotiations, but rather their interpersonal effect. It is worth noting that the intrapersonal effect is examined by those studies in particular that focus on moods and affect. The cluster exhibits the rather high density of 0.44, along with an average degree centrality of 7.06 – by far the highest in this period. Consequently, this cluster is interconnected quite strongly and exhibits a considerable amount of publications that are defining for it (see Figure 4-8). Van Kleef (2004a), van Kleef (2004b), Friedman (2004), Sinaceur (2006), Keltner (1999) and Parkinson (1996) are especially worthy of mention as defining publications. To begin with Keltner (1999) and Parkinson (1996), both consider emotions from a social rather than from an individual perspective, and present corresponding literature overviews. Keltner (1999) examines the social functions of emotions and delivers four distinct levels at which emotions can appear: an individual, a dyadic, a group and a cultural level, whereby the publication demonstrates how scholars differ in their insights concerning the consequences of emotions. Parkinson (1996) shows how emotions are oftentimes interpersonally or culturally defined, how they have consequences for others, and how they are in many cases communicative phenomena. It is based on these general papers on social functions of emotions that the other defining publications build up and conduct their own empirical analyses of the role of emotions in negotiation settings. Van Kleef (2004a), van Kleef (2004b), Friedman (2004) and Sinaceur (2006) all examine the interpersonal effect of emotions in negotiations. As the most defining publication for this cluster, van Kleef (2004a) delivers the following insights: Bargainers with an angry opponent concede more and have a lower demand level. This is due to the fact that the negotiator construes the opponent's emotional state as information about his or her limits – which in an anger-driven opponent are greater – and concludes according to this information that higher concessions need to be made. However, the emotional effect is absent if no need is perceived to take the emotion of the opponent into consideration (e.g., in the case of large concessions made by the expresser, the negotiator is not affected by the counterpart's emotion). Van Kleef (2004b) confirms these findings and complements them with the insight that negotiators are only affected by the opponent's emotion if no time pressure is present, since time pressure hinders the information processing, thus rendering the receiver incapable of processing the emotion. Moreover, receivers are only influenced by the other's emotion in the predicted direction inasmuch as their power is limited accordingly. Sinaceur (2006) confirms this by working out that negotiators concede more to angry opponents if they have poor alternatives, which can be equated with a low degree of power. Lastly, Friedman's (2004) investigation delivers a perspective that stands in contradiction to the others presented.[39] This publication finds that the expression of anger lowers the resolution rate because the expression of anger generates an angry response, which the publication designates as emotional reciprocity. However, as was found by the preceding publications, if the negotiator is vulnerable on account of his or her low reputation, an angry opponent does not hinder settlement.

The largest cluster of period 3 (21 publications) is cluster XI, with a thematic focus on *social motives*. This topic appears for the first time in period 2 in a small cluster of three publications and within a two-tiered cluster, both with barely dominating publications. Consequently, the impact of this topic increases from period 2 to period 3. Unlike period 2, the publications of the focal cluster change completely. The most defining publications are McClintock (1977) and De Dreu (1995). However, these defining publications hardly deliver new insights; rather, they confirm the findings of this topic in the preceding period. In this way, McClintock (1977) adds to the previous findings a behaviour of a bargainer can serve more than one social motive (e.g., a tit-for-tat strategy can serve both cooperation and individualism). The author De Dreu is highly defining for this topic, as he is first author of six publications. With respect to the key figures of social networks, the level of density is rather low (e.g., compared to the largest cluster in period 4 with a size of 37 publications). Therefore, the proximity and the interconnectedness of the publications with respect to their content is smaller. Besides, the average degree

[39] The contradictory findings shown here are revealed on the basis of the most defining publications. It is also possible that contradictory findings can be revealed from the other, less defining publications within a cluster.

centrality amounts to 4.67, which mirrors a lesser importance of the single publications in contrast to the high value of 7.06 for the *emotions* cluster.

Co-Citation Network Period 4 (2010-2018)

Period 4 is the largest network of the four time frames with 90 publications. The best solution of the Girvan-Newman algorithm lies in the division of the network into eight distinctive clusters (Y=0.57). As was already the case for period 3, the 2010-2018 time frame is once more characterized by a lack of interconnections between the different clusters. Moreover, a first look at the network structure indicates that one topic particularly stands out from the other clusters with regard to its size and average degree centrality, namely cluster VIII focusing on *emotions* (see Figure 4-9). Besides, the investigation of the topics identified in this period yields the finding that nearly all topics have appeared in one of the preceding periods (see Table 4-9). Especially the topics of period 3 recur in period 4. Accordingly, a saturation of negotiation topics is visible. The only newly identified topic is *deception*.

Like the two preceding periods, time period 2010-2018 exhibits several clusters which are quite small: cluster I with three publications, cluster II with four, and clusters III and IV, each with a size of five. Their topics are *power*, *automated negotiations*, *fairness in ultimatum games* and *culture*. The topic of *power* was worked out for the first time in period 3. Its size has barely changed from the preceding period, thus its importance is reliably low. Moreover, the publications of that small cluster have hardly changed. However, a further aspect is added, which is that *power* provoked by strong alternatives on the part of a negotiator makes it more likely that he or she will make the first offer, which would lead to a bargaining advantage (Magee 2007). The topic of *automated negotiations* already exhibited only a slight influence within the networks of periods 2 and 3. The publications of that topic have changed completely in contrast to period 3, which holds also for its topical focus. The publications of this cluster focus on the development of algorithms for multi-issue negotiations. The main attention here is on the learning capabilities of the autonomous agents for the purpose of deducing the opponent's preference structure (e.g., Hindriks 2008). For an efficient and effective negotiation strategy, the autonomous agent should infer the preferences of the other party by relying on the offers made by this negotiation party (Coehoorn 2004).

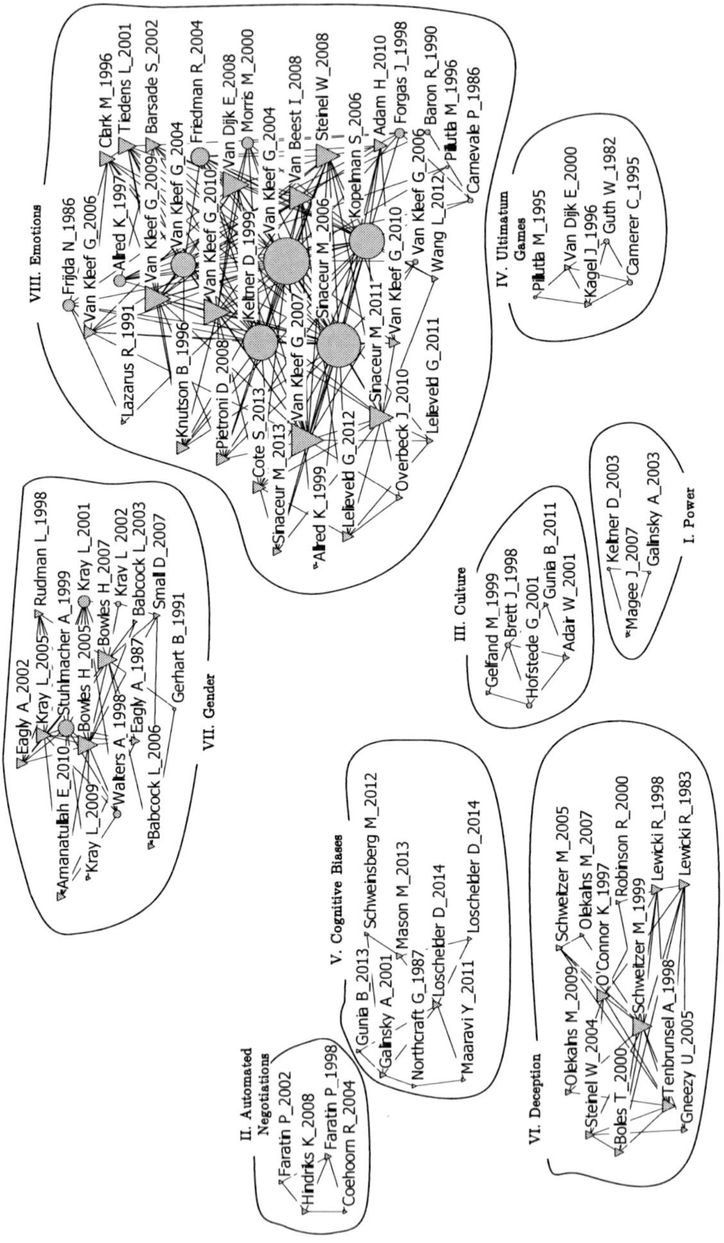

Figure 4-9: Network of Period 4

Source: Author's own illustration.

On the other hand, the publications on the topics of *fairness in ultimatum games* and *culture* are a mixture of the publications of periods 2 and 3, in which the topics appeared for the first time, and newly appearing publications in period 4. Both topics barely changed in size, which is a reflection of their consistently low impact. With regard to the average publication year, the topic of *fairness in ultimatum games* is the cluster with the oldest publications, which is demonstrative of maturation of this topic (average publication year 1994) (see Table 4-9). This goes along with the fact that the topical emphasis hardly changes over against the preceding period. The publications of the topic of *culture* investigate such multiple aspects of cultural differences as distinct trust levels or different social motives between cultures leading to miscellaneous effects on consideration of interests or joint gains. Besides, tests are conducted for behaviour adaptations of the opposing party's behaviour in intercultural negotiations. However, the adaptation does not lead to more integrative solutions, and hence an adaptation is not sufficient; rather, a real motivation to reach integrated agreements is necessary (Adair 2001).

Period 4

Cluster	Size	Density	Avg. Degree Centrality	Avg. Year	Topic
I	3	1.00	2.00	2004	Power
II	4	0.83	2.50	2003	Automated Negotiations
III	5	0.80	3.20	1994	Ultimatum Games
IV	5	0.70	2.80	2002	Culture
V	8	0.39	2.75	2008	Cognitive Biases
VI	12	0.47	5.17	2000	Deception
VII	16	0.38	5.75	2002	Gender
VIII	37	0.28	10.16	2004	Emotions (esp. Anger)

Table 4-9: Results of Period 4
Source: Author's own illustration.

The topic of *cognitive biases* reappears in period 4. However, since its first appearance in period 2, the number of its publications has decreased slightly, to eight publications. Moreover, since period 2 the publications and authors have changed completely. While publications by Tversky and Kahneman dominated the cluster in period 2, in period 4 outstanding publications are rare on this topic. In addition, this cluster exhibits by far the highest average publication year in the network (2008), thus making it a recent theme in 2010-2018 (see Table 4-9). Since the topic had already been covered around 20 years earlier, one can say that it was rethought in period 4. While in period 2 the cluster's main emphasis was on framing, the focus of period 4 is clearly on the first offer as an anchor, whereby the effects of first offers are examined from different perspectives. On the one hand, it is worked out that a precise value influences the counterpart more strongly than an approximate value, as it serves better as an anchor and underlines the competence of the negotiator (e.g., 10,959 vs. 11,000) (e.g., Loschelder 2014). On the other hand, the influence factors on first offers are investigated. For instance, first offers are only helpful if the provider does not reveal his/her preferences, or if the first offer is not too high in cases in which one is faced with a rather powerful counterpart (Loschelder 2014; Schweinsberg 2012).

The only new topic to emerge in period 4 is the cluster *deception*, with 12 publications and a rather high density of 0.47. Deception describes a transmission of information that encourages the counterpart to draw wrong conclusions (Boles, Croson, and Murnighan 2000). The main focus of the cluster lies on the influence of deceptive behaviour along the lines of misrepresentation or lying in negotiations. A few publications stand out slightly from the others. These are the publications by Schweitzer (1999), Tenbrunsel (1998) and O'Connor (1997). Schweitzer (1999) works out that a broad range of bargainers draws on deception while negotiating. For this reason, the publication deals with solution options for dealing with deceptive behaviour. It is revealed that direct questioning lowers lies of omission, but can by the same token increase lies of commission. In order to prevent the latter, the recommendation is made to continuously validate relevant claims and receive written guarantees. These two basic forms of misrepresentation (lies of (c)omission) are also revealed by O'Connor (1997), which examines deceptive behaviour in negotiation settings in which both parties actually have the same interests, but fake opposing interests for one negotiation object to gain an advantage for other negotiation objects. Two factors that serve to lower the effect of deception are the sharing of information and cooperative rather than individualistic motives. Be that as it may, emphasis is also placed on the fact that deception leads to a higher individual negotiation outcome for the user, so that it is not always harmful. Lastly, Tenbrunsel (1998) examines how incentives and expectations with respect to the counterpart's deception effect one's own deceptive behaviour. The publication works out the fact that higher incentives (such as rewards for being deceptive) lead to more frequent deceptive behaviour and increase the expectation that the counterpart's misrepresentation will increase, as well.

The second largest cluster in period 4 deals with the topic of *gender*. In period 4, this topic has significantly increased its influence as compared to its first appearance in period 3, as its size has risen from seven publications in period 3 to 16 publications in period 4. The publications within the cluster are a combination of publications from the previous period and new ones. Since the average publication year has increased from 1997 in period 3 to 2002 in period 4, the cluster is defined by more recent publications. The publications by Stuhlmacher (1999), Bowles (2005), Bowles (2007) and Kray (2005) can be deduced as most defining publications due to their slightly larger node sizes and thus to their higher influence on the topic. Of these defining publications, only Stuhlmacher (1999) reappears from period 3, while the other publications are new in this period. As was worked out in period 3, the defining publications of period 4 underline that the role of stereotypes exerts an especially strong influence on the gender gap in negotiation settings. Bowles (2007) stresses that females are stereotypically more linked to characteristics such as niceness and lower levels of salary, for which reason the counter-stereotypical negotiation for more compensation leads to negative evaluations. Kray (2005) delivers approaches for how these stereotypes can be weakened. One such approach can be via the provocation of reactions to the stereotypes; another avenue of approach could be emphasis placed on how helpful stereotypical female traits are for negotiations. Bowles (2005) reveals that women can let go of their stereotypes and thus significantly enhance their negotiation results when not negotiating for themselves, but as the representative of a group. Gender differences do not always influence the negotiation outcome, though. No difference between males and females is visible in cases of high structural ambiguity involving a great deal of uncertainty both about one's own limits and about those of the other party.

By far the largest cluster of period 4 is the topic of *emotions*. With its 37 publications this cluster stands out in size from the other clusters not only in period 4 but also in all other clusters within the time frames investigated. The topic of *emotions* was already a quite influential topic in period 3, with 17 publications, which made it the second largest cluster in the 2000-2009 time frame. Since 20 additional publications emerge in the cluster, its impact increases significantly from period 2000-2009 to 2010-2018. Nearly all the publications of period 3 reappear among the 37 publications of period 4, some of them with increased influence (e.g., Kopelman 2006), and others with decreased influence (e.g., Parkinson 1996). The newly added publications have increased this cluster's recentness to an average publication year of 2004, compared to period 3 with its average year of 1999. The most frequently examined emotion remains anger, followed by happiness as the second most investigated emotion, whereby anger is compared mainly to either happiness or emotional neutrality. Moreover, once again most of the studies examine the interpersonal effect of emotions. As is the case with period 3, the added publications predominantly represent empirical studies. Only a few of them are conceptual in nature (e.g., van Kleef (2009), who introduces a framework to account for the effect of an opponent's emotional expression on the observer's behaviour, see chapter 5.2.1). As shown in Figure 4-9 and Table 4-9, the cluster is characterized by a high level of interconnectedness, owing to its density of 0.28, which is for its size of 37 publications still on a substantial level. Consequently, the contentual proximity of this research topic is quite high, a feature that was already present in period 3. A closer look at the dominating authors within this topic shows that van Kleef clearly stands out, represented as he is in the cluster with eight publications as first author. Besides, the average degree centrality of 10.16 is particularly outstanding in period 4 as in all other periods.

As the visual analysis of this cluster already reveals, multiple publications have a quite high degree centrality and are thus defining for the cluster. Of major relevance are the publications by van Kleef (2004a), Sinaceur (2006), Kopelman (2006), Keltner (1999) and van Kleef (2007). Of these, van Kleef (2004a), Sinaceur (2006) and Keltner (1999) were also among the most defining publications in period 3. By contrast, the influence of Kopelman (2006) has increased from the prior period to period 4, while van Kleef (2007) emerges for the first time in this period. Van Kleef (2007) confirms the results of the publications in period 3 by working out that low power negotiators claim less value when confronted with an angry counterpart. At the same time, the publication complements its investigation by assessing the appropriateness of the expressed anger. This yields the finding that in the case of high power emotion receivers, an inappropriate expression of anger on the part of the opponent causes a higher claim of value by the former. Thus the effect of anger is influenced by the power structure and the appropriateness of the anger. Unlike to the other defining publications, Kopelman (2006) investigates the effect of strategic displays of emotions on bargaining situations. While the other publications lay their emphasis on genuine emotions, Kopelman (2006) considers faked emotions. The publication finds out that the display of positive emotions leads on the one hand to a greater deal-closing likelihood and on the other hand to a greater likelihood that a future business relationship will be striven for. Moreover, the display of positive emotions is effective in a distributive negotiation, as the opponent makes larger concessions. These findings contradict the other publications, in which the expression of anger is more effective than the expression of happiness.

4.4 Discussion

The aim of the first study was to reveal the intellectual structure of negotiation research and its evolution over time, in order to strengthen the shoulders of giants. For this reason, a citation analysis was conducted as a first step toward determining the most-cited publications. Based on this citation analysis, a co-citation analysis ensued, which worked out the defining themes of negotiation research along with their interconnectedness. In the following, the main findings are provided along with their discussion.

Citation Analysis

The top cited publications were determined, whereby the books of Fisher (2011) *Getting to YES* and Raiffa (1982) *The Art and Science of Negotiation* were found to be particularly outstanding. In addition, the analysis revealed that nearly all highly influential publications are books representing standard works for the field of negotiation research. This also explains the increasing average age of references from period to period. Moreover, it was possible to work out periodical developments in which the publications of period 1 barely reappeared, and the ones of period 2 proved quite defining for the two following time frames. Consequently, period 1 was shown not to be defining for the remaining time periods. This finding supports the insights of several scholars who also concluded that there is a great deal of research prior to 1980 that is not characteristic for negotiation research (e.g., Backhaus, Hummelsberger, and Wilken 2005; Bazerman, Curhan, and Moore 2001; Herbst 2007) (see chapter 2.2).

Co-Citation Analysis

For each of the four time frames, a network was created that allowed for the deduction of topic clusters. The peculiarity of these topics was that their level of detail was quite high. Unlike other co-citation analyses that frequently work out more general topics, it was possible to deduce quite specific topics (e.g., Backhaus, Lügger and Koch (2011) work out more superordinate topics such as pricing strategies in business-to-business marketing research). This allows for a great depth of analysis. As time periods progressed, the degree of interconnectedness between the identified topics diminished. While several interconnections existed in period 1, these relationships were barely present in period 2, and the topics of period 3 and 4 were not connected at all. This underscores the fact that negotiation publications refer exclusively to each other in terms of their direct relatedness on the level of content, but that they do not consider other publications more comprehensively. Thus, interconnections between topics is negligible for negotiation research, and the different topics do not influence each other. With regard to the topics of the time periods and their evolution over time, Figure 4-10 summarizes the results. The font size of the topics proportionately represents the number of publications within the respective cluster, and thus the impact of that cluster. The asterisk highlights topics reappearing among the time frames.

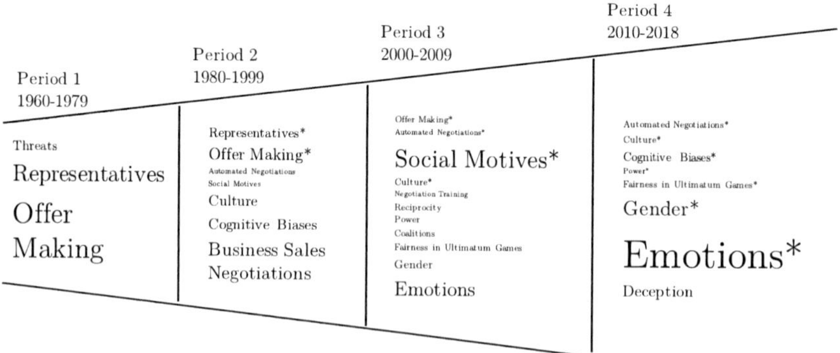

Figure 4-10: Evolution of Negotiation Research Topics
Source: Author's own illustration adapted from Backhaus, Lügger and Koch (2011).

With respect to the evolution of the negotiation topics, the analysis reveals that multiple topics reappear across the time periods (see here and in the following Figure 4-10). In this manner, for instance, the topic of *offer making* appears in periods 1, 2 and 3, albeit with diminishing influence. Also, the topic of *automated negotiations* appears with only small influence in periods 2, 3 and 4 – each time with different publications and a different focus with respect to content. While each period encompasses several new topics beyond that, period 4 represents an exception. Nearly all topics of period 4 are ones that reappear from the preceding periods, such that only one new topic emerges in that time frame, which reflects a saturation of main negotiation topics.

By assessing each of the four time periods separately, the topics of *offer making* and *group representatives* are especially relevant for period 1, *social motives* and *emotions* for period 3, and *gender* and *emotions* for period 4. Only period 2 displays no topic that stands out clearly. Evaluation of the topics across all time frames shows that the topic of *emotions* in particular stands out from all other topics, making up the second largest topic in period 3 and by far the largest topic across all periods in period 4. Next to its dominating size, the topic of *emotions* exhibits by far the highest average degree centrality, so that especially for this topic quite influential publications can be determined (e.g., van Kleef 2004a, Sinaceur 2006, Keltner 1999). The author van Kleef stands out especially here, as multiple publications by him are among the clusters. The analysis of this topic reveals contradictory findings pertaining to the effects that emotions have on the negotiated outcomes.[40] While several publications work out that the expression of anger is helpful in negotiations, as it occasions the opposing party to concede more and has a lower demand level, other publications underline the downside of anger stemming from the fact that an angry opponent elicits an angry response, which in turn increases the risk of an impasse. Moreover, it is worked out that to the extent that anger is a genuine emotion, an angry negotiator performs better, while in case of strategic emotions happiness is superior. Thus the assessment of an emotion's effectiveness depends primarily on the individual moderators. In a large number of cases, the level of power influences the emotion's effectiveness; but also the distributiveness/integrativeness of the negotiation situation or the underlying time pressure have an influence on the relationship.

[40] Contradictory findings could also be found for other topics, like *offer making* or *culture*; however, due to their diminishing or small influence across the periods, their further investigation is not seen to be as promising as an investigation of the topic of *emotions*.

By comparing the main topics detected to the generic negotiation framework introduced in chapter 2.3, it becomes apparent that most of the themes can be assigned to the factors influencing the negotiation process, whereby most topics relate to the personal influence factors (e.g., gender, culture, cognitive biases, emotions, etc.), which serves to confirm their high relevance, worked out in chapter 3.2.1. With regard to the negotiation process itself consisting of the phases negotiation preparation and negotiation implementation, fewer topics are identified by the co-citation analysis that can be assigned to these phases. In this way, only *offer making* appears within the themes and tactics such as *deception* and *threats*. However, further impactful topics that deliver more aspects both for the preparation and the implementation phase are not characteristic for the negotiation research field to date.

The outcome of the first empirical study closed the research gaps presented in chapter 3.1.2. First of all, the qualitative nature of narrative literature reviews was criticized, along with their limited human processing capacity; thus the call arose for a quantitative approach to investigate negotiation research that would be capable of processing very large amounts of publications for the purpose of working out main topics. The application of the quantitative approaches citation and co-citation analysis ensured the objectivity and the broad inclusion of negotiation publications (with 66,294 references entering the analysis) called for, thus eliminating the lack of such a consideration. Moreover, since multiple research areas were investigated and no limitations were set on the negotiation topics and journals, the lack of a comprehensive approach that would include research areas relevant to business negotiations, negotiation topics and journals broadly was remedied. Lastly, a contemplation of the evolution of negotiation research was called for, as most simple literature reviews do not examine the dynamic structure of the research field. Through the investigation of four time periods and their comparison, this lack was also alleviated.

Nevertheless, the first study is not free of limitations. While chapter 6.3 deals with the limitations of the studies in more detail, one limitation should be already mentioned here. Study I delivered the intellectual structure of the negotiation research field by focusing on the most influential publications. Because of this focus, the claim of the study was not (nor could it be) to depict the negotiation research field in its entirety, but rather to detect impactful and strongly researched areas. Consequently, Study I did not come up with all research topics examined within the last 60 years and with all publications that can be assigned to them, but with the most defining ones. Thus, researchers and practitioners should see the study's results as a starting point from which they can get an impression of the research field indeed, but from which on they have to do research for more publications and more themes on their own.

With the insights gained by the co-citation analysis, the shoulders of the negotiation giants are strengthened and the foundation is laid for the second study of this thesis, which aims at growing the dwarf. Due to the high investigation of the topic of *emotions*, its high actuality through the presence in recent time periods, and its strong positive development from time period 3 to 4, the importance and recentness of that topic was worked out. Moreover, a large share of empirical studies that provide contradictory results was detected by the co-citation analysis. Based on the relevance of the topic, an aggregation of these contradictory findings to form one overall valid result seems quite conceivable for the purpose of deducing a consistent picture of the findings. To this end, relevant moderators that influence the overall effect should be examined, in order to gain a unified impression. A meta-analysis is best suited to this purpose, as it addresses these aspects. Consequently, in the following Study II, such a meta-analysis will be conducted to examine the effect of emotions on negotiation outcomes and its moderators.

5 Study II: Growing the Dwarf – A Meta-Analysis on the Effect of Emotions on Negotiation Outcomes

5.1 Research Objectives and Study Overview

In its structuring of the complex negotiation research field within the last 60 years, Study I enables the organization of knowledge generated by prior negotiation scholars. In this way, it reinforces the foundation of negotiation research – or in the words of the metaphor accompanying the thesis at hand: the shoulders of the negotiation scholars – by determining key research topics based on the relations between the most relevant publications of the research giants. Along with the giants' shoulders, the dwarfs, which represent current scholars building on the findings of the preceding researchers, are the second component of the metaphor introduced in chapter 1.2. By referring back to the prior work of the giants, the dwarfs are able to gain a wider perspective. It is exactly this point where Study II wants to position itself, by relying on the strengthened shoulders and becoming a rising dwarf itself.

The co-citation analysis has revealed the topic of *emotions* as the most frequently examined theme in negotiation research in the last decades (see here and in the following chapter 4.3.2). It has also been shown that the predominant emotion dealt with by researchers is *anger*, which is usually compared to *emotional neutrality* or *happiness*, as the second most frequently investigated emotion. Moreover, most studies within the topic clusters focus on the interpersonal effects of these emotions, which means that their effect on the behaviour of the person receiving the emotions is what is examined, rather than the behaviour of the person expressing them. The results of the co-citation analysis have furthermore illustrated that the findings of empirical studies within the clusters show contradictory results regarding how helpful the emotions are with respect to negotiation outcomes. Consequently, a lack of knowledge is present with regard to the usability of emotions: Is anger helpful or harmful in comparison to emotional neutrality or displays of happiness? What are the factors that influence the (dis-)advantageousness of anger? Owing to the lack of any satisfactory approach in negotiation research that would account for this lack of clarity, it has been impossible to date to reliably answer any of these questions (see chapter 3.2). It is this gap that Study II proposes to close.

All previous studies addressing the contradictory findings of the interpersonal effects of anger (over against happiness or neutrality) in negotiations are conceptual in nature and have consisted in verbal descriptions of their insights (see chapter 3.2). Study II clearly distances itself from such subjective presentations and relies on a meta-analysis as the most adequate quantitative method recommended in cases of contradictory findings arising from many empirical studies (Eisend 2014; Lipsey and Wilson 2001). The main strength of a meta-analysis is that it quantitatively aggregates various findings of empirical studies by relying on statistical methods in order to obtain one overall effect size (Bijmolt and Pieters 2001; Chow 1987). In this way, it is not restricted to a specific quantity of studies able to be included in the analysis; it is much rather unlimited as to the number of possibilities it provides to account for them. Thus, meta-analyses are endowed with great explanatory power, owing to the large number of studies and individuals they build on (Eden 2002; Ellis 2010). In addition to the overall main effect a meta-analysis helps to achieve, it also permits the integration of moderator variables, which deliver explanations for variability within this main effect (Stanley 2001; Stanley, Doucouliagos, and Jarrell 2008).

For the establishment of a profound conceptual foundation, the subsequent meta-analysis builds on the most popular model for determining the interpersonal effects of emotions in social interactions: the Emotions as Social Information (EASI) model (van Kleef 2009). This meta-analysis is furthermore characterized by an extensive literature search, which profits from the search conducted in Study I. As a result of this literature search, the meta-analysis is based on a sample consisting of 94 studies, including 267 effect sizes and encompassing 16,795 individuals. It thus is among the largest meta-analyses to have been published in the field of personal factors in negotiations to date (see chapter 3.2.1). Besides, the meta-analysis splits negotiation outcomes in economic outcomes (EOs) and subjective outcomes (SOs), as both are highly relevant (but quite different!) figures for negotiations – whereby the latter are frequently overlooked in studies.

Study II is structured in the following way: First, the conceptual framework is introduced, based on which the hypotheses for the meta-analysis are derived (chapter 5.2). In this process, the EASI model from which the hypotheses are derived is presented first (chapter 5.2.1), then the resulting conceptual model is set up (chapter 5.2.2). After that, the methodological foundation and research design of the study are presented in chapter 5.3, which consists first of an introduction to the principles of and the procedure involved in a meta-analysis (chapter 5.3.1). Subsequently, the procedure for collecting the articles and the coding process of the studies are described in chapter 5.3.2. Chapter 5.4 presents the results of Study II, starting with a short descriptive presentation of the study and the moderator characteristics (chapter 5.4.1). Then follows a presentation of the main effects of the meta-analysis (chapter 5.4.2), and a depiction of the results of the meta-regression model, which includes the moderators, in chapter 5.4.3. Finally, Study II closes with a discussion of the results in chapter 5.5.

It is with this purpose in mind that the author stands on the shoulders of the negotiation giants, as she relies on their most frequently examined research theme and also integrates their empirical study results. Without this input this approach would not be possible. On the strength of it, though, the author of this thesis is able to stand on the giants' shoulders and see further, ascending as she does thanks to their findings. In this way, she enriches negotiation research and practice with new and valuable insights based on giants' prior work. The outcomes lead to clarification in negotiation research with respect to the appropriateness of the expression of the emotions of anger and happiness. Additionally, the meta-analysis is able to detect explanatory factors that account for the deviations occurring in the effects observed in the studies. Professionals can gain considerable benefit from this analysis, as it provides them with clear, reliable instructions for the use of emotions in their negotiation practice.

5.2 Conceptual Framework

5.2.1 Emotions as Social Information (EASI) Model

As described in chapter 2.3.1.2, emotions can be classified as either intrapersonal or interpersonal. Intrapersonal effects refer to the emotions felt and (partly) expressed by an affected individual that in turn influence his/her own insights and behaviour (Barry, Fulmer, and van Kleef 2004; Forgas 1995). Until the millennium, most negotiation studies on emotions focused on the intrapersonal effect or on diffuse mood states (van Kleef 2009). However, the intrapersonal experience of emotions is usually expressed in many ways – whether as a reflection in the individuals' faces or bodies, in their choice of words or in the active sharing of their emotional experiences (here and in the following van Kleef and Sinaceur 2013). Thus expressions of emotions can be observed by others who in turn respond to. This constitutes the interpersonal effect of emotions. Consequently, emotions shape not only the affected individuals, but their counterparts as well, which makes emotions an eminently social matter (e.g., Frijda 1986; Keltner and Haidt 1999; Parkinson 1996).

While in its early times the predominant focus of negotiation research on emotions was on intrapersonal effects, in recent decades researchers have called for a much higher focus on interpersonal effects (e.g., Barry, Fulmer, and van Kleef 2004; Barry and Oliver 1996; Morris and Keltner 2000). One result of this more recent trend has been the EASI model, which has become quite popular within the last years (e.g., Bhattacharjee and Moreno 2017; Kung and Chao 2019). The aim of the EASI model is to explain the interpersonal effects of emotions in social interactions (van Kleef 2007). The EASI model explains the effect that emotional expression on the part of an individual has on the behaviour of another person so that it describes emotions from a social-functional perspective (van Kleef 2009). In strong contrast to prior models, which merely considered diffuse moods (e.g., Forgas 1995; Schwarz and Clore 1983), the EASI model seeks to explain the effects of discrete emotions such as sadness, anger and happiness (van Kleef, De Dreu, and Manstead 2010). Thus, it becomes more specific by deducing concrete effects for specific emotions (van Kleef 2009). As negotiations represent prime examples of social interaction (see chapter 2.1), the EASI model is frequently applied to negotiation contexts (e.g., van Kleef 2009, 2014), which makes it an appropriate model for the meta-analysis at hand.

The EASI model consists of two main paths and two moderating forces (see here and in the following Figure 5-1). The two main paths describe two distinct processes involved in the effect of the expressed emotion on the observer's behaviour: the observer's affective reactions and the observer's inferential processes (here and in the following van Kleef 2007). Both paths are moderated by the observer's information processing and social-relational factors. Each component will be described more in detail in the following.

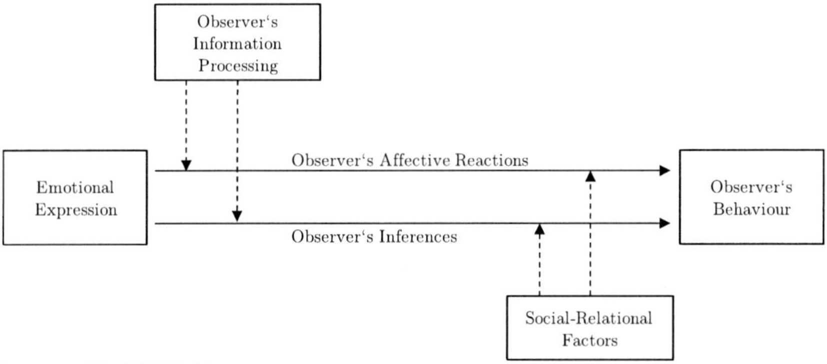

Figure 5-1: The EASI Model
Source: Author's own illustration adapted from van Kleef (2009).

First, the observer's *inferential processes* relate to any inferences the observer can draw concerning the expresser based on the latter's emotional expression (van Kleef 2009). Various information can be deduced from the expression of emotions, such as social intentions, feelings, or the expresser's appraisal of the situation (Clore 1994; Keltner and Haidt 1999; Schwarz 1990). Different behaviours on the part of the observer can be predicted based on the emotion expressed by the opponent. Thus since the informational value of discrete emotions can be generalized, it is possible to deduce the expected direction of the reactions (van Kleef 2009). In this way anger, for instance, arises in a social context if the goals of an individual are not accomplished and the expresser blames the other person for it (Smith et al. 1993). For this reason, observers of outbursts of anger may infer from the anger expressed by their counterpart that they have done something wrong and need to compensate for it through reactive behaviour (here and in the following van Kleef 2009). Moreover, observers may deduce from an anger expression that their opponent has reached his/her limits which in turn can lead to strategic behaviour by the former (e.g., making concessions in a negotiation in order to prevent an impasse). By contrast, in the case of a happy opponent, observers infer that expectations have been met and everything is going well (Smith et al. 1993). Based on this information, observers stick to their course and infer that no need exists for any change in behaviour (van Kleef 2009).

Second, *affective reactions* relate to a number of possible affective processes that emerge within the observer on account of the partner's emotion (here and in the following van Kleef 2009). There are two types of such affective reactions that can be distinguished. First, emotions can spread directly from the individual expressing emotion to the one observing it. This effect is known as emotional contagion; it represents the tendency to adopt the emotion of the opposing person (e.g., an observer confronted with an angry opponent turns angry as well). This process can take place by mimicking the nonverbal emotional expression of the expresser (e.g., via face or voice), leading to a feeling of this emotion due to physiological feedback. Mimicry, however, is not mandatory for emotional contagion, as this process is also possible through conditioning or perspective-taking (here and in the following Hatfield, Cacioppo, and Rapson 1992; van Kleef, Homan, and Cheshin 2012). Therefore, no face-to-face (FTF) interaction is necessary for emotional contagion to occur, and it can also be reached for instance through a computer-mediated interaction. The second type of affective reaction relates to the effect the expression of emotion has on impressions and interpersonal liking (van Kleef 2009). In this manner, expressions of happiness were found to increase the degree to which the observer likes the person expressing happiness, the wish to see the latter again, and the observer's satisfaction with the interaction; whereas manifestations of anger led to diminished willingness to meet again, less satisfaction and to a dislike of the counterpart (e.g., Clark and Taraban 1991; van Kleef, De Dreu, and Manstead 2004b).

The affective reactions and inferential processes can lead to identical but also to divergent behaviour on the part of the observer (here and in the following van Kleef 2009). Inferential processes and affective reactions may converge by eliciting the same behaviour. For instance if a friend of yours is sad due to a death in the family, this leads to affective reactions of compassion and to the inference that he/she needs your help, leading you behave accordingly in support of your friend. Nevertheless, affective reactions and inferences can also motivate diverging behaviours. For instance, a negotiator perceives the anger of the opposing party, which may elicit his/her own anger, which in turn leads to competitive behaviour (affective reaction), or which may lead to the inference that the other party has reached his or her limit, thus leading to strategic cooperation (van Kleef 2009; van Kleef, De Dreu, and Manstead 2004a). In these cases, it is the underlying situation that determines which path is the more dominating.

Both processes are moderated by the observer's information processing and by social-relational factors (here and in the following van Kleef 2009). As the moderating forces increase the relative strength of the paths, they determine which path is experienced more strongly by the observer. The first moderator, *information processing*, relies on the core assumption of the EASI model: that emotions contain information. Consequently, the interpersonal effects of discrete emotions are dependent on the observer's ability to process information, and on how motivated he or she is to do so. Given a large amount of information processing, the behaviour of the observer is more strongly influenced by the inferences process than by the affective reactions, and vice versa. The amount of information processing is mainly determined by individual characteristics and situational factors (De Dreu and Carnevale 2003; van Kleef 2007). Personality factors that enhance information processing are factors such as the need for cognitive load, whereas need for cognitive closure (the need for closure of events and actions) decreases information processing on the part of the observer (Cacioppo and Petty 1982; Kruglanski and Webster 1996). With regard to situational factors that enhance the information processing motivation, factors such as an individual's involvement in the task and his or her accountability for a situation play a role (Eagly and Chaiken 1993; Petty and Cacioppo 1986). Lastly, situational factors that cause a decrease in the motivation or, as the case may be, the capability to process information are factors such as mental fatigue (Webster, Richter, and Kruglanski 1996), time pressure (Heaton and Kruglanski 1991) and a high power position held by the emotion receiver (Fiske 1993; Goodwin et al. 2000).

The second moderator, *social-relational factors*, encompasses variables such as status relations between the emotion expresser and the observer (equal vs. different), their interdependence (who is more dependent on whom) or "display rules" of an organizational or cultural nature, which control the adequacy of an expression of emotion (van Kleef 2007; Matsumoto 1993). Thus the most significant underlying question pertaining to the social-relational factors is in how far the emotion expressed is seen by the observer as appropriate, i.e. whether the emotional reaction is appropriate given the underlying situation (van Kleef 2007, 2009). As an example, social interactions in which a negative emotion targets the observer personally rather than his/her behaviour are usually seen as inappropriate (Steinel, van Kleef, and Harinck 2008). In cases in which an expression of emotion is highly inappropriate, the strength of the affective reactions path becomes greater, thus increasing the expression of affective responses by the observer and decreasing his or her likelihood to act on behalf of the counterpart's intentions (van Kleef 2007).

5.2.2 Conceptual Model and Hypotheses

In the following, the EASI model introduced in the preceding chapter will be used as the basis for deriving the hypotheses and the conceptual model for the subsequent analysis. According to the EASI model, interpersonal emotions can influence the observer's behaviour either along the path of the observer's affective reactions or along that of the observer's inferences.

Main Effect on Economic Outcomes

The pivotal factor for determining the main path in the EASI model is the assessment of the competitiveness of the social interaction (here and in the following van Kleef, De Dreu, and Manstead 2010). Individuals are more likely to feel the counterpart's emotion in situations in which the goals of parties are cooperatively linked. One example of this is a study that has shown how the moods of cricket players were more strongly correlated with their team members' feelings (cooperative relationship), than with the feelings of players from competing teams (Totterdell 2000). Nevertheless, the social interactions examined in the thesis at hand are negotiations that are typical competitive settings (van Kleef, De Dreu, and Manstead 2010). Much as negotiations may have integrative components in which it is possible for bargainers to form cooperative behaviour, the *nature* of negotiations is competitive, as for the most part they involve divergent interests and incongruent goals (such as the price in a sales negotiation) (see chapter 2.1). Along with the lower likelihood of emotional contagion in competitive settings, competitiveness is usually associated with a lesser degree of trust than with cooperation (Deutsch 1973). It therefore becomes increasingly important that the intentions of the opponent be detected, as any such information is valuable (here and in the following van Kleef, De Dreu, and Manstead 2010). It is for these reasons that a competitive environment is associated with higher relevancy of the observer's inferences than with affective reactions.

Since the emotion of *anger* is compared to *happiness* and *emotional neutrality* in the meta-analysis at hand, as is done in the primary studies as well, the meaning of the dominance of the path of the observer's inferences needs to be assessed for all three of them. With regard to an expression of anger, the inferential path means the following: The inferential processes foster the derivation of information contained in the anger expressed, which in turn makes observers more likely to adapt their behaviour in favour of the persons expressing anger (here and in the following van Kleef 2009). This is due to the fact that the observers either assume the opposing party to be close to its limits, or else they think that they have done something wrong that they need to correct in order to avoid an impasse. Accordingly, the observers make concessions towards the persons expressing the emotion. Consequently, it is assumed that an expression of anger has an advantageous economic effect for the person expressing the emotion, as concessions lead to an increase in the expresser's EOs. By contrast, when happiness is expressed, the inferences made by the observer are that the counterpart is content with the course of the negotiation and that the observer does not need to change anything, since an impasse is quite unlikely to occur (van Kleef 2009). And in cases in which the opponent shows no emotion at all, the observer' behaviour is assumed not really to be adaptive in favour of the counterpart, since no information and thus no limits are recognizable. However, in terms of changes it causes in observers' behaviour toward the opposing party, happiness is assumed to be even weaker than emotional neutrality, as the observers of happiness are able to infer more strongly the information that everything is fine. This leads to the first two hypotheses (H):

H_{1a}: *Compared to emotional neutrality and happiness, anger increases the EOs of the expresser of anger.*

H_{1b}: *The positive influence of anger on the EOs of the expresser of anger is stronger compared to happiness than compared to emotional neutrality.*

In addition to the hypotheses of the main influence of anger compared to that of happiness and neutrality on the EOs of the person expressing anger, the EASI model also delivers two categories of moderating variables – motivation and ability to process information, and social-relational factors – that may influence the strength of the main effect for which hypotheses are deduced in the following.

Information Processing

The EASI model proposes the *power* level of the emotion recipient as a relevant moderating factor regarding the recipient's information processing motivation (van Kleef 2014). As a rule, the presence of power allows the party in possession of power to control both his or her own and the others' resources, which leads to a decreased dependency on the decisions made by the counterpart (Magee, Galinsky, and Gruenfeld 2007). Thus usually those individuals in a position of power are the ones making decisions (here and in the following Galinsky, Gruenfeld, and Magee 2003). In line with this, it was found that individuals in possession of power exhibit an increased tendency toward action-oriented behaviour. Due to their power position, they engage less in cognitively complex thinking and have a stronger tendency to implement their plans (Gruenfeld 1995). By contrast, in the absence of power there is an increase in individuals' propensity to deliberate (Galinsky, Gruenfeld, and Magee 2003). For these reasons, information processing motivation is higher for individuals in a position of little power, as they are more dependent on the information they can get from the opposing party in order to infer his or her intentions (De Dreu and van Kleef 2004). As the information processing motivation increases, the EASI model's path of inferential processes becomes stronger, which leads to a stronger positive effect of an expression of anger on the EOs of the person expressing it. This leads to the following hypothesis:

H_2: *If the power level of the anger recipient is low, the positive effect of anger on the EOs of the expresser is stronger than if the power level is high.*

A second moderating factor of the EASI model which affects the information processing motivation of the counterpart is the degree of *time pressure* present in a negotiation (van Kleef 2014). The effect of time pressure on information processing is commonly described by an individual's need for cognitive closure (De Dreu 2003). In cases where the need for cognitive closure is strong, individuals are characterized by considerable cognitive impatience, which leads to rash judgments based on inconclusive knowledge and rigid thoughts (here and in the following De Dreu 2003; Kruglanski 1989). By contrast, if there is only little need for closure, individuals are more prone to seek information in order to create a profound knowledge base for a situation. In the literature it is considered to be common sense that when time pressure is high, the need for cognitive closure is stronger than when it is only slight (here and in the following De Dreu 2003; Kruglanski and Freund 1983). This assessment is justified by the insight that time pressure leads to a closed-mindedness that in turn motivates individuals to seek cognitive closure. Thus, individuals cease considering alternatives and do not engage in thorough information processing. Consequently, as the information processing motivation increases in cases of low time pressure, the positive effect of expressions of anger on the EOs of the expresser also increases. This leads to the following hypothesis:

H_3: *If the degree of time pressure is low, the positive effect of anger on the EOs of the expresser is stronger than in the presence of high time pressure.*

Along with the hypotheses on information processing deduced directly from the EASI model, the thesis at hand aims to take a closer look at two additional moderators, which are closely linked to the set of information processing moderators presented by the EASI model. These moderators are the competitiveness of the negotiation setting and the negotiation medium applied.

First, the *competitiveness* of the given negotiation influences the effect of the expression of anger. As has been worked out for the main effect, the nature of negotiations is a competitive setting in which the parties have incongruent goals, whereby the level of this competitiveness varies from negotiation to negotiation. Owing to their distributive nature – the selling party always seeks a higher price and the buying party a lower one – price negotiations, for example, are strongly competitive (here and in the following Halevy 2008). However, various other negotiation settings are thinkable, in which integrative potential exists, which would increase the cooperative behaviour of bargainers (e.g., a job negotiation over vacation days and working hours). In such situations, the competitiveness of the parties decreases, since both sides reach better outcomes when they cooperate. As was stated for the main hypothesis H_1, competitiveness increases the information processing motivation of the party receiving the emotion. Consequently, if due to distributive negotiation issues the underlying negotiation is even more competitive, the information processing motivation increases accordingly. This amplifies the positive influence of the expression of anger on the EOs of the expresser:

H₄: *If the negotiation is strongly competitive, the positive effect of anger on the EOs of the expresser is stronger than for cooperative negotiations.*

Second, the negotiation *medium* through which the expression of emotion is transmitted influences the degree to which an individual is capable of processing information. Negotiation media can be classified according to their media richness, which refers to the amount of details and information that can be transferred via a communication medium (Daft and Lengel 1986) (see chapter 2.3.1.1). The more strongly a communication medium enables the processing of multiple information cues simultaneously, the richer it is considered to be (Li 2007). Cues can be visual, vocal and written; they incorporate such criteria as facial expressions, the sound of a person's voice, or body language (Daft and Lengel 1986). As richer communication media are able to transfer more information, they enhance an individual's information processing ability. As worked out, the richest communication medium – the one that offers the most cues and enables direct feedback – is FTF. From this richest medium downwards, one can observe a decrease in the amount of available cues, from cases in which there is no FTF interaction (e.g., interactions via webcam), to ones where there are no visual cues (e.g., an interaction via phone call), and, finally, to ones in which only written cues are available (e.g., an interaction via e-mail) (Swaab et al. 2012) (see also chapter 2.3.1.1 for a more detailed consideration). Consequently, if a negotiation is based solely on written cues, for instance in computer-mediated negotiations, the amount of information which can be transferred from the text is much lower than in FTF negotiations. Accordingly, the ability to process information decreases, which in turn weakens the inferential processes path of the EASI model. This leads to the following hypothesis:

H_5: *If anger is expressed in purely text-mediated negotiations, the positive effect of anger on the EOs of the expresser is reduced than when negotiation takes place via FTF.*

Social-relational Factors

With regard to moderating social-relational factors, the deciding criterion for the relative strength of the main paths is the appropriateness of the expressed emotion within a specific given context (see chapter 5.2.1). If an observer perceives an expressed emotion as inappropriate, negative emotions are triggered within that observer (Bucy 2000). In this way it holds that the higher the degree of inappropriateness of an emotion, the higher the relative strength of the affective reaction paths in the EASI model (here and in the following van Kleef 2014). Accordingly, strongly negative emotions felt by the emotion recipient outweigh his or her motivation to process information.

A common factor which determines the appropriateness of an emotion is its *authenticity* as perceived by the recipient. In cases in which a communicated emotion seems inauthentic, the appropriateness of the expression of emotion decreases (Rafaeli and Sutton 1989). Due to its inauthenticity, the emotional expression is perceived as dishonest and as a manipulative tactic, which in turn leads to negative emotions in the observer (van Kleef 2014; Rafaeli and Sutton 1989). An emotion is perceived as inauthentic if the negotiator gets the impression that it is only expressed for tactical reasons and thus represents a faked emotion rather than a genuine one (Côté, Hideg, and van Kleef 2013). The negative affective reactions in turn trigger competitive behaviour in the perceiver (here and in the following van Kleef 2009). By contrast, if the emotion is perceived as authentic, the affective reactions path becomes weaker, as no negative reactions are felt; thus information processing stays in the foreground, and consequently the EOs for the person expressing anger are higher in the presence of an authentic emotion than in cases of a mere tactical expression of it. This leads to the following hypothesis:

H_6: *If the anger expression is perceived as authentic by the receiver, the positive effect of anger on the EOs of the expresser is stronger than for perceptions of inauthentic anger.*

A second factor for determining the appropriateness of an expression of anger is the *target of the emotion*. In this way, one can distinguish whether anger is directed toward the behaviour of the recipient of anger or toward this recipient personally (van Kleef 2009). Depending on what or whom the target of the expression of anger is, the perceiver may behave cooperatively or competitively. In cases in which the expression of anger is directed at the recipient personally, he/she usually appraises it as inappropriate, which triggers negative emotions in the recipient and leads to competitive behaviour on his or her part (here and in the following Steinel, van Kleef, and Harinck 2008). Nevertheless, if the expression of anger is directed toward the recipient's behaviour, negative reactions are less intense, the information processing motivation is stronger, and cooperative behaviour is more plausible. Thus, the positive influence of anger on the EOs of the expresser is greater if anger targets the recipient's behaviour, than if it targets his or her person; which is mirrored in the following hypothesis:

H_7: *If the anger expression is directed at the behaviour of the recipient rather than at the person, the positive effect of anger on the EOs of the expresser is stronger.*

In addition to the main effect and moderating effects derived from the EASI model, there are three control variables that enter the meta-regression of the effect on EOs, namely the author of the studies, the publication status (grey literature vs. peer reviewed), and the type of negotiation (e.g., sales negotiation or job negotiation). These control variables analyse whether the results are dependent on one of these factors.

Main Effect on Subjective Outcomes

The above-derived hypotheses all target the EOs attainable by the person expressing anger. Nevertheless, the EOs are not the only thing influenced by the interpersonal effects in a negotiation. Next to economic values, such SOs as the desire for retaliation or the desire for future interaction are quite important in negotiations (see chapter 2.3). For the effect of the emotion expressions on the SOs, the affective reactions path becomes more relevant. As previously described, the assessment of a person's impression of the partner and of how well-liked or how disliked he or she is (which is precisely what SOs are supposed to measure) brings affective reactions to the fore, which leads to the following consequences for an expression of anger, happiness or neutrality on SOs (here and in the following van Kleef 2007): If observers are confronted with an angry counterpart, they have a worse impression of and a lower level of trust toward their counterpart, and do not want to cooperate again. In contrast, observers of a happy counterpart like this opponent and receive a better impression of him or her. Observers of emotionally neutral counterparts lie somewhere in between, but should exhibit higher SOs than observers with an angry counterpart. This leads to the following hypothesis:

H$_8$: *Compared to emotional neutrality and happiness, anger decreases the SOs of the recipient of anger.*

Moderators of the effect of anger on SOs cannot be deduced from the EASI model, as the SOs are elements of the affective reactions path, and thus do not represent behaviours on the part of the emotion observer. As the EASI model provides moderators that affect the relationship between IV (emotion expression) and DV (behaviour of the observer), it does not deliver moderators for the main paths themselves.

Taking all hypotheses and the control variables together, one gets the conceptual model visualized in Figure 5-2, on which the conducted meta-analysis builds.

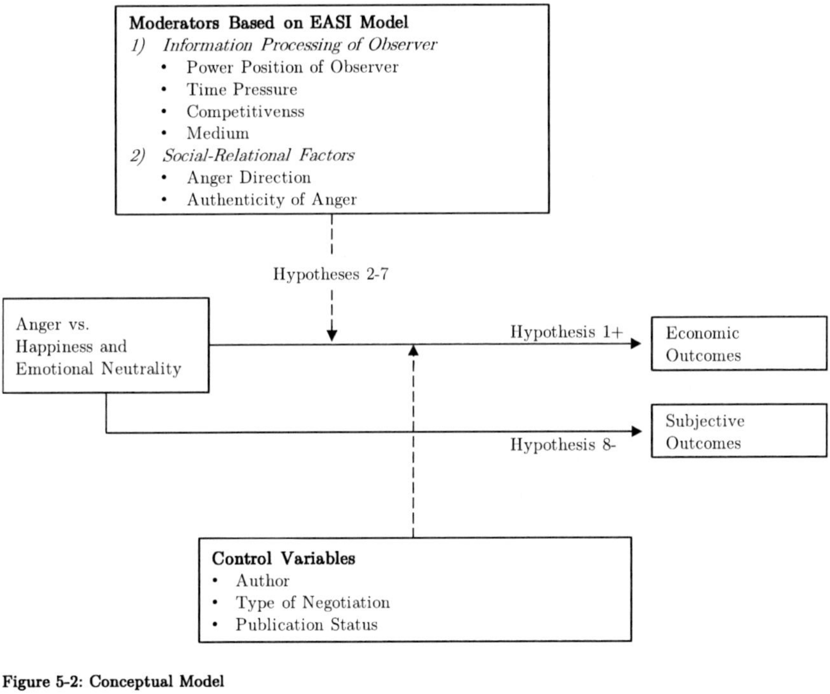

Figure 5-2: Conceptual Model

Source: Author's own illustration.

Prior to the implementation of the meta-analysis, the following sections of this dissertation provide the reader with relevant methodological information, as well as with key figures, statistical procedures and assumptions of meta-analyses, all of which are prerequisite to a proper understanding not only of the results of the analysis but of the procedures for attaining them as well. The next chapter starts by providing all these important explanations and closes with the collection and coding of the data.

5.3 Methodological Foundation and Data Collection

5.3.1 Introduction to Meta-Analysis

From the 1970s onwards, meta-analyses have developed to become a widely applied tool in social sciences (Lipsey and Wilson 2001). The exorbitant increase in the number of published studies has made it difficult for scholars even to keep the overview of all research findings within a single clearly demarcated research topic (here and in the following Eisend 2014). Moreover, these various findings are frequently inconsistent and contradictory. At this point, a meta-analysis represents the adequate method to account for this problem. A meta-analysis can be seen as a form of survey research in which not individuals, but rather research studies are surveyed (here and in the following Lipsey and Wilson 2001). In the same manner as would be done for a survey of participants, a sample of studies investigating the effect of interest is gathered and the studies are "interviewed" in such manner that they are carefully read by the researcher, who then codes all relevant characteristics and quantitative results. Then follows the main aim of a meta-analysis: the quantitative synthesis of the coded empirical study results with the help of statistical methods analogous to those applied in primary data analysis (Denson and Seltzer 2011; Eden 2002). Consequently, meta-analyses are only applicable to the quantitative study results of empirical investigations and do not permit summarization of such publications as theoretical papers (Lipsey and Wilson 2001).

A major advantage of meta-analyses over other review methods (e.g., qualitative reviews) is that they are much more sophisticated, reliable and objective, characterized as they are by a structured and systematic mode of procedure in which each step is documented such that it can be verified by other researchers (Eisend 2014; Lipsey and Wilson 2001). This verifiability is ensured by an organized search strategy for the detection of relevant studies, systematic coding of the variables of interest, and quantitative data analysis (Lipsey and Wilson 2001). Besides, meta-analyses allow for an inclusion of large amounts of studies and their results, so that the relationship under investigation can be tested on a much larger scale (Eden 2002; Eisend 2014; Ellis 2010). Moreover, meta-analyses do not focus solely on significant results of primary studies; rather, they include non-significant relationships as well, so that the effect rather than its significance remains in the foreground (here and in the following Combs et al. 2011; Lipsey and Wilson 2001; Schewe, Hülsheger, and Maier 2014). As insignificance is frequently attributable to high sampling errors which become larger in small sample sizes, meta-analyses can correct this phenomenon by virtue of their ability to create large samples. Lastly, a meta-analysis allows the identification of relevant moderators that represent boundary conditions for the main effect, which means that they provide information concerning those circumstances under which the effect either holds or does not hold (Cortina 2003; Cortina and Folger 1998; Geyskens et al. 2009).

Since a meta-analysis is interested in summarizing the findings of studies investigating the effect of interest, the results of the studies have to be comparable (here and in the following Lipsey and Wilson 2001). This means that on the one hand they have to be conceptually comparable inasmuch as they have the constructs and relationships in common that are the object of their examination (e.g., all studies have to investigate the effect of anger and emotional neutrality or happiness on EOs or SOs). On the other hand, the results have to be present in a similar statistical form. For this purpose, the quantitative results are converted into so-called *effect sizes* (here and in the following Borenstein et al. 2009). Effect sizes reflect the strength of a relationship between two variables, whereby the magnitude of a treatment effect (in the present case: the effect of anger on EOs or SOs) is set into relation to the magnitude of a comparison group (happy or neutral on EOs or SOs). The summary effect aimed for is then simply a weighted mean of all effect sizes gained from the studies, whereby the inverse variance of each effect size is usually recommended as the weighting factor.

In the following sections, first, both the information about the effect size suited for the meta-analysis at hand and its calculation and reconstruction are provided (chapter

5.3.1.1). Then the meta-analytic models for the main effect and for the moderator analysis are presented (chapter 5.3.1.2). Lastly, one of the main and most critical aspects of meta-analyses is illuminated: the so-called publication bias (chapter 5.3.1.3).

5.3.1.1 Computation and Reconstruction of Effect Size

Across studies, effect sizes are calculated in order to deduce the summary effect (Borenstein et al. 2009). Different statistical numbers, such as correlations, elasticities, mean differences/ratios or contingency tables, serve as potential effect sizes, whereby the researcher has to determine which effect size measure he/she wants to rely on consistently (Eisend 2014; Schmidt and Bijmolt 2019). This consistency among effect sizes is quite important, as it enables their summarization (here and in the following Borenstein et al. 2009). Thus, if multiple statistical numbers have to be used for the sake of determining the effect sizes (e.g., correlations and mean differences), as not only one single measure is consistently calculable from the studies, then the effect sizes have to be aligned to one of these statistical numbers (e.g., all effect sizes are transformed into mean differences).

The applied effect size is dependent on the underlying data structure, as it has to be computable from the information provided by the studies (Borenstein et al. 2009). Since the studies of the present meta-analysis mainly report means for the examined effects (see chapter 5.4.1), mean differences are the measure that best suit. Nevertheless, simple mean differences have the disadvantage that they do not account for different scales on which outcomes can be measured (e.g., some studies may report EOs in currencies while others report points) (here and in the following Eisend 2014). Thus the standardized mean difference (ES) is recommended, as it delivers a standardization which enables a comparison between means of different scales. In this way, the standardized mean difference sets the mean of the treatment group (in the present case: anger) in relation to the mean of the comparison group (in the present case: happiness or neutrality) divided by the pooled standard deviation (Lipsey and Wilson 2001) (see Formula 5.1).[41]

[41] Hedges (1981) proposed a correction factor for ES as it tends to be upwardly biased. Nevertheless, this holds for ES based on small sample sizes, especially smaller than 20 (Lipsey and Wilson 2001). As sample sizes of the present meta-analysis by far exceed the value of 20, the application of the correction factor is not seen as necessary.

$$ES = \frac{\bar{X}_1 - \bar{X}_2}{sd_{pooled}} \tag{5.1}$$

ES = standardized mean difference
$\bar{X}_1$ = mean of group 1
$\bar{X}_2$ = mean of group 2
sd_{pooled} = pooled standard deviation

The pooled standard deviation in turn accounts for the sample sizes of each group and the respective standard deviations (Borenstein et al. 2009) (see Formula 5.2). Thus, the mean difference is standardized by the sample sizes and standard deviations of the groups. Any standard errors provided by the studies are transformed to standard deviations by multiplying them with the rooted sample size (Altman and Bland 2005).

$$sd_{pooled} = \sqrt{\frac{(n_1 - 1) * sd_1^2 + (n_2 - 1) * sd_2^2}{n_1 + n_2 - 2}} \tag{5.2}$$

sd_{pooled} = pooled standard deviation
n_1 = sample size of group 1
n_2 = sample size of group 2
sd_1 = standard deviation of group 1
sd_2 = standard deviation of group 2

For the calculation of the overall effect size, it is necessary to determine the sampling error of the respective standardized mean differences as the effect sizes are weighted with their inversed variance when being aggregated (Eisend 2014; Lipsey and Wilson 2001). This sampling error is calculated via Formula 5.3.

$$\varepsilon = \sqrt{\frac{n_1 + n_2}{n_1 * n_2} + \frac{ES^2}{2 * (n_1 + n_2)}} \tag{5.3}$$

ε = sampling error
n_1 = sample size of group 1
n_2 = sample size of group 2
ES = standardized mean difference

Although as a rule means are provided by the studies, they are not present in every case. In such cases, the standardized mean differences need to be determined based on other observed values stated by the respective study. In this way, several effect sizes are calculated based on the correlation coefficient between the treatment group and the outcomes. The correlation coefficients can be transformed into standardized mean differences via Formula 5.4 (Lipsey and Wilson 2001).

$$ES = \frac{2r}{\sqrt{1-r^2}} \tag{5.4}$$

ES = standardized mean difference
r = correlation coefficient

Besides, several studies provide frequencies for specific outcomes (e.g., the desire for future interaction as SO) for treatment and comparison group. These frequencies can be arranged in a 2x2 contingency table – a table which reports the absolute or relative frequencies of combinations of the specific characteristic attributes (Lipsey and Wilson 2001). To remain with the example, a 2x2 contingency table would include the frequencies for anger expression of whether a future interaction is desired or not as well as the frequencies for the comparison group (either happiness or neutrality) pertaining to an endorsement or a rejection of a future interaction. From these frequencies a correlation coefficient can be deduced according to Formula 5.5 (Lipsey and Wilson 2001). The correlation coefficient can in turn be transferred to ES via Formula 5.4.

$$r = \frac{(CF_{11} * CF_{22} - CF_{12} * CF_{21})}{\sqrt{(CF_{11} + CF_{12}) * (CF_{21} + CF_{22}) * (CF_{11} + CF_{21}) * (CF_{12} + CF_{22})}} \tag{5.5}$$

r = correlation coefficient
CF_{11} = cell frequency$_{11}$ in a 2x2 contingency table
CF_{12} = cell frequency$_{12}$ in a 2x2 contingency table
CF_{21} = cell frequency$_{21}$ in a 2x2 contingency table
CF_{22} = cell frequency$_{22}$ in a 2x2 contingency table

Lastly, a few studies report t-values and F-ratios based on their comparison of treatment and comparison group in t-tests or Analysis of Variances (ANOVAs).[42] In order to transfer these values into standardized mean differences, Formulas 5.6 and 5.7 have to be

[42] t-tests and ANOVAs basically examine whether differences between groups of (an) IV(s) with respect to a DV are present or, more generally speaking, in how far a relationship between IV(s) and a DV exists (Backhaus et al. 2016). In the present case, that is the difference between anger and the comparison group regarding their influence on the negotiation outcomes. t- and F-statistics thereby represent the key figures with which the underlying H_0 hypotheses (that no difference is present) are tested, which leads either to their confirmation or to their rejection (Backhaus et al. 2016).

applied, respectively (here and in the following Lipsey and Wilson 2001). One special feature of the ES determination for F-ratios is that this formula will always yield a positive value, since F-ratios cannot become negative. For this reason, the correct sign has to be put on the effect size by the researcher, whose task it is to deduce this information from the underlying experiment.

$$ES = t * \sqrt{\frac{n_1 + n_2}{n_1 * n_2}} \tag{5.6}$$

ES = standardized mean difference
t = t-value of an independent samples t-test
n_1 = sample size of group 1
n_2 = sample size of group 2

$$|ES| = \sqrt{\frac{F * (n_1 + n_2)}{n_1 * n_2}} \tag{5.7}$$

$|ES|$ = absolute value of the standardized mean difference
F = F-value of an one-way ANOVA
n_1 = sample size of group 1
n_2 = sample size of group 2

In most meta-analyses multiple effect sizes are deduced from a single study (e.g., Schmidt and Bijmolt 2019). This holds for the meta-analysis at hand as well (94 studies comprise 267 effect sizes). As an example, it is possible to infer two effect sizes (one based on fake and one based on genuine anger) from a study that manipulates fake and genuine anger and compares the EOs of its participants to the comparison group (Côté, Hideg, and van Kleef 2013). Thus, one gets a nested structure, that is, one in which multiple effect sizes are nested within a single study so that the number of effect sizes exceeds the amount of studies (see Figure 5-3).

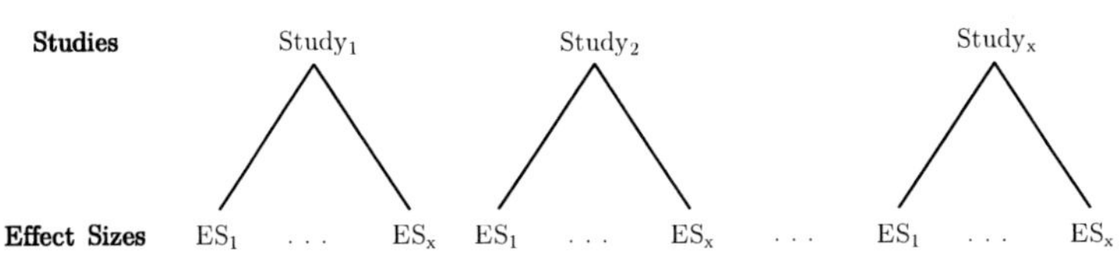

Figure 5-3: Structure of Effect Sizes Nested in Studies
Source: Author's own illustration.

However, the presence of multiple effect sizes in one study leads to dependencies among these effect sizes (Bijmolt and Pieters 2001; Borenstein et al. 2009). These dependencies arise for multiple reasons, such as the same underlying experimental setting or the repeated usage of participants at different points in time (e.g., participants are asked for their demand level after each negotiation round), their usage for multiple comparisons (e.g., a fake and genuine anger group are both compared to the same comparison group), or their usage for the measurement of multiple outcomes (e.g., participants play multiple negotiation games) (Borenstein et al. 2009; Mazei et al. 2015).

Multiple possibilities exist for how to deal with such dependencies, whereby one can differentiate between *single value approaches* and *complete set approaches* (here and in the following Bijmolt and Pieters 2001). Single value approaches imply for instance that an average measurement of the effect sizes of one study is calculated, or that one measurement is randomly selected per study (e.g., the effects of A and B are averaged or one of these effects is randomly selected). Nevertheless, the results of single value procedures are generally quite unsatisfactory with regard to their detection of the true measurements of effects, especially for moderator variables. As they are aggregated at the measurement level, a lot of information is lost which otherwise might have been included. For example, by aggregating fake and genuine anger to a single measure, the important information would get lost as to how the authenticity of the emotion influences the effect of anger on negotiation outcomes. Thus, single value approaches are not recommended for use in meta-analyses.

Complete set approaches outperform single value approaches in all respects (here and in the following Bijmolt and Pieters 2001). The complete set approach most commonly applied in meta-analysis is the treatment of all effect sizes as independent replications, with no attention given to the dependencies. This approach is sound, as long as the dependencies do not lead to biased results. However, most studies that rely on independent replications do not test for whether dependencies of within-study effect sizes actually lead to severe biases. The second complete set approach, dependent replications, should be assumed in the event of a severe dependency problem. This approach statistically accounts for the nested dependency structure of effect sizes in studies via a so-called multilevel or hierarchical linear model (Hox, Moerbeek, and van de Schoot 2017) by splitting the model into multiple levels according to the underlying nested structure (van den Noortgate and Onghena 2003; Raudenbush and Bryk 1985).[43] Thus, in the given case one would obtain a within-study level and a between-study level.

In order to decide which complete set approach to follow, the severity of the underlying dependency structure needs to be evaluated. A key ratio to use for evaluation of the potential dependency problem is the intra-class correlation (ICC) index (Hox, Moerbeek, and van de Schoot 2017; Rosenthal 1979). This index sets the variance of the between-study level in relation to the overall variance (sum of between-level and within level variance) (Hedges 2009) (see Formula 5.8). The ICC can range between 0 (variance at the between-study level equals 0) and 1 (variance at the within-study level equals 0).

$$ICC = \frac{\sigma_{BS}^2}{\sigma_{BS}^2 + \sigma_{WS}^2} \tag{5.8}$$

ICC = intra-class correlation
σ_{BS}^2 = variance at the between-study level
σ_{WS}^2 = variance at the within-study level

When the between-study level variance is large, the studies seem to differ quite strongly (Borenstein et al. 2009; Hedges 2009). Furthermore, a large amount of between-study level variance implies strong correlation between the effect sizes within a study (Hox, Moerbeek, and van de Schoot 2017). In the present case, however, the between-study level variance is close to zero for the main model (the anger effect on EOs), so that the intra-class correlation approaches zero as well (see Appendix A-8). For this reason, although in the underlying dataset multiple effect sizes are nested within studies, the differences across the studies' effect sizes are not severe (Borenstein et al. 2009). In this manner, effect sizes are less correlated within studies than between studies, which leads to a negligible dependence structure among the effect sizes of one and the same study. Thus, complex multilevel models, which are especially essential in cases of present dependencies, are not necessary (Bijmolt and Pieters 2001). Besides, the results of a multilevel model do barely differ from the ones obtained from a traditional meta-analytic

[43] For a better understanding of multilevel models and their results, Appendix A-7 shortly introduces the main aspects and formulas.

random-effects model, which is conducted in the following (for the results of the multilevel models for the main effect on EOs and SOs and the moderating analysis on EOs see Appendix A-8 and A-9, and for the random-effects model see chapter 5.4.2 as well as 5.4.3). This underlines anew that dependencies are not biasing the results. For these reasons, the meta-analysis at hand follows the approach of multiple meta-analyses (e.g., Kugler et al. 2018; Swaab and Swaab 2009) and relies on traditional meta-analytical techniques introduced in the following.

5.3.1.2 Meta-Analytic Models

The simplest traditional model represents a *fixed-effects model* (here and in the following Borenstein et al. 2009). The fixed-effects model assumes that the true effect sizes of the different effect sizes observed are the same. For this reason, deviations which occur from one observed effect size to another must be attributed exclusively to sampling errors. If sample sizes were infinitely large, the sampling error would be zero and the observed effect sizes would equal the true effect sizes. However, the assumption that true effect sizes are actually identical does not take into account the fact that effect size deviations might also be attributable to real differences between effect sizes that occur on account of differences in the experimental setups in the studies or on account of the manipulations of the variables (e.g., in one study the participants negotiate via text, while another study calls for bargaining via FTF). As discussed previously, the underlying dataset is characterized by differences arising from experimental designs and manipulations. Thus, it seems inappropriate to assume that true effect sizes are exactly the same.

Indeed, it is more appropriate to assume that effect sizes are similar but not identical which is the case in *random-effects models* (Borenstein et al. 2009). As these models are much more realistic and less strict, many meta-analyses are conducted using a random-effects model (e.g., Kong, Dirks, and Ferrin 2014; Kugler et al. 2018; Stuhlmacher and Morrissett 2008). Therefore, it will be applied in the meta-analysis at hand as well. Contrary to the fixed-effects model, the random-effects model is based on the assumption that the observed effect size (ES_i) corresponds to the overall effect size together with its true variation across effect sizes and its sampling error (Borenstein et al. 2009) (see Formula 5.9).

$$ES_i = \overline{ES} + \vartheta_i + \varepsilon_i \tag{5.9}$$

i = consecutive number of effect sizes
ES_i = standardized mean difference i
$\overline{ES}$ = weighted mean effect size
ϑ_i = true variation of effect size i
ε_i = sampling error of effect size i

The sampling error can be directly calculated via Formula 5.3 from the sample sizes and effect size measure for each observed effect (Lipsey and Wilson 2001). The sampling error is assumed to be normally distributed with zero mean and variance ϖ_i^2 (Viechtbauer 2010). By contrast, since the error term ϑ_i is unknown, it can only be estimated within the model (Marín-Martínez and Sánchez-Meca 2010). It is also assumed to be normally distributed with zero mean and variance τ^2 (Viechtbauer 2010). The estimation of τ^2 is based on the restricted maximum likelihood estimation in the present case. The regular maximum likelihood method estimates the parameters of a probability distribution under which the underlying likelihood function is maximized in such manner that the effect sizes observed become the most probable ones (Eliason 1993; Hox, Moerbeek, and van de Schoot 2017). Because the regular maximum likelihood estimator tends to underestimate the heterogeneity of the population in finite samples, its restricted version is recommended, as this latter version takes that drawback into account (Viechtbauer 2005).[44]

In the random-effects model, the overall effect size can be determined in the following way: All observed effect sizes are weighted with their inversed variance in a first place (here and in the following Borenstein et al. 2009) (see Formula 5.10). The inversed variance consists of the variance of true variation of the effect sizes (τ^2) and their respective sampling variances (ϖ_i^2). The weighted effect sizes of the studies are then summed up and divided by the sum of all inversed variances (Lipsey and Wilson 2001). As the sampling error decreases in cases in which the sample size increases, effect sizes based on larger sample sizes are therefore more strongly weighted in order to obtain the integrated effect size (Borenstein et al. 2009).

[44] For more details on the maximum likelihood estimator and further possible estimators please refer to Viechtbauer (2005).

$$\overline{ES} = \frac{\sum(\frac{1}{v_i} * ES_i)}{\sum\frac{1}{v_i}};\ whereby\ v_i = \tau^2 + \varpi_i^2 \tag{5.10}$$

$\overline{ES}$ = weighted mean effect size
i = consecutive number of effect sizes
v_i = overall variance of standardized mean difference i
ES_i = standardized mean difference i
τ^2 = variance of true variation of effect sizes
ϖ_i^2 = sampling variance of effect size i

If the overall model exhibits a high degree of true heterogeneity (high values for (τ^2)), moderator analyses or so-called *meta-regressions* or mixed-effects models[45] are conducted (Eisend 2014; Schewe, Hülsheger, and Maier 2014; Stanley, Doucouliagos, and Jarrell 2008). Such moderator analyses aim to reduce the amount of true heterogeneity by accounting for a part of this variability through their regression of relevant moderating factors on the observed effect sizes (here and in the following Eisend 2014). In order to evaluate the degree of real heterogeneity of the observed effect sizes, one can apply two further statistics along with τ^2: the Q statistic and the resulting value of I^2. First, the common test for homogeneity, the Cochran's Q test, proves whether the deviations in observed effect sizes from the weighted mean are higher than would be expected on the basis of sampling error alone (here and in the following Borenstein et al. 2009; Cochran 1954). The Q-value which is tested is thereby calculated via Formula 5.11. The H_0 hypothesis assumes homogeneous effect sizes, i.e. that variability would only occur due to sampling errors. Thus, if the Q statistic becomes significant, the H_0 hypothesis has to be rejected and the presence of heterogeneity can be assumed.

$$Q = \sum \frac{(ES_i - \overline{ES})^2}{v_{\overline{ES}}} \tag{5.11}$$

Q = Q-value of Cochran's Q test
i = consecutive number of effect sizes
ES_i = standardized mean difference i
$\overline{ES}$ = weighted mean effect size
$v_{\overline{ES}}$ = variance of weighted mean effect size

[45] A meta-regression that is based on a random-effects model is also called mixed-effects model, as the estimated coefficients of the moderators represent fixed effects and the variances of the true variation of the effect sizes (τ^2) depict the random effects (Schwarzer, Carpenter, and Rücker 2015; Viechtbauer 2010).

As the Q-statistic only delivers the information, whether heterogeneity is present or not, it is common practice to calculate the I^2 value as well (here and in the following Eisend 2014; Higgins et al. 2003) (see Formula 5.12). This value measures the relative magnitude of true heterogeneity by setting the heterogeneity between the effect sizes in relation to the overall variation. Values of 25%, 50% and 75% can be assigned to levels of low, moderate and high heterogeneity (here and in the following Higgins et al. 2003). In cases in which $I^2=0$, all variability is due to sampling error rather than to real differences between effect sizes. However, if I^2 reaches high levels, it is appropriate to conduct a moderator analysis (Eisend 2014).

$$I^2 = \left(\frac{Q - df}{Q}\right) * 100\% \tag{5.12}$$

I^2 = degree of true heterogeneity between effect sizes
Q = Q-value of Cochran's Q test
df = degrees of freedom

A quite important factor which has to be underlined at this point is that effect sizes already contain the information about the effect of an IV (emotion expression) on a DV (EOs or SOs). This distinguishes a meta-regression from moderator analyses in standard regressions that examine an effect of IVs on a DV while integrating additional moderators (Hair et al. 2014). Contrary to this, the moderators in a meta-regression represent simple IVs which are regressed on the already aggregated effect of emotion expression on negotiation outcomes (here and in the following Eisend 2014). Thus, a meta-regression formulates a regression equation of the kind known from simple regression analyses. As is the case without moderators, the regression equation additionally accounts for true heterogeneity as well as for sampling error (see Formula 5.13).

$$ES_i = \alpha + \sum_k \beta_k * M_k + \vartheta_i + \varepsilon_i \tag{5.13}$$

i = consecutive number of effect sizes
α = intercept of the meta-regression model
k = consecutive number of moderators
β = estimated coefficient of a moderator
M = moderator
ϑ_i = true variation of effect size i
ε_i = sampling error of effect size i

The intercept of the model can be interpreted as the weighted mean effect size for moderators which can be sometimes interpreted but which is not necessarily meaningful (e.g., Hüffmeier et al. 2014; Schmidt and Bijmolt 2019). Since the estimated coefficient of each moderator states the impact and the direction of a moderator on the main effect, it represents the value with the highest informative value (Eisend 2014). A model that includes moderators is a so-called conditional model (here and in the following Mazei et al. 2015; Raudenbush and Bryk 2002). If no moderators are applied, one speaks in turn of an unconditional model. For the unconditional model, the term $\sum\beta_k{*}M_k$ is omitted from Formula 5.13 so that one gets to Formula 5.9, where α represents the intercept of the unconditional model and corresponds to the weighted mean effect size ($\overline{ES}$). Thus, the unconditional model is equivalent to a regression analysis without any moderators from which the mean effect can be deduced.

As the meta-regression depicts a linear regression model, the assumptions for the latter have to be fulfilled in order to receive unbiased and efficient estimates (Backhaus et al. 2016). These assumptions will shortly be presented and subsequently be tested in the meta-regression. One of the most relevant assumptions is that *multicollinearity* among the moderator variables (which represent IVs in a meta-regression) has to be absent (here and in the following Backhaus et al. 2016; Hair et al. 2014). Multicollinearity can be measured by assessing the degree to which one of the IVs of a model can be explained by the remaining IVs. The so-called variance inflation factors (VIFs) are the most popular measure for this relationship. VIF values contain the information about the combined effect of the IVs by treating each IV as a DV and regressing it against the remaining IVs of the model. The obtained R^2 of each of these regression models represents the amount by which an IV can be explained by all the other IVs in the model. From this value, the VIF value is calculated for each of the IVs via Formula 5.14. The higher R^2 is, and thus, the larger the dependence of an IV on the other IVs is, the larger the VIF value becomes. Consequently, higher VIF values indicate higher levels of multicollinearity. A VIF value corresponding to 1 represents the lowest possible and most optimal value mirroring the absence of multicollinearity. Upper boundary values are not clearly specified in the literature; however, values starting at a threshold of 5 or 10 are generally seen as indicators for multicollinearity.

$$VIF = \frac{1}{1 - R^2} \tag{5.14}$$

VIF = variance inflation factor
R^2 = coefficient of determination

Besides, the *residuals* of the regression model have to fulfil several assumptions. First, the residuals have to follow a normal distribution (here and in the following Backhaus et al. 2016; Hair et al. 2014). For this purpose, it is common to apply histograms that plot the frequency or density of the residuals, respectively. If the bars of the histogram approach a normal distribution curve (slight deviations are allowed), the assumption is fulfilled. Next to the visual inspection, statistical tests such as the Shapiro-Wilk test (1965) can be adduced as well, in order to test for normality. Their H_0 hypothesis assumes a normal distribution; thus, in order to retain H_0, the tests should not reach significance. Moreover, a further assumption is that the residuals exhibit a constant variance (homoscedasticity). This assumption can be checked in a scatterplot in which the predicted values of the DV (x-axis) are plotted against the standardized residuals of the model (y-axis). In order to fulfil the assumption, the standardized residuals should strew unsystematically around a residual value of 0 within the diagram, and no shape should be recognizable. In cases of funnel- or diamond-shaped scatterplots, for example, the assumption of constant variance is violated and heteroscedasticity is present. With the help of that plot, two further assumptions can be verified. On the one hand, the residuals should not exhibit a non-linear relationship. If they mirror a quadratic relationship, for example, the assumption of linearity of the regression model is violated. Lastly, residuals should be independent of one another. Thus, if clear trends become visible from the scatterplot, this assumption is violated. That appears for instance if separated point clouds fall around the horizontal line around a residual value of 0, mirroring a potential event-based dependence. All in all, if the residuals spread randomly around the horizontal line of 0 in the scatterplot without showing any pattern at all, the assumptions of constant variance, linearity and independence are fulfilled.

The violation of one or more of these assumptions would lead to different consequences, such as the occurrence of biases in the estimated coefficients of the model, inefficiency of the parameters, or invalidity of the significance tests (here and in the following Backhaus et al. 2016). Remedies for such violations of the assumptions would be for instance a lntransformation of the DV, the dropping or combining of variables with a high level of multicollinearity, or the use of a generalized least squares estimation.

5.3.1.3 Publication Bias

Prior to the presentation of the data collection, a factor which exhibits considerable potential for damaging the results of a meta-analysis needs to be introduced, namely the presence of *publication bias* (Geyskens et al. 2009). Publication bias is the pattern in which mainly those studies are published that report significant results (Ellis 2010; Stanley 2005). Especially in current times of strict review processes for journals, insignificant results become hard to publish (Schneider 2010; Sutton 2009). On the one hand, statistical significance is influenced by the underlying sample size, whereby small samples are at a strong disadvantage over against large samples when it comes to reaching significance (here and in the following Stanley 2005). This is due to the fact that smaller samples are confronted with larger sampling errors, which makes the deduction of a significant effect much harder. On the other hand, large effects tend to be significant more frequently than small effects, irrespective of the underlying sample size (here and in the following Borenstein et al. 2009). Since insignificant results, inasmuch as they are considered for publication at all, tend in most cases to appear in grey literature only, they are less likely to be included in a meta-analysis. This may lead to potential biases consisting in a lack of small effects of small samples (Sutton 2009). The consequence of this is that when null findings are left out due to non-publication, there is an overestimation of the overall population effect size in meta-analysis (Egger et al. 1997; Ueno, Fastrich, and Murayama 2016).

Approaches to dealing with potential publication bias vary among meta-analyses. Some researchers ignore the possible threat of publication bias and do not account for it at all (e.g., Hüffmeier et al. 2014). Furthermore, there are a number of researchers who argue that the exclusion of unpublished and non-peer-reviewed studies is intentional for purposes of quality assurance (e.g., Weisz et al. 1995). Nevertheless, the majority of researchers conducting meta-analyses account for publication bias for the sake of preempting its severe consequences. The most common preventive measure taken against the negative effect of publication bias is the inclusion of a sufficient number of results from grey literature and unpublished works such as proceedings, dissertations, working papers or raw data (Eisend 2014; Sutton 2009). Moreover, exploratory methods exist that enable the researcher to check the underlying dataset for potential publication bias, as do methods that allow for the determination of its extent (Eisend 2014).

A helpful and commonly applied graphic to explore potential publication bias is a *funnel plot* (Viechtbauer 2010). A funnel plot sets the sampling errors (y-axis) in relation to the underlying effect sizes (x-axis) (Sutton 2009) (see Figure 5-4). The assumption underlying the plot is that effect sizes with a larger distance from the overall mean have larger sampling errors due to smaller sample sizes, which leads to an inverted funnel shape (here and in the following Mazei et al. 2015). This inverted funnel shape is visualized by the triangle in Figure 5-4, which represents the 95% confidence interval around the overall mean effect size depicted by the vertical line. Publication bias is seen as absent if the effect sizes are distributed symmetrically around the underlying mean effect size within the inverted funnel and no gaps are visible (here and in the following Eisend 2014). By contrast, potential publication bias arises if the lower part of the inverted funnel exhibits apparent gaps, which is indicative of a lack of small effects based on small sample sizes. Depending on the side on which these data points are missing, the overall mean is biased in the other direction, as publications seem to be favoured in this trend (Borenstein et al. 2009). Besides, if many observed effect sizes fall out of the 95% confidence interval, this fact is also seen as indication that large effect sizes rather than small effect sizes based on small samples get published (Mazei et al. 2015). Consequently, the funnel plot of Figure 5-4 indicates potential publication bias, as the lower left part of the funnel barely includes any effect sizes which leads to an overestimation of the positive estimated overall effect size.

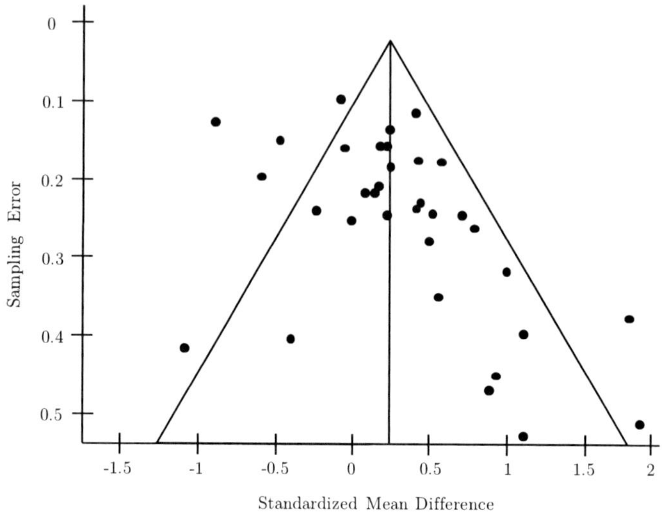

Figure 5-4: Exemplary Funnel Plot
Source: Author's own illustration.

Nevertheless, a funnel plot is based solely on visual inspection by the researcher, thus making it quite subjective (Stanley 2005). Therefore, additional tests are usually conducted to determine the presence of potential publication bias. One of these tests is the *trim and fill method*, which assesses in statistical terms the asymmetry of the funnel plot (Duval and Tweedie 2000a; b).[46] This method has furthermore been established as the most popular approach to adjusting for publication bias (Borenstein 2005). The trim and fill method is a one-sided approach that checks for asymmetry separately on either side of the funnel plot (here and in the following Sutton 2009). For example, when testing for asymmetry on the left side of the plot, the number of effect sizes is determined that would need to be trimmed off the right side in order to attain a symmetric funnel. Once that number is found, the adjusted overall mean effect size is calculated and the trimmed

[46] Further tests for asymmetry are for example the Egger's test, which investigates the presence of potential publication bias via a linear regression, whereby the intercept of the model represents a measure of asymmetry (Egger et al. 1997). In addition, in order to exclude publication bias, effect sizes and their sampling variances should not exhibit any relatedness to one another (here and in the following Begg and Mazumdar 1994). Thus, if both are shown to be positively correlated within a rank test, publication bias is assumed to be present, as small effect sizes with high sampling variance seem to be missing. These tests serve only to confirm asymmetry, whereas the trim and fill method also provides a possibility to adjust for publication bias (Borenstein 2005). Moreover, the methods can be seen as complementary to one another, inasmuch as they usually lead to the same results (e.g., Kugler et al. 2018).

effect sizes are reinstated by inversely complementing the same amount of missing effect sizes on the other side, in order to maintain the adjusted overall mean. In order to receive the respective number of trimmed or missing effect sizes, an iterative procedure is necessary that estimates the two.

This number can be determined by two estimators (in the following R_0 and L_0), whereby the two can – but do not have to – lead to the same result (here and in the following Duval and Tweedie 2000a; b). Consequently, it is recommended that both be calculated, in order to obtain a more reliable estimate of publication bias. For both estimators, the initial mean effect size is subtracted from each observed effect size. The received absolute values are then ranked (with the smallest values at the lowest ranks), whereby those rankings based on a negative difference get a negative sign (e.g., 1, -2, 3, 4, -5, -6, 7, 8, 9, 10, 11 for 11 effect sizes and 3 less than the pooled effect size) (here and in the following Sutton 2009). Afterwards, a vector is built that includes the rankings ordered ascendingly (e.g., -6, -5, -2, 1, 3, 4, 7, 8, 9, 10, 11).

In order to calculate the first estimator (R_0), the rightmost run of consecutive ranks has to be determined from the created vector (here and in the following Duval and Tweedie 2000b; Sutton 2009). For the given example, it is $y^*=5$ as the rightmost series has the length of 5 (7, 8, 9, 10, 11). Therefore, according to Formula 5.15, the first estimator results in 4 missing effect sizes.

$$R_0 = \gamma^* - 1 \tag{5.15}$$

R_0 = first estimator of number of missing effect sizes
γ^* = rightmost series of consecutive ranks

For the second estimator (L_0), the sum of the rankings greater than the overall mean (T_N) is calculated (here and in the following Duval and Tweedie 2000b; Sutton 2009). Thus, those rankings are summed that did not receive a negative sign. The corresponding value in the given example is 53 (1+3+4+7+8+9+10+11=53). According to Formula 5.16, the second estimator (L_0) consequently corresponds to a value of 3.81 with N=11 (the number of effect sizes in the present case). Thus, L_0 corresponds rounded to 4 missing effect sizes as well.

$$L_0 = [4T_N - N(N+1)]/[2N-1] \quad (5.16)$$

L_0 = second estimator of number of missing effect sizes
T_N = sum of the ranks of those observed effect sizes which are greater than the overall mean
N = number of effect sizes

After determining the number of missing effect sizes, the effect sizes with the largest gap from the pooled effect size (L_0) or the largest effect sizes (R_0), respectively, are removed according to that number (four in the present example) (here and in the following Sutton 2009). Thereupon, the pooled effect size is calculated again, an assessment is made as to whether or not the funnel plot is still asymmetric, and if this holds true, the next four effect sizes are trimmed off from the plot. This procedure is repeated until the funnel plot reaches symmetry. By doing so, the trim and fill method has to be seen as an approach that assesses the sensitivity of the results of a meta-analysis due to specific selection mechanisms; it is not an approach aimed at attaining a better or more valid estimate of the overall effect (Viechtbauer 2010).

Lastly, it needs to be emphasized that, much as the underlying assumption of publication bias can hold true for many cases, it does not necessarily do so. The examination of publication bias takes place based on the assumption that small effect sizes from small studies arising from unpublished/grey literature are missing in the meta-analysis. However, for various reasons other than the presence of a biased sample, small studies may indeed find large rather than small effects (here and in the following Borenstein et al. 2009). In this way, smaller studies may conduct their experiments with better quality controls than larger ones; or they may have included participants from which the investigated effect can be more easily deduced. To test adequately for whether the assumption holds true for the present meta-analysis, the common practice is followed to integrate the publication status as a moderator variable in the meta-regression model (e.g., Swaab et al. 2012; Walters, Stuhlmacher, and Meyer 1998). By comparing the effect sizes of unpublished/grey literature to those of peer-reviewed literature, it can be deduced whether they do indeed differ. In cases in which no difference can be detected between the two groups, publication bias does not seem to exert a subjective influence on the results.

5.3.2 Generation and Description of Database

5.3.2.1 Collection of Articles

For the meta-analysis to be conducted, empirical studies are needed that compare the effect of anger on negotiation outcomes to that of happiness or emotional neutrality. In order to collect these relevant studies, this dissertation applies a diverse literature search strategy for articles. This strategy is based on two starting points. In a first step, the strategy relies on the results of the co-citation analysis of Study I. As was emphasized in chapter 3.3., the co-citation analysis not only delivers the topic for the subsequent meta-analysis, but also reduces the search effort involved in identifying pertinent studies for the latter. This is due on the one hand to the fact that the co-citation analysis clusters publications according to their topical proximity so that relevant publications can easily be deduced from the thematic clusters. On the other hand, the co-citation analysis delivers a quite detailed search string along with sets of inclusion and exclusion criteria that can be used for the meta-analysis (in a slightly adapted version) as well. From the two network clusters focusing on the topic of *emotions* (in periods 3 and 4), 18 articles can be identified that fulfil the criteria for the meta-analysis at hand.[47] Thus, these articles directly enter the analysis (see here and in the following Figure 5-5).

[47] The remaining publications revealed by the co-citation analysis are either reviews, books, or they focus on moods or other emotions which are not investigated by the meta-analysis at hand.

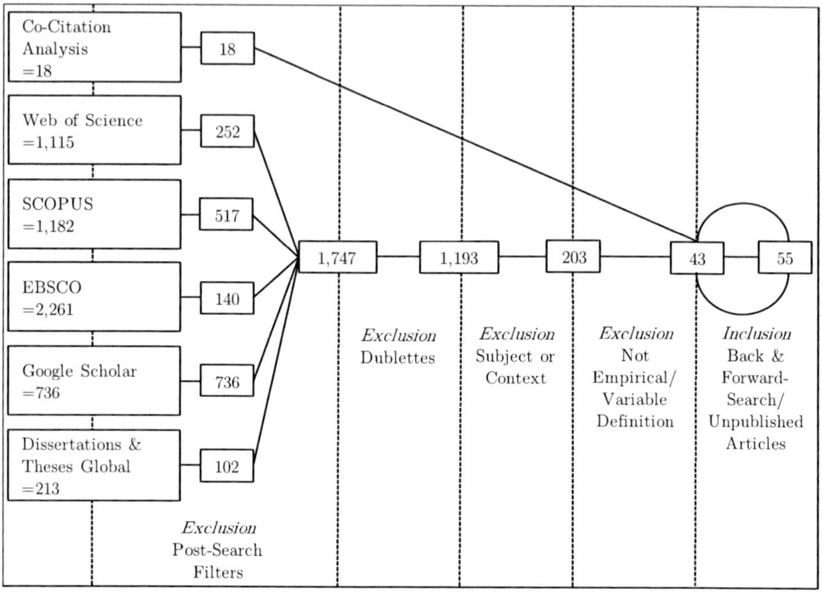

Figure 5-5: Process of Literature Search
Source: Author's own illustration.

Moreover, these articles are used for the determination of the relevant search words to be applied in the second step of the literature search, which consists in an extensive complementary online search. For this online search, the following online databases are adduced: WOS, Google Scholar, EBSCO (focusing on the databases Academic Search Premier, Business Source Premier, PsycArticles and PsycInfo), SCOPUS and ProQuest's Dissertations & Theses Global.[48] The decision to use multiple databases is made because the databases differ in part with regard to the journals they include or the time frames within which they provide information about a specific journal. This decreases the likelihood of relevant articles being overlooked. Journal articles are deduced from all databases, except for Dissertations & Theses Global, which is considered due solely to its

[48] The literature search conducted in Study I focused solely on the WOS as its database, although the study was much broader with regard to the embedded topics. This was done for two reasons. First, as the literature search in Study I was not limited to any specific topic, this factor in itself led to a myriad of potentially relevant articles (see chapter 4.2.3). A broadening to further databases would have caused a overwhelming number of articles from which large parts would not have been relevant. Second, a consistent notation of the cited references of the relevant articles was necessary while processing the former. However, the different databases could not provide this consistency. Therefore, the WOS was used as the sole database on account of its advantages, as presented in chapter 4.2.3.

supply of doctoral theses. Besides, SCOPUS delivers articles in press as well as conference proceedings, and EBSCO accounts for dissertations as well. As was the case for the first study of this thesis, the search for articles is limited to those written in English (as the approved scientific language). However, in contrast to Study I no time frame is determined, so that publications from all years can be included in the analysis. For the search, the keywords deduced from the 18 articles identified by the co-citation analysis are applied in each of the databases in the subsequent combinations: "emotion*" or "happy" or "happiness" or "anger" or "angry" and "negotiat*" or "bargain*" or "conflict", whereby the search words are used in their truncated forms (characterized by the asterisk) if necessary, thus allowing for the inclusion of all inflected forms (e.g. emotions, emotional, negotiations, bargainer, bargaining etc.). Besides, the search terms are applied to the title only, as the extension to keywords and abstracts has proven to lead to imprecise results. This procedure leads to a number of articles ranging between 213 and 2,261 across the databases (in sum 5,507 articles).

From here onward, multiple exclusion steps follow for the identified articles, thus limiting the amount to those relevant for the analysis (see Figure 5-5). First, post-search filters are applied within those databases that allow for such options (WOS, SCOPUS, EBSCO, Dissertations & Theses Global). Such post-search filters determine, among other things, the inclusion/exclusion of research areas, specific key words, or multiple topics. For the search in WOS, the same filters are used as the ones applied in the (co-)citation analysis (see Appendix A-1). For the other three databases, post-search filters are created in a similar pattern depending on which filters can be chosen in the online database (see Appendix A-10 for the search string and inclusion/exclusion criteria in SCOPUS and Appendix A-11 for the ones in EBSCO).[49] This leads to an overall amount of 1,747 articles after the first processing step.

Second, through the use of multiple databases, the dataset contains a specific share of articles which can be found more often. Therefore, articles that show up more than once are only included once in the dataset, thus leading to a reduced number of 1,193 articles.

[49] As the post-search filters applied to Dissertations & Theses Global are negligible, they are not listed in the Appendix.

Third, on the basis of their titles and abstracts, the remaining articles are evaluated according to inclusion and exclusion criteria (determined by the researcher) for their suitability for the analysis. This exclusion step focuses on how well the studies fit with regard to their subject and context. Again, the same criteria are drawn on as were used for Study I (see chapter 4.2.3 or Appendix A-2) supplemented with specific criteria for Study II. For example, one criterion adopted from Study I is that a negotiation setting or a related game situation has to be present if it is to be included in the further processing steps. As negotiations have their origin in research on game-theory (see chapter 2.2), such experimental games are included as well. Thus articles that focus solely on the effects of emotions without addressing the context of negotiations are not included (e.g., Graham et al. 2008; Tiedens 2001). Besides, relevant research areas need to be considered by the articles, and hence negotiations need to take place in a business environment or a comparable setting. Thus political or legal negotiations, for instance, are not relevant for the analysis (e.g., Tagar, Federico, and Halperin 2011).

In addition to the criteria adopted from Study I, the following emotion-specific criterion holds: Studies are excluded that deal with moods rather than discrete emotions (e.g., Baron 1990; Carnevale and Isen 1986; Forgas 1992). As discussed in chapter 2.3.1.2, moods are indeed affective states, as are emotions; nevertheless, they need to be clearly distinguished from emotions, as they are not only more enduring, but also less intensive and more diffuse, as distinctions are generally only drawn between "being in a good mood" or "being in a bad mood" (Ekman 1984; Forgas 1992; Lazarus 1991; Thompson 2015). As the two concepts differ, a merger would bias the results. In most cases, it is possible to deduce clearly whether a discrete emotion is present (e.g., van Kleef, De Dreu, and Manstead 2004b; Lelieveld et al. 2012), or whether it is a matter of moods (e.g., Forgas 1998; Kramer, Newton, and Pommerenke 1993). Be that as it may, in some cases the assignment to either mood or emotion is not clearly possible, as the article in question merely refers to the overall concept of affect. In such cases, it is necessary to adduce the manipulation of the experiments, as this procedure provides insight into the examined affect. For example, Carnevale and Isen (1986) use funny cartoons and a gift to manipulate the participants in their experiment into a condition of positive affect. Alternatively, Baron (1990) chooses pleasant scents emitted by air-fresheners in order to generate positive affect. Both are prime examples of mood manipulations, as they set the participants into a generally "good" feeling. The application of the set of inclusion and exclusion criteria reduces the amount of relevant articles enormously to 203 articles. Thus, the basic direction of the remaining articles fits the context sought for the meta-analysis.

Fourth, the articles are evaluated on the basis of their empirical nature and the appropriateness of their definition of variables. Since a meta-analysis aggregates numerical data, non-empirical articles have to be excluded. Besides, several inclusion and exclusion criteria are determined regarding the variable definition:

- Articles must compare an anger expression with an expression of happiness and/or with a neutral comparison group regarding their effect on EOs and/or SOs (defined in chapter 2.3). Thus, articles that focus solely on other emotions, such as anxiety or disappointment, are not taken into account (e.g., Brooks and Schweitzer 2011; van Kleef, De Dreu, and Manstead 2006). In addition, papers are not considered for the analysis if anger is compared to other emotions which are not part of the meta-analysis at hand (e.g., Lelieveld et al. 2011). Lastly, articles are not considered that do not include information about a negotiation outcome, but rather focus solely on behaviours, for instance the effect of anger on deceptive behaviour (e.g., Yip and Schweitzer 2016).

- Some articles examine an overall variable called "negative emotion" or "positive emotion". In such cases, it is verified which discrete emotions are subsumed under these overall terms, or which instructions are given by the experimenters to participants in the experiments for the expression of a negative or positive emotion. If anger or happiness are among these emotions, and if they dominate the construct; or if the briefing explicitly permits recognition of the respective emotions, the articles are integrated into the meta-analysis (e.g., Bhattacharjee and Moreno 2017; Kopelman and Rosette 2008; Wang, Han, and Su 2017). However, if the overall variable merely subsumes multiple diffuse moods or unsuitable other discrete emotions, the articles are not taken into consideration (e.g., Brett et al. 2007; Mueller and Curhan 2006). In addition, two articles are integrated in the meta-analysis which examine joy next to anger (De Melo, Carnevale, and Gratch 2012; De Melo, Gratch, and Carnevale 2014). As joy is quite related to happiness (e.g., Lyubomirsky, King, and Diener 2005; Tng and Au 2014) and can be seen as synonym in these studies, this approach is seen as appropriate.

- Articles are excluded from the analysis that examine the effect of emotional transitions on negotiation outcomes (e.g., Filipowicz, Barsade, and Melwani 2011). Emotional transitions refer to the circumstance that individuals do not necessarily stay in their currently felt emotion, but instead may change from one emotional state to another during negotiations (e.g. from anger to happiness). However, as these shifts do not reflect one concrete discrete emotion, their consideration is precluded. In addition, studies are precluded from the analysis that deal with the suppression of the respective emotions, i.e., if emotions are felt but not acted out (e.g., Shao et al. 2015). As the suppression of an emotion is not comparable to a neutral (control) negotiator, these studies measure a different outcome than the one sought by the meta-analysis at hand.

All these criteria lead to the exclusion of 160 articles, thus resulting in a total of 43 articles that are integrated into the meta-analysis. Of these 43 publications, 18 articles stem from the co-citation analysis, which amounts to a share of approximately 42%.

Fifth and last, based on these 43 articles, a forward and backward search is conducted in order to identify further relevant publications. For the backward search, the reference lists of the 43 articles are screened for additional pertinent publications. The forward search is conducted via Google Scholar, and it is analysed which (relevant) articles cited the 43 identified publications. The forward-and-backward search lead to 10 additional articles. Furthermore, it is also accounted for the reduction of potential publication bias by searching for unpublished articles as well (in addition to the articles in press and conference proceedings received from the search in the online databases). This is done by sending requests to scholarly online communities and mailing lists, on the one hand. For example, the online community of the Academy of Management is consulted which scholars can use as a platform in order to swap knowledge. Moreover, the mailing lists of the Society for the Psychological Study of Social Issues, the Society of Experimental Social Psychology and the International Association for Conflict Management are queried as well. On the other hand, the past two years of conference proceedings of the International Association for Conflict Management and the past five years of the conference proceedings of the Academy of Management are searched for potentially unpublished relevant publications. Lastly, several of the dominant researchers of negotiations are asked for unpublished negotiation studies, whereby the identification of these major authors takes place based on the results of the citation and co-citation analysis. From this, influential researchers are determined for the field of negotiations in general and for emotions in negotiations in particular. Among the researchers contacted are van Kleef, Thompson, De Dreu and Lewicki. This leads to additional two publications, resulting in a final dataset of 55 articles included in the meta-analysis.

5.3.2.2 Coding Procedure

The next step for deducing the relevant information from the articles identified by means of the previously described process is coding. As articles usually contain more than one study, and since as a rule the objects of investigation, the experimental setup and the samples change across studies of a single article, the coding takes place on the basis of studies rather than on the basis of articles. A single such study can provide multiple effect sizes, as both outcome measures or both comparison groups can be examined within that study (see chapter 5.3.1.1). Nevertheless, not every study of an article is relevant for the underlying meta-analysis. For example, there are a number of studies that do not examine relevant outcome measures and are thus not relevant for further analysis, for which reason they are not coded (e.g., Adam and Brett 2018; Harinck and van Kleef 2012). Thus, the same inclusion/exclusion criteria hold for the studies as the ones applied for the articles (see preceding chapter 5.3.2.1).

A coding sheet is developed in which the relevant information of each study is recorded. To begin with, *descriptive characteristics* of the studies' participants are coded, such as the overall sample size, the gender structure, the average age of the participants and their occupational status (a distinction is made between students and professionals). Furthermore, characteristics of the conducted experiments are coded. This includes the type of experiment (laboratory experiment or field experiment)[50], the type of negotiation conducted in the experiment (e.g., sales negotiation), the negotiation content (e.g., sale of a car), the number of issues negotiated (either single or multiple), the role of the

[50] Laboratory experiments are characterized by their artificial reconstruction of real life settings in the lab (here and in the following Balzert, Schröder, and Schäfer 2011; Gerber and Green 2009). They offer a higher degree of controllability and thus greater internal reliability than field experiments. Field experiments, by contrast, mirror the real behaviour of individuals in the field, for which reason they have a much higher external validity than lab experiments.

emotion receiver in the experiment (e.g., seller), and whether the negotiation is a dyadic or multi-party negotiation. In order to account for potential publication bias, the publication status of the studies is coded as either peer-reviewed or grey literature (including research such as dissertations, working papers, conference proceedings, or raw data). Lastly, the involvement or non-involvement of researcher van Kleef – as first author or as co-author – in the origin of the effect sizes is coded. Van Kleef is the most popular researcher of the interpersonal effects of emotions in negotiations (see Study I), and his share of studies published on the topic of investigation is accordingly large (see Table 5-2 in chapter 5.4.1). As the frequency of the occurrence of this one (co-)author increases the likelihood of similar outcomes, the analysis checks for this variable.

Coding of Main Effects

As stated previously, the standardized mean difference is chosen as the effect size for the meta-analysis at hand. In order to calculate this effect size and the corresponding standard error needed to deduce the aggregated effect, the *means, standard deviations* and *group sizes* need to be coded for both, the treatment group (anger) and the comparison group (happiness/neutrality) (Eisend 2014). In most cases, the information about means and standard deviations is present in numerical values. However, in some instances, means and standard deviations are only provided in bar charts (e.g., Adam, Shirako, and Maddux 2010; Dehghani, Carnevale, and Gratch 2014). Thus, the values need to be read off the bars and are consequently not as precise as exact numbers. Besides, in some cases the respective standard deviations of the means are missing. In such cases authors are contacted for the information (e.g., van Kleef, De Dreu, and Manstead 2004a). In addition, a few studies report only results that are significant. Nevertheless, as a meta-analysis does not distinguish between significant or non-significant outcomes (see chapter 5.3.1), the authors are also asked for the missing insignificant values (e.g., van Kleef and Côté 2007; Sinaceur and Tiedens 2006). In these cases, in which correlations, t- or F-values or frequencies are stated (e.g., Hillebrandt and Barclay 2017; Jäger, Loschelder, and Friese 2016), they are coded and transferred to the standardized mean difference effect size in the coding sheet according to the formulas provided in chapter 5.3.1.1. Lastly, standard errors are transformed into standard deviations, in order to enable them to be used in the present meta-analysis (e.g., Tamir and Ford 2012).

Regarding the sizes of treatment group and comparison group, some studies provide the information directly (e.g., Tng and Au 2014). However, in many cases the exact group sizes are not stated; rather, the studies indicate only the overall sample size (e.g., Khooshabeh et al. 2013). In such cases, it is assumed that the participants are equally distributed across groups, and so the sample size is simply divided by the number of groups in order to arrive at the group sizes.[51] If the group size has to be calculated in this way, it is important to keep the experimental setup in mind. In the event that the participants negotiate via computer (which is frequently the case), the groups can simply be calculated as described above (e.g., Adam and Shirako 2013). However, if they negotiate with another participant, the group size of the respective emotion condition needs to be halved, as only the effect of the emotion expresser (for EOs) or emotion receiver (for SOs) is of interest (e.g., Campagna et al. 2016; Campagna, Mislin, and Bottom 2019).

With regard to the outcome variables, the presence/non-presence of EOs or SOs is coded, in order to be able to distinguish the two. Taking a closer look on the kinds of *EOs* included, it is important to emphasize that the established hypotheses in chapter 5.2.2 target the economic effect of an emotion expression for the expresser. Consequently, joint gains or Pareto efficient[52] results do not represent relevant EO measures, as they affect both negotiators and hence are not included in the meta-analysis (e.g., Overbeck, Neale, and Govan 2010). The most obvious and most frequently examined EOs relevant for the meta-analysis are the concessions made by the emotion recipient to the expresser, as the reason for this being that concessions depict economic values made in favour of the opposing party, thus increasing its ultimate individual gain (e.g., Adam and Shirako 2013). Moreover, the value claimed by the emotion expresser representing his/her individual gain likewise depicts a relevant EO (e.g., Rothman and Northcraft 2015). As well, the demand level of the emotion receiver is frequently provided by the studies. As a higher value claimed by the recipient is (usually) made at the expense of the opposing party, one can deduce by implication that the outcome of the opponent decreases at the

[51] This assumption is quite realistic, as it is observable in studies which provide the exact numbers that the groups are more or less equal in size. This is not surprising, since the studies' purpose is to compare the groups with one another so that it is aimed for a (nearly) equal distribution for the sake of achieving good comparisons.

[52] Pareto efficiency describes the state for which the joint gain of the negotiation parties is maximized (Raiffa 1982; Yang, Singhal, and Xu 2013).

same time (e.g., Rothman and Northcraft 2015). Consequently, the effect size calculated from the demand level of the emotion recipient has to be reversed coded before entering the meta-analysis. The EOs are frequently expressed in such monetary terms as price in a specific currency, or in a negotiated salary (e.g., Campagna, Mislin, and Bottom 2019); however, in many cases the EOs are depicted as chips or points gained in the negotiation (e.g., Butt and Choi 2006; Pietroni et al. 2009). If there are points or chips that can be gained, the underlying outcome can be – but is not necessarily – a financial value. Points can also arise from negotiated outcomes such as days of vacation, the start of employment or the length of a warranty period (e.g., Butt and Choi 2006; Wang, Northcraft, and van Kleef 2012). Lastly, the willingness to pay or the willingness to concede on the part of the emotion recipient are also reported as an EO (e.g., Belkin and Rothman 2017; Kopelman, Rosette, and Thompson 2006).

Regarding the *SOs*, multiple non-economical outcomes are aggregated, which are partly quite different constructs. Unlike EOs, which are usually easy to embrace, SOs are more diverse (see chapter 2.3). Besides, it is important to stress that, unlike the perspective taken for the EOs (the influence on the individual gain of the emotion expresser), the SOs target the socio-emotional outcomes for the emotion receiver. Although different constructs feed into the SOs' variable, the coded factors can be divided best into three classes. The first category relates to the *subjective appraisals of the negotiated EOs*. This includes for instance the perceived unfairness of the outcome or the satisfaction with or disappointment over the result (e.g., Butt and Choi 2006; Harinck and van Kleef 2012; Kopelman, Rosette, and Thompson 2006). The second category includes constructs targeting the *assessment of the emotion expresser*. Such factors are the perceived threat of the opponent (e.g., Sinaceur et al. 2011), the level of trust felt by the recipient (e.g., Jang and Bottom 2020)[53], whether or not one likes the opponent (e.g., Belkin and Rothman 2017), or the opponent's perceived reliability (e.g., Bhattacharjee and Moreno 2017). Moreover, more general constructs such as the impression of the emotion expresser as it echoes within the emotion recipient (e.g., van Kleef and De Dreu 2010), feelings about the relationship (e.g., Adam and Brett 2018) or feelings of being mistreated (e.g., Wang,

[53] In cases in which a subform of the variable trust (e.g., benevolence or integrity) is present in the studies, this form is included.

Northcraft, and van Kleef 2012) belong to this category as well. Lastly, the third category targets *hypothetical behaviour of the emotion recipient* after the negotiation. This includes for instance the desire for future interaction (e.g., Pietroni et al. 2008), the desire for retaliation (e.g., van Kleef and Côté 2007) or potential cooperation after the deal (e.g., Campagna et al. 2016). Most of the SOs constructs are positively valued by means of such criteria as satisfaction, desire for future interaction, or trust. In these cases with a negative valuation, the resulting effect size is reversed coded (e.g., for perceived unfairness or desired retaliation).

Coding of Moderators of the EASI Model

Regarding the coding of the dummy variable *time pressure*, either the studies themselves use the variable as moderator, indicating that a given amount of time is tight (e.g., van Kleef, De Dreu, and Manstead 2004b), or the presence of time pressure has to be assessed on the basis of the experimental design, which holds for most cases. A time limit is seen as an indication of time pressure. However, this indication is only transferred to a real presence of time pressure if the given time might be perceived as too short to reach an agreement. Thus, time pressure is assumed to be present if multiple issues have to be negotiated within a restricted amount of time, thus increasing the complexity of the negotiation task (e.g., Hunsaker 2018). If, for example, for the successful completion of a negotiation experiment the participants are to negotiate eight issues within a 25-minute time period, such a constellation is seen as a case of time pressure (e.g., Campagna, Mislin, and Bottom 2019). Time pressure is also assumed to be present if within the parameters of an experiment the participants are not told how much time they have, but are merely informed that they have multiple issues to negotiate until their time runs out (e.g., van Kleef, De Dreu, and Manstead 2004b).

In addition, the *power* level of the emotion recipient is coded for the moderator analysis (either low, medium or high), whereby the studies either directly manipulate the power position of the emotion recipient or the power level has to be deduced from the context. The power manipulation of studies is manifold, as they integrate different power constructs, all of which are integrated into the power variable for the underlying meta-analysis, including such power measures as the Best Alternative to a Negotiated Agreement (BATNA)[54], the support of the company, legitimate power, the perceived likelihood of finding another partner, or the presence of abundant alternatives (e.g., van Kleef et al. 2006; van Kleef and Côté 2007). A number of times the studies indicate the emotion expresser's level of power, which then has to be transformed into the power position of the recipient (reversed coding). Inasmuch as a study does not manipulate the power position, the power level is deduced from the dependence structure (provided there is one) between the recipient and the expresser of emotion. This transfer is possible because the power level of one party is frequently the other party's dependency (Bacharach and Lawler 1981; Wolfe and McGinn 2005) (see chapter 2.3.1.2). A potential dependence structure is determined based on the roles exhibited by the recipient and expresser of an emotion in each of the studies. In cases where a dependence structure exists, it is possible to determine whether the emotion recipient has greater power or less power than the individual expressing emotion. If for example the emotion recipient is a human resource manager in a job negotiation, he/she is assigned to a position of considerable power, since the emotion expresser (as a job applicant) is dependent on the manager's decision. If no obvious dependencies are present (e.g., in sales negotiations), both negotiators are seen as equally powerful, which leads to a medium power level for the emotion recipient. However, the dependence structure only serves as an indicator for the level of power inasmuch as power has not been manipulated by the study. Thus, the manipulation of power is seen as pivotal.

The assessment of whether the expressed emotion is *genuine* or *fake* (dummy variable) is based on the perception of the person receiving the emotion. If the experiment allows the participant to notice that the expression of emotion takes place for tactical reasons and is thus not genuine, it is coded as fake. This holds, for instance, if an actor or a participant in the experiment is explicitly directed to fake anger, the spuriousness of

[54] The BATNA of a bargaining party reflects the basis for this party's evaluation of the attractivity of a potential negotiation outcome (Fisher, Ury, and Patton 2011). It can be interpreted in terms of the opportunity costs to the negotiation outcome; thus it mirrors the best alternative to this outcome (Backhaus and Pesch 2018). Hence, it would not be meaningful for a bargaining party to accept an agreement which is worse than its BATNA (Fisher, Ury, and Patton 2011). An attractive BATNA offers a negotiator the choice between his/her best alternative and the proposed deal; therefore it significantly improves this negotiator's own power position (Lewicki, Barry, and Saunders 2015).

which is observable for the participant, since he/she perceives cues – such as incongruence between the chosen words and the facial expressions of the counterpart – that would be inappropriate for a truly felt emotion (e.g., Campagna, Mislin, and Bottom 2019). In cases in which the emotion expressed is clearly genuine, or in which the recipient is unable to deduce the authenticity of the emotion, it is assumed that the receiver surmises the emotion to be real, as no reason exists to believe it is being faked (e.g., Hillebrandt and Barclay 2017). If the individual expressing emotion represents an avatar in computer-mediated negotiations and if the participants cannot deduce any other information, it is assumed that they construe the emotion as genuine (e.g., Hareli et al. 2013).

The moderator *direction of the emotion* (dummy coded with the groups "at behaviour" or "at person") is evaluated on the basis of the emotion manipulation. If the manipulation makes explicitly recognizable that the expressed emotion is targeted at the behaviour of the recipient, it is coded as the same (e.g., the emotion expresser phrases that the offer makes him/her really angry (e.g., Sinaceur et al. 2013)). However, if the emotion is explicitly directed at the person, if both behaviour and person are addressed by the emotion, or if no reference to the behaviour of the opponent is observable, the variable is coded as "at person" (e.g., Lelieveld et al. 2013; Steinel, van Kleef, and Harinck 2008).

The moderating variable *competitiveness* of the negotiation is once again either drawn as a moderator from the studies (e.g., Adam and Brett 2015) or deduced from the experimental setting. This variable is divided into three groups: competitive, cooperative and balanced. When deduced from the context, an assessment is made as to which strategy the negotiation dyads follow (either competitive, cooperative or balanced). In addition, since the social value orientation of a negotiator is related to his/her chosen strategy (see chapter 2.3.2), this information is also used to determine the competitiveness of a negotiation (e.g., if a prosocial individual is present, the negotiation is assessed as cooperative (e.g., van Kleef and De Dreu 2010)). Lastly, the specification as to whether a competitive or a cooperative strategy is present or whether the situation is balanced, is evaluated on the basis of the underlying form of the negotiation (either distributive or integrative), since both constructs are related as well (see chapter 2.3.2). Thus, if it is apparent that a negotiation is distributive, it is coded as competitive, while an integrative negotiation is seen as cooperative (e.g., Rothman and Northcraft 2015). In cases in which no indications are present, or in which both distributive and integrative issues are negotiated, the negotiation is assessed as balanced (e.g., Yip and Schweinsberg 2017).

Lastly, the *medium* of the negotiation is coded. The medium variable is classified into four groups, which accounts for the amount of informational cues deducible from the medium. These four groups are FTF, visual channels (e.g., webcam conversations), vocal channels (e.g., telephone calls) and textual channels (e.g., e-mails). Of these four groups, FTF represents the medium with the largest amount of informational cues, followed by visual channels, vocal channels and, lastly, with the fewest informational cues, textual channels (see chapter 5.2.2). In most of the studies, the negotiation medium is fixed and does not vary during the interaction. However, in several cases multiple media are applied; for example, the negotiators bargain via textual messages first and then switch to video messages (which is mostly done in order to manipulate the expression of emotion) (e.g., Hideg and van Kleef 2017). In the cases with multiple channels, the "richest" medium – according to the media richness approach presented in chapter 5.2.2 – is coded (i.e., to stay with the example provided, visual rather than textual cues are coded).

Determination of Inter-Rater Reliability

Most of the coded variables for the meta-analysis at hand can be determined objectively, such as the sample sizes of the groups, the gender structure or the comparison group against which anger is compared. The coding of such variables is called low inference coding as the coder does not have to read between the lines and does not need to make interpretations (here and in the following Eisend 2014). However, some of the variables are characterized by a high degree of subjective latitude (high inference coding). This encompasses all six moderating variables of the EASI model. For these high inference variables, it is recommended in the literature that the quality of the coding be assessed by deploying a second coder. By doing so, one's own coding results can be compared to those of the second coder, thus offsetting their high degree of subjectivity, and delivering the option for adjustments if necessary. This decreases the likelihood of wrong coding decisions distorting the results of the meta-analysis. The second coder should be a person who is unfamiliar with the details of the research problem and design, so as to preclude any biases due to his/her prejudices (Rosenthal 1995; Wilson 2009). A student assistant of the institute who has no knowledge of the hypothesized effects is selected for this purpose. The coding sheet developed by the researcher is extensively explained to the second coder, the variables are clarified and examples are provided.

It is recommended in the literature that half or at least 20 of the studies be coded by the second coder (Hüffmeier et al. 2014; Lipsey and Wilson 2001). The number of effect sizes and studies for the EOs for which the moderating EASI variables are relevant amounts to 168 effect sizes, which arise from 85 studies (see chapter 5.4.1). In order to fulfil the goal that the second coder codes 50% of the studies, she needs to code at least 43 studies, which account for at least half of the effect sizes as well (at least 84 effect sizes). As the second coder has the possibility to leave gaps in the event of great uncertainty concerning a coding decision, the first 50 studies are provided to the coder as a buffer for missing values. After the first 10 studies (20% of the coding sample), emerging questions are resolved so that the coder can finish the remaining 40 studies on her own.

The second coder's coding results are compared to those of the author of this thesis. As all variables represent categorical variables, an adequate measure of agreement needs to be determined. In a number of cases the simple agreement proportion, which sets the agreement between the coded values in relation to all coded values, is prone to chance correspondences (Eisend 2014). For this reason, Cohen's Kappa, the most popular and commonly used interrater reliability for categorical variables, is applied (e.g., Hüffmeier et al. 2014; Kugler et al. 2018; Mazei et al. 2015). Cohen's Kappa determines how much better the proportion agreement is than chance alone (Cohen 1960) (see Formula 5.17). Cohen's Kappa values around 0 indicate very low reliability, whereas values starting from 0.41 can be classified as moderate (here and in the following Landis and Koch 1977). Values ranging from 0.61-0.8 indicate a substantial agreement and values from this range upward to 1 are seen as almost perfect.

$$\kappa = \frac{\Pr(o) - \Pr(x)}{1 - \Pr(x)} \tag{5.17}$$

κ = Cohen's Kappa
$Pr(o)$ = observed percentage agreement
$Pr(x)$ = expected percentage agreement based on chance alone

The average interrater reliability in the present case corresponds to 82.70%, which is a quite satisfying value according to the values presented above. More precisely, the interrater reliability for emotion direction is κ=1; for authenticity of the anger expression κ=0.92; for negotiation medium κ=0.82; for power level κ=0.79; for time pressure κ=0.72; and for competitiveness κ=0.71. Thus, all reliability values represent at least a substantial agreement, whereby the emotion direction reaches perfect agreement, and time pressure and competitiveness are the two variables with the highest coding discrepancies. It is based on these coding results that the author of this thesis and the second coder discuss their coding decisions, along with their disagreements and like-mindedness. In cases of discrepancies, the second coder and the researcher discuss until agreement is reached. The remaining studies are then coded by the author alone.

5.4 Results

Prior to the presentation of the results, it is of particular importance to stress that no separate models will be considered subsequently for a comparison between anger and happiness, and anger and emotional neutrality. Instead, anger will be compared within each model directly to both happiness and emotional neutrality, by building a comparison group which encompasses the two. The reason for the decision to merge emotion neutrality and happiness into one comparison group is that both constructs show a quite similar pattern when compared to anger. In this way, their main effects for both EOs and SOs show the same significant effect directions that do not deviate strongly in their magnitudes (see Appendix A-12). Additionally, the single meta-regression models for happiness and emotional neutrality overlap strongly with regard to significant moderators and their direction (the only relevant difference that can be noted is for the variable *negotiation medium*) (see Appendix A-13 and A-14).

One explanation for their similar pattern with respect to anger is the fact that in both constructs there is a stronger presence of pleasantness than for anger and at the same time both exhibit a similar level of arousal (see the circumplex model in chapter 2.3.1.2). Consequently, the segregated considerations of happiness and emotional neutrality would be accompanied by heavy intersections. By merging the two, the positive effect arises that the overall models exhibit greater explanatory power than the separated ones, since the number of included effect sizes is much larger due to the pooling. Thus, additional significant effects can be revealed with the help of this aggregation. Nevertheless, as happiness still exhibits a higher degree of pleasantness than does neutrality, and as both represent different constructs, the degree of difference between happiness and neutrality over against anger is included as a moderator in the meta-regression model.

5.4.1 Descriptive Characteristics

The search and coding procedure lead to an amount of 55 articles pertinent for the meta-analysis.[55] As a rule, these articles contain at least two studies, most of which measure the main effects examined by this meta-analysis; hence effect sizes are deduced from 94 studies overall. In the following section, the characteristics of the studies are briefly described first, and the characteristics of the moderators for the meta-regression second.[56]

The mean publication year of the 94 studies is 2012, mirroring the up-to-dateness of the research topic also underscored by the results of the co-citation analysis. In addition, nearly all studies represent laboratory experiments (98.92%), whereas only a minimal share conduct a field study (1.08%) (see here and in the following Table 5-1). Furthermore, almost all experiments are conducted as dyadic negotiations (94.68%), as a mere 5.32% of the studies create negotiation settings with more than one counterpart. As regards the number of negotiated issues, the majority of studies include multiple negotiable objects (e.g., price, warranty and delivery date) in the negotiation setups (62.92%). With respect to the characteristics of the studies' participants, the average age is 25.1 years.[57] This low mean age can be traced back to the fact that the studies' samples consist mainly of students (88.16%), while only a small amount (11.84%) base their analyses on professionals with more negotiation experience. Furthermore, the gender structure of the participants is quite balanced: 51.15% females and 48.85% males are investigated.

[55] The bibliographic data of the 55 articles can be found in Appendix A-15.

[56] A more detailed overview of the distribution of moderators across the effect sizes and values from which the effect sizes can be calculated can be found in Appendix A-16 and Appendix A-17.

[57] Both the average age and the gender structure are determined based on the studies that indicate this information.

Variable	Percentage	Variable	Percentage
Kind of Study		Gender Structure	
Laboratory	98.92	Female	51.15
Field	1.08	Male	48.85
Number of Issues to Negotiate		Dyadic Negotiation	
Single	37.08	Yes	94.68
Multiple	62.92	No	5.32
Occupation Status		Outcome Measure	
Students	88.16	Economic	62.92
Professionals	11.84	Subjective	37.08

Table 5-1: Study Characteristics
Source: Author's own illustration.

In total, 267 effect sizes can be calculated from the 94 studies, whereby 168 effect sizes measure EOs (62.92%) and 99 effect sizes SOs (37.08%).[58] A total of 237 effect sizes (88.76%) are calculated on the basis of means and standard deviations, while the remaining 30 effect sizes (11.24%) have to be reconstructed via the formulas presented in chapter 5.3.1.1. The overall sample size of the effect sizes amounts to 20,850 participants[59], 10,451 of which account for the anger variable and 10,399 for the comparison condition (happy or neutral). The division within the comparison group is quite balanced with 42.26% effect sizes comparing the effect of anger to happiness and 57.74% comparing it to emotional neutrality (see Table 5-2). Of the overall sample size, EOs are measured based on 13,080 participants and SO measurements are based on 7,770 participants.

Second, the descriptive characteristics of the moderating variables are considered in greater detail which are added for the sake of more precisely analysing the overall effect

[58] Several researchers recommend that corrections be made for potential biases in the effect sizes (so-called artifacts), such as measurement errors of the variables (e.g., Eisend 2014; Hunter and Schmidt 2004). Measurement errors can be corrected with the help of the reliability coefficient of the measurement scale (Eisend 2014). However, as these reliabilities are frequently missing in the studies included in the present meta-analysis, it foregoes correcting for measurement errors both in IVs and in DVs.

[59] As worked out in chapter 5.3.1.1, the extraction of multiple effect sizes from one study usually leads to the inclusion of participants multiple times for the calculation of these effect sizes (i.e., participant A in study i is counted three times, as he/she is used for the determination of three effect sizes). In this case, the overall sample size based on the determined effect sizes exceeds the overall actual number of the participants of the studies (e.g., Schmidt and Bijmolt 2019). This holds for this meta-analysis as well. Thus, it has to be kept in mind that the overall sample size includes one and the same participant of some studies multiple times, which leads to a higher amount than the pure number of participants which is 16,795 individuals. The pure number was used in chapter 3.2.1 to compare this meta-analysis with other published meta-analyses.

of the 168 effect sizes on EOs (see chapter 5.4.3). The methodological moderators of interest are: the underlying type of negotiation; the author from which the effect sizes arise; and the publication status of the underlying articles. Regarding the type of negotiation, the majority of effect sizes arise from sales negotiations (64.29%) (see here and in the following Table 5-2). By far fewer effect sizes – a mere 13.69% – are generated from game settings; 11.31% from job negotiations; and 10.71% from other situations such as bargaining over rents, negotiations on financial resources for a business project or negotiations on potential expansions. With regard to the author from whose studies the effect sizes are determined, the distinction is made between the researcher van Kleef and other authors. Van Kleef is first or co-author in 40.48% of the effect sizes, thus 59.52% arise from other authors. Lastly, a distinction is drawn between effect sizes from peer reviewed studies and grey literature, in order to account for potential publication bias (see chapter 5.3.1.3). 11.90% represent effect sizes from grey literature, which demonstrates a reasonable amount in meta-analysis (e.g., Bijmolt, van Heerde, and Pieters 2005; Schmidt and Bijmolt 2019; Tully and Winer 2014).

Variable	Percentage	Variable	Percentage
Authors		Communication Medium	
Van Kleef	40.48	Face-to-Face	11.39
Others	59.52	Visual	17.09
		Vocal	15.19
Publication Status		Text	56.33
Peer Reviewed	88.10		
Grey Literature	11.90	Time Pressure	
		Yes	28.57
Type of Negotiation		No	71.43
Sales Negotiation	64.29		
Job Negotiation	11.31	Power of Emotion Receiver	
Negotiation Games	13.69	Low Power	14.29
Other	10.71	Medium Power	64.88
		High Power	20.83
Comparison Group			
Happy	42.26	Emotion Direction	
Neutral	57.74	At Behaviour	65.48
		At Person	34.52
Competitivenss of Situation			
Distributive/Competitive	73.81	Emotion Authenticity	
Integrative/Cooperative	7.74	Genuine	84.52
Balanced	18.45	Fake	15.48

Table 5-2: Characteristics of Moderators

Source: Author's own illustration.

There follows a consideration of the six EASI model-based moderators (see here and in the following Table 5-2). For the competitiveness of the situation, a competitive situation is present in most cases (73.81%). A mere 7.74% of all cases account for cooperation. Thus, 18.45% of the effect sizes reflect a balanced negotiation with distributive as well as integrative parts. The predominant medium of communication in the effect sizes are textual messages (56.33%). In 15.19% of the effect sizes, the negotiators have auditory contact and in 17.09% a visual interaction that includes pictures or videos. Only 11.39% of the effect sizes rely on FTF negotiations. The presence of time pressure is existent in 28.57% of the effect sizes. The power position of the emotion receiver is low 14.29% of the time and high 20.83% of the time. Consequently, the power position of the emotion receiver is for the most part – 64.88% of the time – medium. With reference to the emotion direction, the majority of effect sizes are based on an expression of anger directed at the behaviour of the negotiation counterpart (65.48%), while in 34.52% of effect sizes the expression of anger is directed at the opponent personally. Lastly, in views of the emotion authenticity, genuine emotion expressions predominate (84.52%), whereas (strategically) faked expressions are ascertained in 15.48% of the effect sizes.

The results of the meta-analytical models will be presented in the following two sections, starting with the main effect on both negotiation outcomes (chapter 5.4.2), and proceeding in chapter 5.4.3 with the meta-regression on the anger effect on EOs.

5.4.2 Main Effect on Economic and Subjective Outcomes

In the following, the main effect of an anger expression in negotiations compared to the comparison group will be determined both for EOs and for SOs.[60] Prior to the presentation of both overall effect sizes and the assessment of the quality of the models, the potential presence of publication bias will be addressed for both outcomes, and an outlier analysis will be conducted. An outlier analysis is sensible for the following reason: As meta-analyses aim to provide a reasonable summary of quantitative findings, effect sizes that deviate noticeably from the majority of the other research findings are not representative of the results, and may well even be spurious (here and in the following Lipsey

[60] All analyses implemented in the following are conducted using R, whereby recourse is mainly taken to the package "metafor" created for meta-analyses in R (Viechtbauer 2010).

and Wilson 2001). Consequently, due to their disproportionate impact on the overall effect, the inclusion of such extreme values could distort the results, for which reason outliers are subsequently excluded prior to the analysis.

Economic Outcomes

First, the analysis of the main effect of anger on EOs is examined, beginning with a visual assessment of potential publication bias on the basis of a funnel plot (see here and in the following Figure 5-6). The observed effect sizes of the studies range from nearly -2 to almost 2. The sampling errors in turn exhibit a range of 0.08 to 0.57. The inversed funnel shape is built around the overall mean, which lies on the right side of 0, thus indicating a positive effect on EOs by an anger expression. In the upper part of the funnel (with smaller sampling errors) the distribution of the data points in the funnel plot is relatively symmetric. Nevertheless, the lower part of the funnel plot (with greater sampling errors) clearly shows gaps, as far fewer data points lie in this area than in the upper part, which makes the plot asymmetric. Moreover, many effect sizes fall out of the inversed funnel, which also points to the fact that rather large effect sizes from large studies get published. Since small effect sizes from small studies (with high sampling errors) seem to be missing, the visual inspection indicates potential publication bias. Nevertheless, there is no absolute clarity as to the side on which data points might be missing.

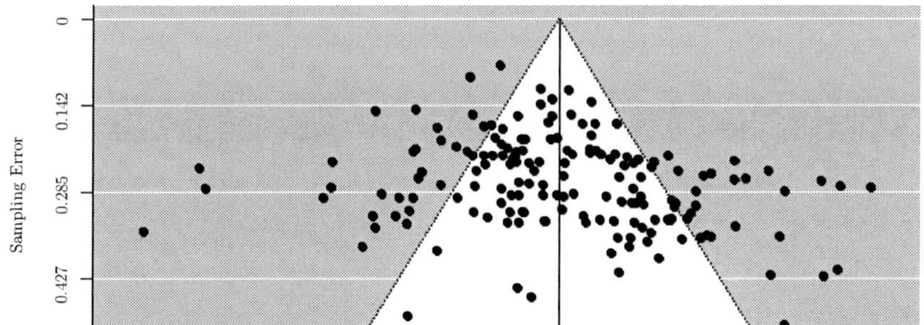

Figure 5-6: Funnel Plot of Effect Sizes Measuring Economic Outcomes
Source: Author's own illustration.

Therefore, a statistical validation is adduced in addition. The trim and fill method introduced in chapter 5.3.1.3 is used for this purpose. As recommended, both estimators R_0 and L_0 are applied. The results, however, are quite different. While the test based on L_0 indicates that 30 studies are missing on the left side of the funnel, R_0 results in 2 missing studies on the right side of the funnel (see here and in the following the example of Figure 5-7 for the results of the estimator L_0 and Appendix A-18 for the estimator R_0; the missing values are indicated by the unfilled dots). Consequently, the statistical assessment offers no clear indication of the side on which effect sizes might be missing; thus it delivers no direction in which the overall effect might be biased.

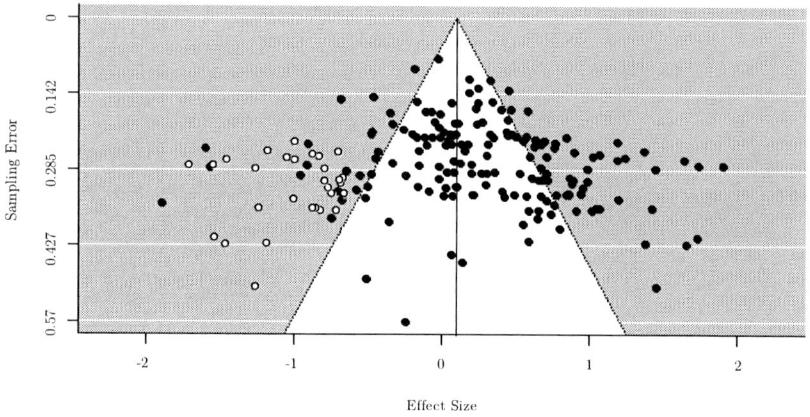

Figure 5-7: Funnel Plot of Economic Outcomes Corrected with Estimator L_0
Source: Author's own illustration.

Moreover, the results do not solve the problem of the lack of small studies with small effect sizes, as both estimators fill the plot outside the funnel in order to attain symmetry. Lastly and most importantly, both estimators lead to the result that the estimate of the overall effect size still remains positive and significant (see Appendix A-19). Thus the overall effect maintains the direction hypothesized for it irrespectively of which side of the funnel effect sizes might be missing on. Accordingly, publication bias does not seem to be much of a problem in the present case, as despite the fact that small studies with small effect sizes are indeed missing, this lack does not seem to bias the results. Nevertheless, the meta-regression model in chapter 5.4.3 accounts for potential publication bias by addressing the publication status of the included studies as a moderator in order to confirm or refute the insights gained from the funnel plot and the trim and fill method.

There follows the outlier analysis based on which potential outliers are determined with the help of a boxplot (see Appendix A-20). The box within the boxplot mirrors the area in which 50% of the data points can be located, whereby its horizontal line represents the median (here and in the following Backhaus et al. 2016). The antennas of the box – or so-called whiskers – mark the range of the data (minimal and maximal data values) excepting outliers. Thus, values outside the whiskers are candidates for outliers. One observed effect size becomes visible, that falls out of the upper whisker and three effect sizes fall out of the lower whisker. Sinaceur et al. (2013) measure an effect size of 1.9, which lies almost 10% above the last value included in the upper whisker. By contrast, the distances of the negative extreme values from the lower whisker are much higher, with a difference of at least 64%. Adam, Shirako and Maddux (2010) deliver two large negative effect sizes of -1.57 and -1.6. Pietroni et al. (2009) exceed these effect sizes even further by indicating a value of -1.9. As the extreme values could distort the results, they are excluded from further analysis. Thus, 4 effect sizes out of 168 (2.38%) are precluded, which represents a reasonable amount (Cohen et al. 2003; Schmidt and Bijmolt 2019).

Based on the remaining 164 effect sizes, the overall effect of an anger expression compared to happiness and neutrality on EOs is calculated. Therefore, an unconditional regression model is calculated with the intercept representing the weighted mean effect size. The determination of the significance of the effect takes place on the basis of the z-statistic, which uses the standard normal distribution (here and in the following Malhotra and Birks 2007). According to the determined level of the z-value, the corresponding p-value can be deduced.[61] The confidence interval is provided on the 95% level. The weighted mean effect size corresponds to 0.2986, which is highly significant owing to its p-value of less than 0.0001 (see here and in the following Table 5-3). The magnitude of this effect size is small to medium (Cohen 1992). The corresponding 95% confidence interval reaches from 0.2152 to 0.3820. Accordingly, negotiators reach higher EOs if they express anger than if they express happiness or emotion neutrality. Thus, the usage of anger is more

[61] For more details on the z-statistic and the relevant z-value table from which the significance level can be obtained, please refer to Malhotra and Birks (2007).

promising than happiness or neutrality. Consequently, H_{1a} can be confirmed by the unconditional model.

τ^2	0.2228
I^2	80.58%

Test for Heterogeneity:

Q(df=163) = 724.0995, p-value < 0.0001

Model Results:

Estimate	s.e.	z-value	p-value	ci.lb	ci.ub	
0.2986	0.0425	7.0188	<0.0001	0.2152	0.3820	***

Significance codes: 0.001(***) , 0.01(**) , 0.05(*) , 0.1(.)

Legend: df=degrees of freedom, s.e.=standard error, ci.lb=confidence interval lower boud, ci.up=confidence interval upper bound

Table 5-3: Main Effect of Anger Expression on Economic Outcomes
Source: Author's own illustration.

With regard to the degree of heterogeneity of the unconditional model, the Q statistic is highly significant (see here and in the following Table 5-3). Thus it is necessary to reject the H_0 hypothesis, which assumes that the effect sizes are homogeneous and thus indicates heterogeneity between the effect sizes. The value of I^2 is able to determine the degree of heterogeneity as it mirrors the relative proportion of the true heterogeneity to all variability in the model. It amounts to 80.58% in the present case, which is evaluated as a high degree of true heterogeneity (Higgins et al. 2003). For this reason, the inclusion of moderators in the model is requested, for the purpose of explaining parts of the heterogeneity. Accordingly, a meta-regression is conducted in chapter 5.4.3 that accounts for relevant moderators.

Subjective Outcomes

Second, the main effect is determined for the SOs. To begin with, a further visual assessment of potential publication bias is made, based on a funnel plot (see here and in the following Figure 5-8). On the x-axis, the observed effect sizes of the studies range from nearly -3 to almost 1. On the y-axis in turn, the sampling errors exhibit a range from 0.13 to 0.42. The inversed funnel shape is formed around the overall mean, which lies on the left side of 0, thus indicating a negative effect on SOs exerted by an expression of anger. The visual inspection of the funnel plot reveals several gaps of data points (upper and lower part) and some effect sizes that fall outside the 95% confidence interval. Nevertheless, the visual inspection is not sufficiently resilient to be able to assess the degree of asymmetry, so that both the Egger's test for funnel plot asymmetry (Egger et al. 1997) and the rank test based on Begg and Mazumdar (1994) are conducted, as well. Both tests show non-significant results, which is confirmation of the underlying assumption of funnel plot symmetry (Egger's test: z-value=-1.1082, p-value=0.2678; rank test: Kendall's τ=-0.0855, p-value=0.2113). Accordingly, publication bias does not seem to be present and likely does not bias the results of the meta-analysis on SOs.

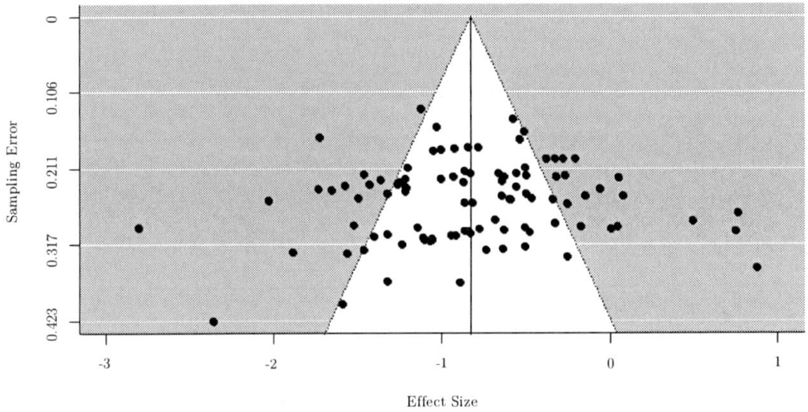

Figure 5-8: Funnel Plot of Effect Sizes Measuring Subjective Outcomes
Source: Author's own illustration.

In addition, an outlier analysis is conducted for which a boxplot is adduced anew (see here and in the following Appendix A-21). Two values fall below the lower whisker of the boxplot, whilst three values exceed the upper whisker, representing potential outliers of the effect sizes. The two lower effect sizes stem from Sinaceur et al. (2011), with a value of -2.36, and from Belkin and Rothman (2017), with -2.81. The upper outliers arise from Tng and Au (2014) (0.75), Kopelman, Rosette and Thompson (2006) (0.76), and De Melo, Gratch and Carnevale (2014) (0.88). As was done for the analysis of the EOs, the extreme values are excluded from further analysis, as they potentially distort the results. Thus, 5 effect sizes of 99 are excluded, which equals 5.05% and is a reasonable value in meta-analysis as well (Cohen et al. 2003; Schmidt and Bijmolt 2019).

The overall effect of anger on SOs as compared to that of happiness and neutrality is determined on the basis of the remaining 94 effect sizes. As was done for EOs, an unconditional regression model without moderators is conducted, for which the intercept represents the weighted mean effect size (see chapter 5.3.1.2). The overall effect size corresponds to -0.8399 and is highly significant, with a p-value smaller than 0.0001 (see here and in the following Table 5-4). The corresponding 95% confidence interval extends from -0.9382 to -0.7416. Consequently, the SOs of emotion receivers are much worse in cases where the counterpart expresses anger rather than happiness or neutrality. According to Cohen (1992), the magnitude of this effect size is large. Thus in contrast to happiness and neutrality, the usage of anger is harmful when considering the SOs of the counterpart. Consequently, H_8 is supported in the unconditional model.[62]

Model Results:

Estimate	s.e.	z-value	p-value	ci.lb	ci.ub	
-0.8399	0.0502	-16.7432	<0.0001	-0.9382	-0.7416	***

Significance codes: 0.001(***) , 0.01(**) , 0.05(*) , 0.1(.)

Legend: s.e.=standard error, ci.lb=confidence interval lower boud, ci.up=confidence interval upper bound

Table 5-4: Main Effect of Anger Expression on Subjective Outcomes
Source: Author's own illustration.

5.4.3 Meta-Regression on Economic Outcomes

A meta-regression is conducted on the basis of the 164 effect sizes investigating the effect of anger on EOs. In total, ten moderators are entered into the model. Of these ten moderators, six represent binary dummy variables. The manifestation of 0 represents the reference group to which the values of 1 are compared (here and in the following Hair et al. 2014). In cases where categorical variables consist of more than two categories, the manifestations have to be converted into dummy variables, whereby the number of needed dummy variables equals the amount of categories minus 1. This is because one

[62] The heterogeneity of the model is not assessed, as no meta-regression (moderator analyses) will be conducted for SOs (see chapter 5.2.2).

of the manifestations is chosen as a reference category against which the other manifestations are compared. This reference category does not enter the regression model. The four moderators *negotiation medium*, *competitiveness of the situation*, *power level of the recipient* and *type of negotiation* exhibit three to four different categories so that they have to be transformed into two or three dummy variables, respectively (medium: visual, vocal and text; competitiveness: competitive and balanced; power level: medium and low; type of negotiation: job, game and others). Thus, for medium, *FTF* is set as reference category, for competitiveness of the situation *cooperative*, for power *high power* and for negotiation type *sales negotiations*. With all independent variables being dummy coded, a regression analysis can be conducted. However, prior to the presentation of the model results, the assumptions presented in chapter 5.3.1.2 are tested, as the meta-regression has to fulfil these assumptions in order to deliver unbiased and efficient results.

First, the assumptions of regression models are tested, targeting the distribution of the residuals. As was stated in chapter 5.3.1.2, they can be visually inspected by plotting the standardized residuals on the y-axis with the corresponding predicted values of the DV on the x-axis. First, the assumption of linearity can be proven via this scatterplot. If no non-linear effect can be clearly deduced from this plot (e.g., curvilinear distributed dots), one can conclude that there is no non-linearity. Figure 5-9 confirms that linearity can be assumed, as no other pattern becomes visible. Second, it is tested for homoscedasticity, which assumes that the variance of the error terms is constant (here and in the following Hair et al. 2014). If heteroscedasticity was present, the residuals of the plot would show a consistent pattern (e.g., they would spread more strongly for higher predicted values of the DV than for lower ones). For the current case, no such pattern can be found, as the dots fall symmetrically around the horizontal line of 0 for the standardized residuals and thus do not show diverging values for lower/higher values of the predicted variable. Consequently, homoscedasticity is present and the assumption is fulfilled. Third, the last assumption which can be proven by the scatterplot is the independence of the error terms (here and in the following Hair et al. 2014). If the error terms were dependent on any sequencing variable (e.g., different points in time of measurements), one could observe this phenomenon from Figure 5-9, as the points would show a clear pattern (e.g., segregated point clouds). However, the residuals are randomly distributed across the scatterplot, so that independency of the error terms can be assumed as well.

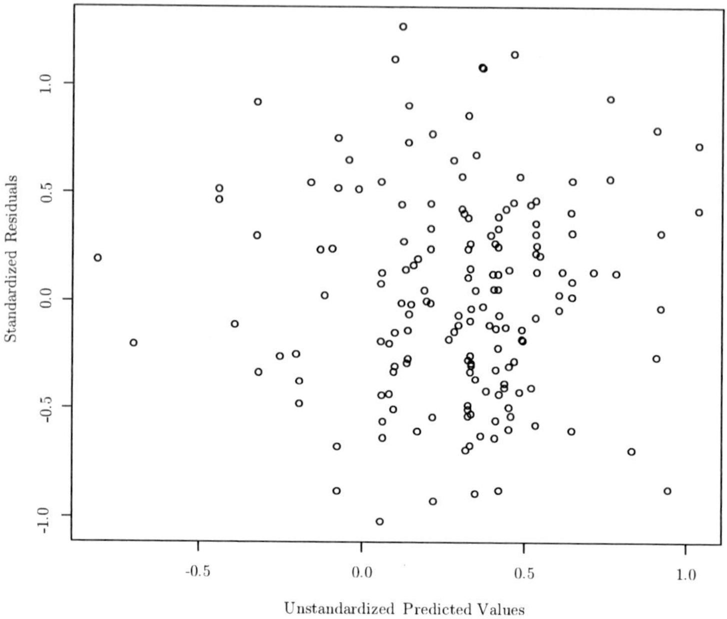

Figure 5-9: Scatterplot for Assumption Testing
Source: Author's own illustration.

Fourth, the assumption has to be fulfilled that the error terms of the model are distributed normally. In order to prove this assumption, the standardized residuals on the x-axis are plotted against their density on the y-axis in a histogram. The resultant distribution of the bars indicates whether a normal distribution is present. As Figure 5-10 shows, the standardized residuals are quite close to a normal distribution laid over the histogram. Besides, the Shapiro-Wilk test of normality is not significant, displaying as it does a p-value of 0.3613. Thus, the H_0 hypothesis of normally distributed error terms can be retained. Consequently, this assumption is fulfilled as well.

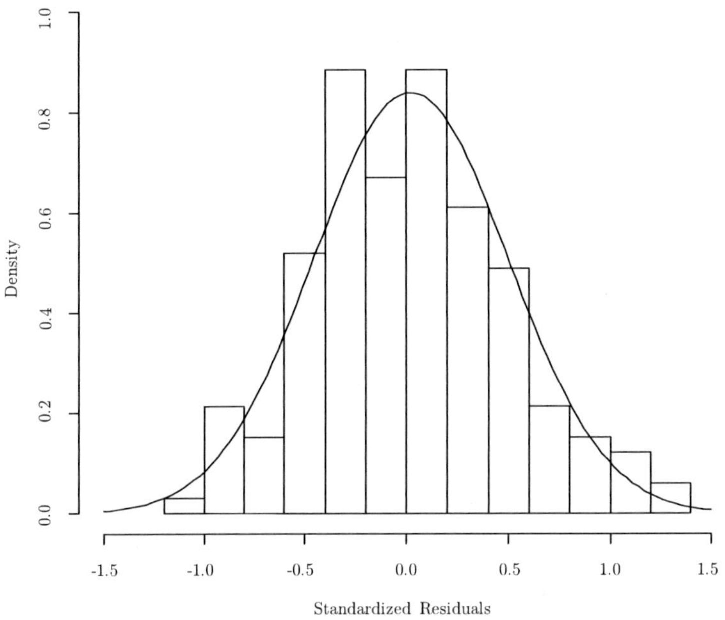

Figure 5-10: Histogram for Assumption Testing of Normally Distributed Errors
Source: Author's own illustration.

Lastly, the assumption must be fulfilled that there is no linear dependency among the moderators of the meta-regression model (here and in the following Backhaus et al. 2016). This assumption is tested with the help of the VIF values introduced in chapter 5.3.1.2. Table 5-5 summarizes the VIF values for the IVs at hand. It is conspicuous that the VIF values of the categorical variables with more than two categories are mostly higher than the VIF values of those variables consisting of only one dummy variable. This is a quite obvious observation, as the manifestations of those variables with more than two categories are related within their variable (e.g., the presence of low power is dependent on the presence or absence of medium or high power). Nevertheless, neither of these VIF values crosses the threshold value of 5. Thus, although the variables with more than two categories are more strongly related within the variable, they do not exhibit a collinear relationship with the remaining variables not belonging to their category. All remaining VIF values are close to 1. Accordingly, violations due to multicollinearity can be precluded.

Variable	VIF	Variable	VIF
Comparison Group	1.1288	Power	
Medium		*Low*	1.6485
Visual	2.4353	*Medium*	2.3643
Vocal	2.4533	Direction	1.9058
Text	3.6962	Negotiation Type	
Competitiveness		*Job*	2.3498
Competitive	3.5796	*Game*	1.7336
Balanced	4.1319	*Others*	1.3940
Time Pressure	1.4565	Publication Status	1.2629
Authenticity	1.3339	Author	1.5511

Table 5-5: Assumption Testing for Multicollinearity
Source: Author's own illustration.

As all assumptions were shown to have been fulfilled, no adaptations and transformations have to be made in the data, hence they can directly enter the meta-regression model. The confirmed assumptions allow the researcher to rely on the significance tests of the model, and to interpret the unbiased estimates (Backhaus et al. 2016).

Table 5-6 summarizes the results of the meta-regression. A positive sign for the estimates stands for a strengthening of the positive anger effect on the EOs of the anger expresser. By contrast, negative signs indicate a reduction in the positive anger effect. The discussion of the moderators will be split up into three parts: first, the comparison group of the main effect is examined in more detail; second, the moderators deduced from the EASI model are presented; and third, the methodological moderators are examined. Effects are interpreted up to a significance level of 10%.

τ^2	0.1554
I^2	73.00%
R^2	30.25%

Test for Heterogeneity:

Q(df=147) = 486.2525, p-value < 0.0001

Model Results:

	Estimate	s.e.	z-value	p-value	ci.lb	ci.ub	
Intercept	-0.6162	0.2801	-2.1998	0.0278	-1.1652	-0.0672	*
Comparison Group	0.1291	0.0806	1.6019	0.1092	-0.0289	0.2872	
Power							
Low	0.4854	0.1387	3.4986	0.0005	0.2135	0.7574	***
Medium	0.3224	0.1223	2.6373	0.0084	0.0828	0.5621	**
Time Pressure	0.0492	0.1001	0.4918	0.6228	-0.1469	0.2453	
Competitiveness							
Competitive	0.2637	0.1573	1.6768	0.0936	-0.0445	0.5720	°
Balanced	0.1883	0.1897	0.9927	0.3209	-0.1835	0.5602	
Medium							
Visual	-0.2486	0.1564	-1.5893	0.1120	-0.5553	0.0580	
Vocal	-0.0206	0.1642	-0.1254	0.9002	-0.3424	0.3012	
Text	-0.3624	0.1434	-2.5267	0.0115	-0.6435	-0.0813	*
Authenticity	0.3827	0.1150	3.3276	0.0009	0.1573	0.6081	***
Direction	0.3233	0.1073	3.0119	0.0026	0.1129	0.5337	**
Negotiation Type							
Job	0.1794	0.1822	0.9847	0.3247	-0.1777	0.5365	
Game	-0.2545	0.1466	-1.7363	0.0825	-0.5418	0.0328	°
Others	0.2000	0.1403	1.4251	0.1541	-0.0751	0.4751	
Publication Status	0.0140	0.1223	0.1141	0.9091	-0.2257	0.2536	
Author	0.0426	0.0959	0.4441	0.6570	-0.1453	0.2305	

Significance codes: 0.001(***) , 0.01(**) , 0.05(*) , 0.1(°)

Legend: df=degrees of freedom, s.e.=standard error, ci.lb=confidence interval lower boud, ci.up=confidence interval upper bound

Table 5-6: Meta-Regression on Economic Outcomes
Source: Author's own illustration.

The *comparison group* variable is dummy coded, including either happiness or emotional neutrality. For that variable no significant difference is detected between happiness and neutrality when comparing them to an anger expression (p=0.1092). However, this insignificance comes close to a significant effect at a 10% level. With happiness coded as 1 and neutrality as 0, the estimate of 0.1291 shows that, compared to happiness, the EOs of the anger expresser increase more strongly than compared to an emotion neutral negotiator. However, due to lacking significance this finding is not resilient. Thus, an increase on the valence dimension (from neutrality to happiness) does not significantly change the effect in comparison to an anger expression. Accordingly, hypothesis H_{1b} is not supported by the model.

With respect to the moderators of the EASI model, five of these six moderators reach significance. The only moderator not to reach significance is *time pressure* (p=0.6228). H_3 hypothesized that a lower degree of time pressure increases the positive anger effect. Due to a lower need for cognitive closure in cases of low time pressure, individuals were assumed to be more prone to seek information in order to gain a better understanding of the situation, which in turn favours the expression of anger. Due to the lacking significance, however, no resilient interpretations can be made. Thus, H_3 is not confirmed by the model.

With regard to the *power* variable, it was assumed that a low power level of the emotion recipient increases his/her motivation to process information. As a stronger motivation to process information strengthens the inference path, it was hypothesized that a low power level of the emotion receiver amplifies the positive anger effect on the EOs of the expresser of anger (H_2). The *power* variable shows a highly significant positive effect. With high power as reference category, low power significantly increases the EOs of the anger expresser even more (coefficient=0.4854, p=0.0005), whereby the coefficient of 0.4854 represents a medium effect size bordering on high magnitude (Cohen 1992). In this way, the positive anger effect is even more pronounced if the opponent has low power rather than high power. Thus, H_2 is confirmed by the model. The positive effect also holds for emotion receivers with medium power; however, it is slightly smaller (coefficient=0.3224, p=0.0084). Consequently, the lower the degree of power of the emotion recipient is, the higher are the EOs for the anger expresser over against happiness and emotional neutrality.

For the *competitiveness* of the situation, H_4 hypothesized that a competitive negotiation is more likely to foster the positive anger effect on the EOs of an anger expresser than a cooperative one is. The meta-regression model reveals a significant positive effect at the

10% level (coefficient=0.2637, p=0.0936). Accordingly, the positive anger effect is more strongly amplified in purely competitive negotiations than in cooperative settings. Thus, H_4 is supported by the conditional model. This significant moderator mirrors that it is more advantageous to express anger in purely distributive settings than in cooperative ones. Nevertheless, for a balanced negotiation situation in which both distributive and integrative components are present, no significant effect is found (p=0.3209).

For the *negotiation medium*, H_5 hypothesized that the positive anger effect is reduced if the negotiation takes place via text only rather than via FTF. This is due to the fact that in contrast to written channels, which exhibit the least possible amount of cues from which information can be deduced, FTF presents the "richest" medium by which most cues are available. H_5 is supported by the conditional model, as textual channels exhibit a significant negative effect on EOs of the anger expresser (coefficient=-0.3624, p=0.0115). Consequently, unlike FTF, textual channels significantly reduce the positive anger effect on EOs for the expresser of anger. Thus, the usage of anger expressions should be avoided in text-mediated negotiations, and should much rather be expressed FTF. The negative effect direction of visual and vocal channels ($coefficient_{visual}$= -0.2486; $coefficient_{vocal}$=-0.0206) somehow supports the finding that fewer informational cues favour a reduction in the positive anger effect; however, as they do not reach significance they are not resilient.

With respect to the *authenticity* of the anger expression, H_6 hypothesized that if anger is seen as genuine and not as a tactical manoeuvre, the positive anger effect on the EOs of the expresser is strengthened, as the affective reactions path is not triggered and the informational inferences stay in the foreground. For this variable, a highly significant positive effect is found (coefficient=0.3827, p=0.0009). As a dummy variable with 1 indicating genuine anger and 0 a (strategically) faked anger expression, the genuineness of anger therefore amplifies the positive anger effect on EOs. Thus, H_6 is supported by the conditional model. Accordingly, bargainers should avoid using anger expressions as a means of manipulating their counterpart. Instead, they should be encouraged to express their genuinely felt resentment.

With regard to the last moderating variable, *emotion direction*, H_7 hypothesized that if the anger expression is not directed at the counterpart personally, but rather at his/her behaviour in the negotiation (e.g., the offer is seen as too high), the positive anger effect on the EOs is strengthened. The variable shows a significant positive effect (coefficient=0.3233, p=0.0026). With 1 standing for an anger expression directed at the behaviour of the negotiation opponent and 0 representing anger targeted at the counterpart personally, the positive anger effect is amplified when directed at the behaviour. Accordingly, the meta-regression model confirms H_7. Therefore, negotiators should take care how they express anger. They should stick to objective aspects and take care that the expression of their anger is not a personal attack on the opponent.

Finally, the methodological moderators are tested. Table 5-6 shows that job negotiations and negotiations from the "others" category do not differ significantly from the anger effect in sales negotiations (p_{job}=0.3247, p_{others}=0.1541). The anger effect examined in negotiation game settings, however, reaches significance and shows a negative effect (coefficient=-0.2545, p=0.0825). Consequently, the positive anger effect on EOs is weaker in game settings, as opposed to experimental sales negotiations. Thus, the results from the game settings somehow cancel out the overall results, as they differ in the negative direction from the vast majority of sales negotiations examined. Consequently, the exclusion of results from game settings should cause the resulting overall positive effect of anger to become much higher. As experimental sales negotiations are very much closer to real life negotiations than are game settings, this moderator shows that the anger effect seems to be promising indeed in real life negotiations. The two other methodological moderators show no significant effect. In this manner, the anger effect is independent of whether the *author* van Kleef is among the authors from whose studies the effect sizes arise (p=0.6570). Last and quite importantly, the variable *publication status* does not become significant (p=0.9091). Consequently, effect sizes from grey literature do not deviate significantly from those of peer reviewed studies. This allows the conclusion that the estimates of the model are not biased due to publication bias.

With respect to the meta-regression model's quality, the τ^2 decreased from 0.2228 in the unconditional model on EOs, to 0.1554. In order to allow the reader to understand the meaning of this decrease even better, an analogous R^2 index is determined. This analogous R^2 index mirrors the variance that is explained by the moderators in the conditional model (here and in the following Rustenbach 2003). It is calculated by setting the τ^2 of the conditional model in relation to the τ^2 of the unconditional model (0.1554/0.2228). This ratio is then subtracted from 1. The R^2 in the present case amounts to 30.25% explained variance through the integration of the moderating variables (see Table 5-6). Thus, the conditional model explains the data much better than the unconditional model presented in chapter 5.4.2. When this value is compared with R^2 coefficients of other meta-analyses, one notices that this amount is quite good (e.g., Balliet et al. (2011) report a R^2 of 10% and Mazei et al. (2015) an R^2 of 15.16%). The better explanation in the model is also visible in the reduced value of I^2, which equals 73.00% and thus has decreased by more than seven percentage points. According to Higgins et al. (2003), this amount is equivalent to moderate heterogeneity, so that high heterogeneity is no longer present. Nevertheless, as the Q statistic shows by its significance as well: there is variance which remains unexplained after the inclusion of the moderating variables, so that further variables might exist that are not included in the model and that could account for the remaining variance.

The next section summarizes and discusses the results of the meta-analysis. There, Study II is assessed with regard to its reliability and the added value it creates for negotiation research and practice.

5.5 Discussion

The meta-analysis at hand has built on the results of the (co-)citation analysis conducted in Study I, in which the umbrella topic of "emotions" has been worked out as the most highly examined research topic within the last decades. With this topic at hand, the aim of Study II was to build on the strengthened shoulders of negotiation research giants and to generate new knowledge from their research findings as a new researcher (dwarf). In particular, the meta-analysis conducted here has sought to shed light on the interpersonal effect of the discrete emotion *anger* as compared to that of *happiness* and *emotional neutrality* in negotiations, the reason being that the influence of anger in bargaining situations has been rather unclear until now. While several studies had found the expression of anger to be of value for the person expressing it, other studies had stressed that anger is a detrimental factor in negotiations, and that emotional neutrality or happiness should be preferred over it (e.g., Adam, Shirako, and Maddux 2010; Sinaceur and Tiedens 2006). Besides, it had been emphasized that these ambiguities most likely arose due to the inclusion of certain moderators in the empirical studies. Thus it was not possible to infer from these results how best to adapt bargaining behaviour with respect to an expression of anger. Hence, a meta-analysis was needed that would bring clarity in the midst of these ambiguities.

This necessity was addressed by Study II. The anger effect was tested on the basis of the EASI model introduced in chapter 5.2.1, a rather new model which aims at explaining the interpersonal effect of emotions in social interactions (van Kleef 2009). Table 5-7 summarizes the results of the hypotheses testing. The checks on the right side indicate whether it was possible to confirm the respective hypothesis, whereby crosses indicate that the hypothesis did not reach significance. A total of six hypotheses could be confirmed, while it was found that two could not find support.

Hypothesis		**Confirmation**
H_{1a}	Compared to higher pleasantness, anger increases the EOs of the expresser	✓
H_{1b}	The anger effect is stronger when compared to happiness than to neutrality	✗
H_2	If the anger recipient has low power, the positive anger effect is stronger	✓
H_3	If time pressure is low, the positive anger effect is stronger	✗
H_4	If the level of competitiveness is high, the positive anger effect is stronger	✓
H_5	If only textual cues are present, the positive anger effect is reduced	✓
H_6	If the anger expression is genuine, the positive anger effect is stronger	✓
H_7	If anger is not personally directed, the positive anger effect is stronger	✓
H_8	Compared to higher pleasantness, anger decreases the SOs of the recipient	✓

Table 5-7: Hypotheses Testing Results of the Meta-Analysis
Source: Author's own illustration.

In this way, it was confirmed that anger increases the EOs of the anger expresser compared to an expression of happiness and emotional neutrality. This result underlines that the inference path dominates the affective reactions path of the EASI model in negotiations. The positive anger effect arises as negotiators infer higher limits from angry counterparts than from emotionally neutral or happy ones, so that they adapt their behaviour in favour for the anger expressers. Thus, it seems to be valuable in negotiations to confront the counterpart with the fact that one feels angry. The opponent will conclude that he/she needs to change behaviour if the negotiation is to continue. However, it was also confirmed that SOs are much lower for receivers of anger than for the observers of emotional neutrality and happiness. Accordingly, although anger has a positive effect on the EOs of the expresser, it should be kept in mind that the SOs of the emotion receiver decrease at the same time. Consequently, if the underlying negotiation is a recurring phenomenon as it arises for example from continuing business relations, or if such relations are strived for, any anger that might be brought to expression should be well thought through.

There were multiple significant moderators that fostered or weakened the positive effect of anger on the EOs of the expresser. With regard to the information processing moderators, it was found that if the power level of the emotion recipient is low, the positive anger effect is strengthened. This holds as well if the underlying negotiation is purely competitive. However, in cases where the negotiation is implemented via text-mediated channels, the positive anger effect is reduced. With the help of the social-relational moderators, the importance of the appropriateness of the anger expression was confirmed. In this way, the positive anger effect is stronger in cases where the anger expression is directed at the opponent's behaviour rather than at the opponent personally. Moreover, the effect increases if the anger expression is perceived as genuine by the receiver and not as a tactical attempt to influence the latter. To summarize: in order to exert a fostering of the positive anger effect on EOs, the emotion receiver needs to be both capable of processing the information hidden in the anger expression and motivated to do so; and the receiver must perceive the anger as appropriate.

The results of Study II are highly resilient, as they were assessed out of multiple perspectives. First, potential publication bias was accounted for, as evaluated for SOs and EOs alike. The verification via diverse graphical and statistical tests led to the result that publication bias has no negative influence on the results of the analyses. This was additionally proven within the meta-regression model for EOs, as it included the publication status as a moderator. Since no significant effect was found, it was underscored once more that publication bias does not distort the results. Second, the assumptions underlying regression models were tested for the meta-regression model. As violations of the assumptions would lead to biased and inefficient results, tests were run to ascertain whether data transformations were necessary in order to correct for violations. As all assumptions were fulfilled and thus no adaptations were needed, the outcomes of the meta-regression were proven to be both unbiased and efficient. Third and last, tests for dependencies among effect sizes within one and the same study were conducted for the sake of determining the adequate meta-analytical model. As no severe dependencies were detected, a random-effects model was conducted. Nevertheless, a multilevel model accounting for dependencies was implemented in addition, thus verifying the results of the random-effects model (see Appendix A-8 and A-9).

To conclude, Study II contributes substantially to existing research on negotiations and closes the research gaps presented in chapter 3.2.2. Study II is the first meta-analysis in the negotiation field to have compared anger with emotional neutrality and happiness; even more: it is the first meta-analysis to consider emotions at all. With its total of 16,795 individuals and 267 effect sizes from 94 studies, this meta-analysis has joined the higher ranks of other meta-analyses in the area of personal factors in negotiations presented in chapter 3.2.1. With the help of several significant moderators and the distinction between the influence of anger on EOs on the one hand and SOs on the other, it provides multiple explanatory approaches that reveal why anger is helpful in some cases and detrimental in others. Thereby, the meta-analysis of Study II allows for large confirmation and support of the EASI model. From the results of the meta-analysis, multiple practical and theoretical implications can be derived which can be found in chapter 6.2. Nevertheless, also Study II does have a number of limitations, which will be illuminated in chapter 6.3.

Thus Study II fulfilled its aim to build on the strengthened shoulders of negotiation giants, which enabled the researcher grow herself. For this purpose, Study II has built on the shoulders of the giants from two different perspectives. First, it has relied on the research topic most frequently examined by negotiation scholars. Having determined emotions, and especially anger, as most frequently investigated topic in the last two decades, the meta-analysis committed itself to a large subtopic in negotiations that is highly relevant and recent, and to which a great deal of research effort has been devoted. Second, Study II has relied on the giants' shoulders by aggregating these researchers' findings. Accordingly, it is important to emphasize that it was only possible to attain the insights gained from Study II based on existing empirical results of emotion research in the field of negotiations. Or in terms of the metaphor: the outcomes of the meta-analysis at hand are present thanks to the prior work of negotiation giants. It was based on this work that the author of this thesis found new and valuable insights which enabled her to look further from her position on the shoulders of these giants.

6 Closing Remarks

6.1 Summary

Negotiations are omnipresent. Next to professional settings in which negotiations are indispensable, such as business sales negotiations or job negotiations, they shape our private lives as well (e.g., bargaining about the sales price of a car or with one's partner about the next travel destination). For this reason, Neale and Bazerman (1992) claimed that everyone is a negotiator, as every individual negotiates many times in his/her life. In addition to their high practical relevance or even because of their great importance in real life, negotiations have gained a great deal of attention from a variety of researchers in different epochs and diverse research areas. In this way, negotiation research is characterized by a long history with multiple scientific approaches (e.g., game-theoretic approaches, social-psychological studies or behaviouristic approaches, see chapter 2.2). Moreover, the many research areas that deal with the topic make it a large research field. Whether politics, legal sciences, psychology, computer science or (business) economics, all these disciplines have spawned manifold research studies that cover negotiations from their own specific perspective. Due to the long history and the manifold research areas, there exists a continuously growing bulk of negotiation publications. This growth is strongly fostered by the current age of digitization, with its vastly improved distribution and accessibility of publications. For this reason, the aim of the thesis at hand has been to work with existing negotiation research publications by conducting two secondary research studies, rather than by joining the large amount of primary research.

The guideline for this purpose has been the quite popular metaphor of dwarfs standing on the shoulders of giants introduced by Bernard of Chartres in the 12th century, which mirrors how research works. This metaphor sets the latest research in relation to past research, with giants representing the large amount of scientific predecessors of past epochs and dwarfs being the current researchers that build on the findings of these predecessors. Thus, the metaphor assumes that current research can only gain knowledge and new insights if it relies on the previous discoveries of scholarly pioneers. By relying on the findings of the predecessors, the dwarfs (current researchers) can look further than the giants, – quite importantly – not because they are wiser; but rather because they are elevated by the gigantic stature of the preceding scholars. With this metaphor in mind, the thesis at hand assessed the current negotiation research situation.

The assessment showed on the one hand that the large amount of negotiation research is present in the form of various single research findings. On the other hand, large sets of studies dealing with quite specific negotiation topics have yielded a mass of insights for (partly) strongly related topics (e.g., is it helpful or detrimental to make the first offer?). It was further underscored that, on account of the isolation of the individual contributions in this research mass, it has become quite difficult to gain an overview of the topics covered by negotiation research. Faced by this clutter, it is something of a challenge to position oneself as a researcher. Thus, the claim advanced by this thesis that negotiation research needs a structure that brings related research insights together in order to give current researchers a reference point. An overview not only of research topics extensively researched in the past, but also of as yet uncharted territory in the field of negotiation research needs to be made available to current researchers, if the latter are to be able properly to position themselves. In terms of the metaphor: the shoulders of the negotiation giants require strengthening so that current research can stand firmly and reliably on them.

Chapter 3.1.1 presented a literature review that proofed what structuring approaches are present in the literature. It was emphasized that a reliable structuring approach should fulfill three requirements. First, reliability is especially present in quantitative procedures. As subjective appraisals and biases are reduced in quantitative approaches, a quantitative proceeding should be preferred over qualitative ones. Second, a reliable approach should broadly account for research topics, research areas, publications, journals and time range. Third and last, such an approach should be differentiating between different time periods in order to allow for more precise conclusions. Chapter 3.1 has shown that a structuring approach that accounts for all three of these aspects is not present in literature so that this gap has been addressed by Study I.

In Study I, the two bibliometric methods citation and co-citation analysis were applied to the field of negotiation research, the former constituting the basis for the latter. This procedure permitted quantitative processing and assessment of the data. Furthermore, it enabled the researcher of this thesis to analyse a huge amount of 66,294 references arising from 1,557 publications, a far greater quantity than would have been possible within a qualitative approach, limited as it is to human processing capability. In the citation analysis, the publications within the dataset of the 66,294 references were sorted according to their citation counts for four consecutive time frames. The most frequently cited of these, thus the most influential publications, were deduced (top x-lists) per period (in bibliometrics, higher citation counts are equated with a higher influence of a publication). The above mentioned co-citation analysis was conducted based on a specified share of the most cited publications, and pointed out the intellectual structure of the research field by identifying frequently examined topics and their interconnections. As co-citation analysis assumes that publications which are more frequently co-cited (thus, occur more often together in reference lists) are more closely related as to content, these publications were merged to topical clusters in a cluster analysis.

The co-citation analysis revealed 16 different topical clusters with different degrees of investigation (more intense investigation and higher impact for larger clusters) across the four time frames. Some topics reappeared within different periods (e.g., automated negotiations), whereas others were less frequently examined than in previous periods (e.g., representative negotiations), or vanished entirely. Interconnections between the different topics were barely present, which is indicative of the fact that researchers more or less “live in their own world”. The influence factors of the negotiation framework presented in chapter 2.3 – in particular, personal factors – were among the topics to a large extent, which confirms their relevance as depicted in chapter 3.2.1. Thus, the preparation and implementation phase, in which concrete tactics, goals and strategies are examined, are less present in existing negotiation literature than are personal influencing factors. It was possible to work out in a clear manner the most important topic, that of “emotions”, which over the last 20 years has developed into a highly examined and current research topic. In the process it was shown that the prime affect examined by the studies focusing on emotions is anger, followed by happiness, and that anger is predominantly compared to emotional neutrality or happiness. Moreover, most of the studies investigated the interpersonal effect of these emotions. Last and quite important, it was revealed that these comparisons are marked by several contradictory and inconsistent results, as the studies find different effects for the suitability of the emotion expressions on negotiation outcomes.

These findings made up the starting point for Study II. In the same manner in which giants form the basis for the dwarfs, the first study, in which the giants' shoulders were strengthened, constituted the baseline for the second study, which is devoted to the second part of the metaphor, i.e., the dwarf. The most frequently examined topic of "emotions" specified the umbrella theme for the second study, while the concrete topic for Study II was determined by inconsistent findings pertaining to a comparison of the influence of the specific emotion of anger on negotiation outcomes with the influence of emotionally neutral or happy negotiators. Thus, the aim of the second study was to aggregate these inconsistent findings in order to deduce a reliable overall effect on which both scholars and practitioners can rely. The most adequate method for this purpose was a meta-analysis. As chapter 3.2 has shown, multiple meta-analyses exist in the overall field of personal factors in negotiations. However, it has also been shown that no meta-analysis is among these studies that would account for a comparison of the effect of anger on negotiation outcomes to those of emotional neutrality or happiness. This is the gap addressed in Study II.

Based on the most popular model, which accounts for interpersonal effects of emotions in social interactions (the EASI model), eight hypotheses were derived that were to be proven by the meta-analysis in Study II. Two hypotheses targeted the main effect of anger on negotiation outcomes, compared to the main effect of the comparison group (emotional neutrality and happiness), whereby one of these hypotheses focused on the EOs of the emotion expresser and one focused on the SOs for the emotion receiver. The remaining six hypotheses included potential moderating factors derived from the EASI model that might influence the main effect of anger on EOs (which are the power level of the emotion receiver, the degree of time pressure in the negotiation, the competitiveness of the situation, the negotiation medium, the emotion direction and the genuineness of the emotion expression). Based on the results of the co-citation analysis and by means of an additional online search, a total of 94 studies entered the meta-analysis, including 16,795 individuals and accounting for 267 effect sizes. There followed the calculation of two unconditional random-effects models for the main effect on EOs (164 effect sizes after the exclusion of outliers) and the main effect on SOs (94 effect sizes after the exclusion of outliers). Additionally, a meta-regression was conducted, which accounted for the moderators potentially influencing the relationship on the EOs.

The results of Study II have shown that an expression of anger increases the EOs of the anger expresser significantly compared to the expression of emotional neutrality or happiness, which confirms the observer's inferences path of the EASI model. Moreover, most moderators that influence this main effect also reached significance. In this way, it was confirmed that the lower the power level of the recipient is, the higher the EOs for the anger expresser are, as low power individuals are more highly motivated to process information. In addition, the positive anger effect occurred in competitive negotiation settings rather than in cooperative ones, as information processing was more important in the former. The expresser of anger reached lower EOs if he/she negotiated via textual channels rather than negotiating FTF, the former being a by far inferior transmitter of informational cues. For the last two moderators, emotion direction and emotion genuineness, it was shown that the EOs for the anger expresser increased in cases in which the anger was not directed at the opponent personally and was perceived as being genuine rather than strategically faked. However, next to the positive effects of anger in negotiations, Study II also revealed that anger is detrimental with regard to the SOs of the emotion receiver. Thus, it was advised that prior to an expression of anger the underlying negotiation situation be evaluated, and that expressions of anger be avoided altogether in cases in which long-lasting (business) relationships are being sought.

To conclude, this thesis represents a prime example of an application of the metaphor of the dwarfs standing on the shoulders of giants to negotiation research. The researcher has brought clarity to the mass of negotiation research by applying bibliometric methods to 66,294 references, and has by means of a meta-analysis aggregated the contradictory findings of the effect of anger on negotiation outcomes for 267 effect sizes arising from 16,795 individuals. On the one hand, the thesis has delivered a guideline for negotiation scholars in order to position themselves with new research projects. On the other hand, it has deduced the aggregated effect of anger expressions and its moderating forces on negotiations, which stands to have multiple implications for scholars and practitioners. In this way the author has strengthened the giants' shoulders, on which she has been raised up in order to extend the knowledge relevant to emotions in negotiations.

6.2 Implications

On the basis of the two empirical studies, multiple implications can be derived for negotiation scholars and practitioners. In the following, these implications will be presented consecutively for science and professionals, respectively.

Implications for Academia

Starting with implications for negotiation science, one can underline first that – unlike the earlier obscure research situation – this thesis offers scholars a resilient overview of chief research topics; an overview on which they can rely and build. This dissertation has found that, in the words of Brett and Thompson (2016), negotiation research is mature but by far not dead, as there is still room left for research. Thus the starting points the thesis provides for future research by negotiation scholars are one of its main implications. At one of these starting points this thesis drew on itself and brought clarity into the contradictory findings of anger (as compared to happiness and emotional neutrality) on negotiation outcomes. It aligned the inconsistent findings and found explanatory variables that account for the observed variations, on which researchers can henceforth rely when making statements on the effects of emotions in negotiations.

Thus, this thesis deduces main research spots as well as under-researched areas, which reveal manifold potential for further research. In this way, it was shown that especially personal factors are examined quite exhaustively by negotiation research. Particularly the last decades were characterised by a strong focus on diverse personal factors, thus rendering more difficult the placement of further primary research studies in this field. As this thesis has shown, an adequate way of dealing with such highly examined topics is to position oneself with a meta-analysis that aims to combine the various research findings of empirical studies. Nevertheless, as chapter 3.2.1 has exposed, for other personal factors next to *emotions* there already exists several meta-analyses that integrate the diverse knowledge of these personal factors (e.g., gender, culture, trust). However, as meta-analyses are characterized by the fact that an update and/or extension of older versions is not unusual (e.g., Mazei et al. 2015 and Stuhlmacher and Walters 1999), there might still be room for meta-analyses focusing on gender, culture or even emotions. After all, it is certainly conceivable that other focal points could be set within these topics. Moreover, the co-citation analysis of Study I has revealed negotiation topics that shaped the research field strongly in early times but lost their dominance in more recent times. For such research themes, scholars could come up with new studies (e.g., negotiations with representatives or the usage of threats in negotiations). Lastly, research themes of relatively slight influence on the research field, such as automated negotiations or negotiation training, could profit from further investigations as well. This also includes research gaps disclosed for entire segments of the negotiation framework presented in chapter 2.3. As Study I has exposed, the phases of negotiation preparation and implementation are relatively under-researched by negotiation scholars. This provides researchers the possibility to extend the field by conducting more studies by focusing on topics such as negotiation goals or tactics.

In addition, the co-citation analysis has shown that the research topics remained more or less unconnected within the different time frames, indicating that they are characterized by high isolation. Thus, the thesis emphasized that negotiation scholars are particularly active within their specific negotiation research topics, but without interacting with others outside their familiar theme. This was also criticized by Herbst's, Voeth's and Meister's literature overview in 2011, and has not changed since then. Thus, the thesis at hand suggests combined research designs and more complex research models integrating multiple topics to be implemented by negotiation scholars. The interplay of multiple negotiation topics, for instance, could intend to analyse in how far emotions can be transmitted in automated negotiations and the consequences they might have. A further example could be to examine in how far negotiation training reduces the usage of cognitive heuristics in negotiations. Consequently, scholars could deduce various possible combinations of topics from the research networks of the co-citation analysis in order to develop cross-cutting research designs.

Yet another quite relevant implication of this thesis can be drawn from the interplay of Studies I and II. In this way, the thesis at hand was one of the first scientific works to use the methods of citation and co-citation analysis as a basis conducting a meta-analysis. It was shown that a co-citation analysis is perfectly suited to finding a relevant, recent and frequently explored research topic on which a meta-analysis can build. It was further demonstrated that the effort for the literature search of the meta-analysis can be significantly lowered by prior application of a (co-)citation analysis. This is on the one hand due to the fact that search strings applied to the (co-)citation analysis can be used (in slightly adapted forms) for the meta-analysis as well. On the other hand, the publications within the thematic clusters of the networks can be used for the meta-analysis, if only to a certain extent (for example, non-empirical studies have to be excluded as they cannot enter a meta-analysis). By doing so, the online search for the meta-analysis at hand could be reduced by approximately 42%, as this relative amount of publications could be adopted from the co-citation analysis. Thus, researchers should be aware of these advantages and should be encouraged to combine both methods in their studies.

Lastly, the dissertation has demonstrated strong confirmation of the EASI model for interpersonal emotions in social interactions by van Kleef (2009). In this way, it has been demonstrated that the observer's inferences path dominates the behaviour of emotion receivers in negotiation settings, as emotion receivers ascribe greater weight to the information processing than to potential affective reactions. Likewise, support has been found for nearly all moderating factors deduced from the EASI model, which underpins the influence of the two moderating forces of the observer's information processing and of social-relational factors. It was confirmed that the observer's inferences path or the affective reactions path weighs stronger, depending on which moderating force is affecting the main effect.

Implications for Professionals

Along with the various theoretical implications drawn from the studies, the latter also deliver implications for professionals. Although the implications of Study I tend to target negotiation academia more than they do negotiation practitioners (as Study I delivered the intellectual structure of the negotiation research field from which negotiation scholars benefit the most), professionals can also benefit from the results of the (co-)citation analysis. In this way, practitioners are encouraged to rely on the detected research networks, inasmuch as they are willing to improve either their own negotiation skills or for instance those of their employees. As the networks show highly relevant research themes and the dominating publications handling them, focal points are set for professionals in order to inform them of highly relevant topics and/or publications.

The practical implications of Study II are manifold. Professionals in particular learn from the results that they can enhance their EOs in a negotiation if they express anger, which deviates from the intuitive opinion that to express anger is definitively harmful for the expresser, as the counterpart feels affronted and the transmission of emotional neutrality or happiness might be much better alternatives. But far from it! Anger does not represent an emotion that should be suppressed in any event. Quite the contrary: in a variety of settings it should be even preferred over a happy or emotionally neutral manifestation. Consequently, in such settings professional negotiators should be encouraged to express the anger they feel, as it enhances their own EOs or those of the party for whom they are conducting the negotiation (e.g., for a company). It is critical, however, to bear in mind that the advantageousness of an expression of anger in negotiations has been found to be dependent on several factors that can be classified as situational and social-relational factors.

Starting with the situational factors, negotiators should analyse the present negotiation situation according to these factors prior to their expression of anger. First, they should keep in mind that the power position of their opponent should be low in order to increase the positive anger effect. As a good negotiation preparation consists of the identification of the counterpart's negotiation goals and thus of his or her alternatives (see chapter 2.3.2), the opponent's power position could be deduced from this information (as alternatives and power are related constructs). Should it be possible to assess the counterpart's power level as low, an expression of anger obtains the desired effect. It is also important to underline at this point that a person's own power position is only partially suitable for the assessment of the power position of the counterpart, as being in control of a power position of one's own does not necessarily mean that the other party has little power, or vice versa (e.g., a recruiter in a job negotiation may, as the person making the hiring decision, have considerable power over a candidate; however, if the candidate has multiple advantageous alternatives, his/her power is not low at all).

Second, the advantageousness of anger depends on the negotiation medium by way of which it is communicated. In this way, negotiators should keep in mind that anger should only be communicated via media in which as many informational cues as possible can be processed. Thus, if your counterpart is able to see your facial expressions and gestures, is able to hear your tone of voice and to process words you utter without any physical distance or asynchronism in the transmission, anger can much better be expressed. In these cases, the counterpart can evaluate and interpret the anger expression better than in textual messages. Consequently, misunderstandings and incomprehension become less likely. However, anger should definitely be avoided in negotiation settings with sparsely informational cues (e.g., text-mediated negotiations), as the anger effect is worsened in these cases. Thus, negotiators should wait with their anger expression until a FTF or videoconference meeting, and should refrain from manifesting it in other situations.

Third, negotiators should be aware of the negotiation context in order to assess the use of anger. In this way, anger should be especially expressed in the context of distributive or competitive negotiations, respectively. The prospects of concessions being made are more promising in these settings than in cooperative negotiations.

Regarding the social-relational factors, professionals should pay attention to the appropriateness of the anger expression. If the counterpart gets the impression that the anger expression is inappropriate, the anger effect worsens. Accordingly, professionals should mind that they are only allowed to express anger if it is not directed at the opponent personally (but rather at his/her offer, for instance) and if the anger expression is perceived by the other party as genuine. Thus, negotiators should acquire the habit of remaining impersonal in negotiations and place their words accurately prior to their pronouncement. Anger should not be communicated in the form of a hate speech directed at the other party, but rather as a reflection that the person expressing anger is dissatisfied with specific circumstances of the negotiation. Therefore, negotiators should definitely receive training in handling the expression of anger properly. The same holds for the genuineness of the anger expression. If negotiators want to profit from the positive anger effect even though they do not feel angry at all, they have to be quite careful with their strategically faked expression. The moment the emotion receiver recognizes that the anger is not real, he/she questions the emotion expresser, which leads to a reduced effect. Thus, if such tactical expressions of anger are desired, training sessions should be attended by the ones who aim to profit from them.

Finally, this thesis has shown as well that SOs of emotion receivers are dramatically impaired by an expression of anger. The variable SOs encompassed multiple constructs, namely the subjective appraisal of the negotiated outcome, the assessment of the emotion expresser, and the hypothetical behaviour of the emotion receiver after the negotiation, so that all three constructs are negatively influenced by an anger expression. The emotion receiver will not be content with his/her outcome, as he/she will have made large concessions to the counterpart. Moreover, the image of the anger expresser is worsened, as the liking and the level of trust felt for the counterpart decreases, thus negatively influencing the relationship between the parties. This leads in the end to a decrease in the likelihood of a recurring negotiation for the emotion receiver, and the latter becomes more likely to report negatively about that expresser. Consequently, despite all positive effects that an expression of anger can have on the EOs of a negotiator, the degree to which such an expression could negatively influence the relationship between the negotiators should be precisely evaluated. If the bonds are already quite strong or if the negotiation is a one-time event, anger can be expressed by keeping the previously described basic conditions in mind. However, if the formation of new (business) relationships is the aim, negotiators should be more cautious and better forego an expression of anger.

6.3 Limitations and Directions for Future Research

It is obvious that the previously presented implications underscore the contributions of the thesis at hand. Nevertheless, the implications should be considered against the background that the conducted studies are not free of a number of limitations, which will be considered separately and in greater detail in the following for each of the studies. Some of these shortcomings result in potential for future research on which other researchers can build.

Study I

Initially, as stated in chapter 4.4, a (co-)citation analysis delivers mainly researched and quite impactful research topics for a research field, however, is not able and does not claim to deliver all topics and all publications that have shaped the research field in its entire existence. This is not a particular limitation of bibliometric analyses; rather, it is something that is not feasible with any approach. Thus the aim at this point is to emphasize once more the expectations with which the results of the (co-)citation analysis should be regarded. The results mirror the intellectual structure of the negotiation research field of the past 60 years by focusing on the most impactful publications to have given rise to topical clusters. Consequently, researchers may wonder why topics are missing in the networks. Such a lack is due to the fact that the missing topics do not result from the most impactful publications, for which reason they turn out not to be as defining for the research field as the ones detected by Study I. Besides, Study I functions as a structuring approach and thus as an overview of research topics. Accordingly, in order to gain more detailed knowledge on the specific topics (or the topics not covered by the analysis), summarizing text books or qualitative literature reviews (with a special focus on one specific topic) are recommended by the author of this thesis (e.g., see Voeth and Herbst 2015b or the literature reviews presented in chapter 3.1.1 as starting points).

Thus, coming to the first real limitation of the first study, it has to be underlined that the results of every (co-)citation analysis are only as good as the database that underlies it (Fernandez-Alles and Ramos-Rodriguez 2009). In order to ensure that the analysis takes place on the basis of a comprehensive data pool, Study I relied on the databases SSCI and SCIE, as do many other (co-)citation analyses due to their large inclusion of journals and research areas since 1945 (see chapter 4.2.3). Nevertheless, despite their broad inclusion, SSCI and SCIE are not all-encompassing as, for example, they do not cover every journal (from their first issue onward), which is a lack that has been criticized by other researchers as well (e.g., Marsilio, Cappellaro, and Cuccurullo 2011). As the databases are not all-embracing, publications might be missing that would have been relevant for the current analysis. The inclusion of potentially missing publications might have changed the results of the citation analysis and the topical clusters derived by the co-citation analysis. A remedy for this shortcoming would be to rely on a greater number of databases than the two applied (Marsilio, Cappellaro, and Cuccurullo 2011), as was done in the meta-analysis of Study II. However, the reason why Study I did not follow this approach lies in the fact that databases provided from different suppliers (e.g., Google Scholar, SCOPUS, WOS with SSCI and SCIE) are not compatible with regard to their output of bibliographic data (e.g., they provide bibliographic data in different formats or use different abbreviations). As a mass of bibliographic data is processed in bibliometric analyses, the databases have to be compatible, as necessary adjustments would otherwise involve too much effort. Thus, if databases should become more aligned in the future, or computerized tools are available that are able to create such an alignment, the thesis at hand recommends rerunning the analysis with the application of a larger number of databases.

Second, SSCI and SCIE provide only the information of first and co-authors for the publications from which the bibliographic data is drawn, while for the bibliographic data itself only the first author is indicated. Accordingly, in cases in which an author of a publication (no matter whether first or co-author) is the co-author of a publication cited in the former's reference list, this is not traceable in the underlying analysis, as this information is not provided. Consequently, the self-reference values determined for the citation analysis might be underestimated for several publications. The citation analysis forewent a different handling of self-citations than of normal citations. However, the supplement of missing self-references due to co-authorship might have changed this decision if the results would had deviated strongly from those not accounting for self-references. This might have altered the results of the citation analysis and with them the networks of the co-citation analysis. Nevertheless, as the self-references which could be determined based on the data provided showed hardly any influence on the rankings of the citation analysis (see chapter 4.3 or Appendix A-3), it is assumed that the missing self-citations, if included, would not influence the results of the analysis strongly.

Third, although bibliometrics have the huge advantage that they are capable of quantitatively processing and analysing large amounts of numeric bibliographic publication data, the whole process is not completely free of subjective interventions on the part of the researcher. In this way, the choice of the publications from which the references were drawn, the final topic labelling of the clusters, the selection of the relevant research areas and the data cleaning were (partly) performed by the researcher, which necessarily involved a degree of subjective influence on these parts of the study. Nevertheless, the choice of publications from which the references were taken took place on the basis of a detailed catalogue of criteria, which reduced the scope for interpretation for the publications and with it the negative influence of human biases. Likewise, the topic labelling was unambiguous in most cases, thus reducing the likelihood of subjective influences. Moreover, the determination of those research areas that are defining for business economics took place on the basis of the assessments by several scholars (however, other classifications could have been conceivable as well, as the classification is not always consistent (e.g., the inclusion of macroeconomics)). Lastly, the possibility cannot be excluded that in the data cleaning process of the 66,294 references mistakes or inconsistencies have gone undetected and thus were not adapted; however, as two individuals – the author of this thesis and an assistant – corrected the data, this amount should be as low as humanly possible.

Fourth, the focus on business-relevant research areas and journals limited the analysis to those areas and other key topics could have been revealed by including or focusing for instance on legal or political sciences. Thus, scholars seeking to shed light on the research field from another perspective (e.g., negotiations in legal sciences), should replicate the present (co-)citation analysis with research areas relevant for their emphasis.

Lastly, in addition to the identification of the most frequently examined research topic of *emotions* in Study I, which was directly revisited in Study II, Study I's main implication was the revelation of many potentials for future research (see chapter 6.2). Multiple further relevant topics (e.g., culture or gender) or under-researched areas (e.g., automated negotiations or negotiation training) have been discovered on which future research can build.

Study II

First, although the endeavour was made to account for all studies relevant to the analysis by an extensive literature search, it cannot be precluded that relevant studies are missing in Study II. There could be a number of reasons for this. On the one hand, further search terms could have led to further relevant studies capable of serving as the points of departure for back and forward searches that might bring to light relevant studies as well. On the other hand, unpublished articles or other grey literature might be missing in the dataset. Although Study II tried to account for these publications, for example by contacting researchers or screening conference presentations for running research projects, there might still be grey literature not integrated in the analysis. This becomes especially likely in light of the fact that a number of the scholars contacted did not answer who might have provided data for unpublished works. Besides, as a few researchers only provided values for significant results and did not share the missing values for the insignificant effects, several effect sizes are definitely missing.

Second, the coding information of several moderating variables could not in all cases be directly derived from the studies, but rather had to be deduced from other variables, for the sake of avoiding missing data. This was the case multiple times for the variable *time pressure*, which was built from the variable *time limit* combined with the variable *complex negotiation (multiple issues)*. The same was true for power, which was deduced on the basis of the variable *role of the emotion receiver* in several cases. Thus, the determination of these values took place on the basis of (profoundly derived) assumptions made by the author of this thesis, which thus displays degree of subjectivity. Besides, the coding of other moderating variables of the EASI model was also partly accompanied by subjective decisions. For example, the decision as to whether an anger expression is perceived as genuine or fake by the emotion receiver or whether the negotiation situation is competitive rather than cooperative was not free from several subjective appraisals. Nevertheless, based on the uniformity of the coding rules applied to all studies and the verification by a second coder, the degree of subjectivity is assumed to be justifiable and a consistent coding is present. Be that as it may, the one or the other inadvertent mistakes in the coding of the moderating variables, in the deduction of the study characteristics, and/or in the determination of the effect sizes cannot be excluded with absolute certainty.

Third, although the I^2 of the conditional model of the anger effect on EOs was smaller than the unconditional model without moderators, it nevertheless remained on a high level, thus indicating potential for further explanatory factors. The EASI model proposes further moderators such as personality variables, cultural display rules, or involvement in the task (van Kleef 2007). However, Study II already tried to integrate moderators from the EASI model in the meta-regression for which information can be deduced from the studies. In this way, for instance, hardly any information on personality factors of the participants was provided by the studies, which made it impossible to include such factors into the meta-regression. Nevertheless, one could try to create variables out of the given information that come close to those constructs of the EASI model, as was also done for several variables in Study II. However, as the basis is predominantly missing for these variables, this thesis claims that further primary studies are needed that consider such variables in order to integrate them into further meta-analyses. Moreover, a meta-regression accounting for potential moderators for the anger effect on SOs displays potential for further research, as this was not covered by this thesis, but would be helpful for negotiators nonetheless. However, the EASI model would not be the correct conceptual baseline for this purpose. Since the EASI model considers not attitudes and feelings, but rather the behaviour of individuals as DV, the model provides no potential moderators for the latter. Thus, further theories or models would be needed to hypothesize moderating effects for the anger effect on SOs.

A fourth limitation which offers perfect room for further research is that Study II has been limited to the emotions of anger and happiness as the most frequently examined emotions. Nevertheless, other emotions (if only far fewer of them) are explored by negotiation publications, whose aggregated effect on negotiation outcomes (influenced by several moderating factors) would be helpful for negotiators as well. As the EASI model is not limited to specific emotions (van Kleef, De Dreu, and Manstead 2010), it also delivers approaches for other emotions besides anger and happiness, so that further research could also build on this model as a theoretical foundation. Other discrete emotions examined by negotiation studies are for example guilt, disappointment or fear (e.g., van Dijk et al. 2018; van Kleef, De Dreu, and Manstead 2006; Lelieveld et al. 2012). Negotiation researchers and practitioners would benefit if these emotions were to be integrated into a meta-analysis alongside anger and happiness, in which the diverse emotions' influences on negotiation outcomes would be compared in a more complex model. By doing so, negotiators could assess the appropriateness of emotions much better in bargaining situations, as they would have information about a great deal of different emotions, which would allow for a more general statement about emotions in negotiations.

Finally, Study II revealed that most of the studies included in the meta-analysis are based on laboratory experiments with students as participants. Aside from the several advantages laboratory experiments offer (such as a higher controllability compared to field experiments), laboratory studies also have the main shortcoming that real-life settings are only artificially reconstructed, which make their external validity inferior to that of field experiments (Balzert, Schröder, and Schäfer 2011; Gerber and Green 2009). Besides, students are frequently used as participants, as they are easier to recruit than practitioners. However, students have only limited or no negotiation experience at all, which makes inferences from such individuals more disputable (Agndal, Åge, and Eklinder-Frick 2017). These facts represent shortcomings attributable to the data present in the primary studies, and these shortcomings have to be kept in mind while relying on the results of this meta-analysis. Further research should strongly focus on research studies with professionals as participants in more real-life settings (e.g., Voeth et al. 2020; Voeth and Lenzing 2017). That would make the transferability of study results much easier.

Appendix

TS=("negotiat*" OR "bargain*") NOT TS=("law*" OR "politic*" OR "medic*" OR " NEUROLOG*" OR " PHILOSOPH*" OR "ECOLOG*" OR "NUTRITION*" OR "CRIMINOLOG*" OR "ENERGY FUEL*" OR "RELIGIO*" OR "TRANSPORTATION*" OR "VETERINAR*" OR "PUBLIC ADMINISTRATION*" OR "SPORT*" OR "HEALTH" OR "GERONTOLOG*" OR "CARDIOLOG*" OR "METEOROLOG*" OR "DENTISTR*" OR "GENETIC*" OR "ZOOLOG*" OR "ANTHROPOLOG*" OR "SURGER*" OR "FORESTR*" OR "ARCHAEOLOG*" OR "PEDIATRIC*" OR " PHARMAC*" OR "OCEANOGRAPH*" OR "ORTHOPEDIC*" OR "CHEMISTR* " OR "NURS*" OR "FOOD*" OR "FISHER*" OR "ONCOLOG*" OR "GYNECOLOG*" OR "GEOLOG*" OR "BIOLOG*" OR "GASTROENTEROLOG*" OR "AGRICULTUR*" OR "PSYCHIATR*" OR "RHEUMATOLOG*" OR "ANATOM*" OR "ANESTHESIOLOG*" OR "OTORHINOLARYNGOLOG*" OR "PATHOLOG*" OR "PHYSIOLOG*" OR "ENTOMOLOG*" OR "HEMATOLOG*" OR "PALEONTOLOG*" OR "TOXICOLOG*" OR "MUSIC*" OR "ALLERG*" OR "PLANT*" OR "touris*" OR "PHYSICS" OR " PHOTOGRAPHIC*" OR "MINERALOG*" OR "UROLOG*" OR "PARASITOLOG*" OR "TRANSPLANTATION*" OR "VIROLOG*" OR "CRYSTALLOGRAPH*" OR "OPHTHALMOLOG*" OR "ASTRO*" OR "DERMATOLOG*") NOT SU=("COMMUNICATION" OR "CONSTRUCTION BUILDING TECHNOLOGY" OR "SOCIAL WORK" OR "PHYSICS" OR "NEUROSCIENCES NEUROLOGY" OR "ASIAN STUDIES" OR "HISTORY PHILOSOPHY OF SCIENCE" OR "MEDICAL ETHICS OR RESPIRATORY SYSTEM "OR "ENVIRONMENTAL SCIENCES ECOLOGY" OR "WATER RESOURCES" OR "NUTRITION DIETETICS" OR "EDUCATION EDUCATIONAL RESEARCH" OR "REHABILITATION OR MATERIALS SCIENCE" OR "CRIMINOLOGY PENOLOGY" OR "ROBOTICS" OR "ENERGY FUELS" OR "RELIGION" OR "ENGINEERING" OR "TRANSPORTATION" OR "VETERINARY SCIENCES" OR "PUBLIC ADMINISTRATION" OR "SPORT SCIENCES" OR "PHILOSOPHY" OR "PUBLIC ENVIRONMENTAL OCCUPATIONAL HEALTH" OR "GERIATRICS GERONTOLOGY" OR "CARDIOVASCULAR SYSTEM CARDIOLOGY" OR "GEOGRAPHY" OR "PHYSICAL GEOGRAPHY" OR "AREA STUDIES" OR "AUTOMATION CONTROL SYSTEMS" OR "MARINE FRESHWATER BIOLOGY" OR "LINGUISTICS" OR "BIOTECHNOLOGY APPLIED MICROBIOLOGY" OR "METEOROLOGY ATMOSPHERIC SCIENCES" OR "DENTISTRY ORAL SURGERY MEDICINE" OR "HEALTH CARE SCIENCES SERVICES" OR "INFECTIOUS DISEASES" OR "MEDICAL INFORMATICS" OR "BIOMEDICAL SOCIAL SCIENCES" OR "ARTS HUMANITIES OTHER TOPICS" OR "GENETICS HEREDITY" OR "ZOOLOGY" OR "RESEARCH EXPERIMENTAL MEDICINE" OR "ANTHROPOLOGY" OR "SURGERY" OR "EVOLUTIONARY BIOLOGY" OR "WOMEN S STUDIES" OR "FORESTRY" OR "ARCHAEOLOGY" OR "HISTORY" OR "PEDIATRICS" OR "RADIOLOGY NUCLEAR MEDICINE MEDICAL IMAGING" OR "GENERAL INTERNAL MEDICINE" OR "PHARMACOLOGY PHARMACY" OR "OCEANOGRAPHY" OR "INFORMATION SCIENCE LIBRARY SCIENCE" OR "ORTHOPEDICS" OR "FILM RADIO TELEVISION" OR "LIFE SCIENCES BIOMEDICINE OTHER TOPICS" OR "BIOCHEMISTRY MOLECULAR BIOLOGY" OR "NURSING" OR "FOOD SCIENCE TECHNOLOGY" OR "FISHERIES" OR "ONCOLOGY" OR "MECHANICS" OR "URBAN STUDIES" OR "CHEMISTRY" OR "LEGAL MEDICINE" OR "IMMUNOLOGY" OR "INSTRUMENTS INSTRUMENTATION" OR "OBSTETRICS GYNECOLOGY" OR "NUCLEAR SCIENCE TECHNOLOGY" OR "GEOLOGY" OR "CELL BIOLOGY" OR "FAMILY STUDIES" OR "BIODIVERSITY CONSERVATION" OR "GASTROENTEROLOGY HEPATOLOGY" OR "AGRICULTURE" OR "SUBSTANCE ABUSE" OR "EMERGENCY MEDICINE" OR "PSYCHIATRY" OR

"RHEUMATOLOGY" OR "ANATOMY MORPHOLOGY" OR "ART" OR "ANESTHESIOLOGY" OR "OTORHINOLARYNGOLOGY" OR "OPTICS" OR "REMOTE SENSING" OR "INTEGRATIVE COMPLEMENTARY MEDICINE" OR "DEVELOPMENTAL BIOLOGY" OR "MICROBIOLOGY" OR "GEOCHEMISTRY GEOPHYSICS" OR "PATHOLOGY" OR "PHYSIOLOGY" OR "ENTOMOLOGY" OR "POLYMER SCIENCE" OR "HEMATOLOGY" OR "ACOUSTICS" OR "LITERATURE" OR "PALEONTOLOGY" OR "MEDICAL LABORATORY TECHNOLOGY" OR "TOXICOLOGY" OR "MUSIC" OR "ALLERGY" OR "PLANT SCIENCES" OR "BIOPHYSICS" OR "IMAGING SCIENCE PHOTOGRAPHIC TECHNOLOGY" OR "MINERALOGY" OR "UROLOGY NEPHROLOGY" OR "PARASITOLOGY" OR "TRANSPLANTATION" OR "METALLURGY METALLURGICAL ENGINEERING" OR "VIROLOGY" OR "CRYSTALLOGRAPHY" OR "OPHTHALMOLOGY" OR "ASTRONOMY ASTROPHYSICS" OR "MATHEMATICAL COMPUTATIONAL BIOLOGY" OR "ENDOCRINOLOGY METABOLISM" OR "DERMATOLOGY" OR "TROPICAL MEDICINE" OR "MINING MINERAL PROCESSING" OR "SCIENCE TECHNOLOGY - OTHER TOPICS" OR "GOVERNMENT LAW" OR "REHABILITATION" OR "MATERIALS SCIENCE" OR "MEDICAL ETHICS" OR "RESPIRATORY SYSTEM") NOT WC=("ETHNIC STUDIES" OR "PSYCHOLOGY CLINICAL" OR "DEMOGRAPHY" OR "HOSPITALITY LEISURE SPORT TOURISM" OR "HISTORY OF SOCIAL SCIENCES" OR "INTERNATIONAL RELATIONS" OR "TELECOMMUNICATIONS" OR "MATHEMATICS APPLIED" OR "BUSINESS FINANCE" OR "ECONOMICS" OR "INDUSTRIAL RELATIONS LABOR" OR "ETHICS" OR "MATHEMATICS INTERDISCIPLINARY APPLICATIONS" OR "PSYCHOLOGY EDUCATIONAL" OR "SOCIOLOGY" OR "STATISTICS PROBABILITY" OR "PSYCHOLOGY PSYCHOANALYSIS" OR "PSYCHOLOGY DEVELOPMENTAL" OR "PSYCHOLOGY MATHEMATICAL" OR "SOCIAL SCIENCES MATHEMATICAL METHODS" OR "OPERATIONS RESEARCH MANAGEMENT SCIENCE" OR "CULTURAL STUDIES" OR "MATHEMATICS" OR "PSYCHOLOGY BIOLOGICAL") *AND* **LANGUAGE:** (English) *AND* **DOCUMENT TYPES:** (Article)

Appendix A-1: Search String in WOS for the Co-Citation Analysis
Source: Author's own illustration.

Inclusion of...	**Exclusion of...**
• Articles in the Application Field of Business Negotiations (e.g., Sales Negotiations, Salary Negotiations,...)	• Articles Dealing with Economical Topics (e.g., Trading between Countries,...)
• Publications with Negotiation Objects such as Prices, Salaries, Delivery Terms, Purchase Quantities, i.e. Objects Relevant for Business Negotiations	• Labour Union/Labour Conflict Negotiations
• Articles Dealing with Negotiations in General, which can be Transferred to Business Negotiations (e.g., how to Negotiate Successfully, which Tools should be used, how does the Negotiation Process look like,...)	• Political or Legal Negotiations (e.g., no Israel-Palestine Conflict, no Wars, no Climate Protection,...)
• Publications Dealing with Negotiation Learning (however the Group of Investigation should be (Young) Adults and not Children)	• Articles Dealing Marginally or not at all with Negotiations (e.g., an Article solely Focusing on Decision Making without having a Negotiation Background,...)
	• Articles Honouring the Life Work of Researchers
	• Articles Focusing on Auctions
	• Publications in which Bargain is meant as Steal and not as Negotiation
	• Publications which are Focusing Predominantly or Exclusively on Mathematics or Informatics (e.g., Service Level Agreements, Cloud Computing,...)
	• Articles Focusing on Trust or Identity Negotiations

Appendix A-2: Set of Inclusion and Exclusion Criteria for the Dataset Creation
Source: Author's own illustration.

Period 1			
Accounted for Self-References		Not Accounted for Self-References	
Publication	**CV**	**Publication**	**CV**
Siegel (1960)	41.24%	Siegel (1960)	41.24%
Schelling (1963)	25.77%	Schelling (1963)	25.77%
Walton (1965)	19.59%	Walton (1965)	19.59%
Deutsch (1960)	14.95%	Deutsch (1960)	15.46%
Thibaut (1959)	14.95%	Thibaut (1959)	15.46%
Liebert (1968)	14.43%	Kelley (1967)	14.43%
Kelley (1967)	13.92%	Liebert (1968)	14.43%
Pruitt (1969)	13.92%	Pruitt (1969)	14.43%
Homans (1974)	12.37%	Pruitt (1970)	12.37%
Pruitt (1970)	11.86%	Deutsch (1962)	12.37%
Deutsch (1962)	11.86%	Homans (1974)	12.37%
Chertkoff (1967)	11.34%	Chertkoff (1967)	11.34%
Komorita (1968)	10.31%	Komorita (1968)	11.34%
Komorita (1969)	10.31%	Komorita (1969)	11.34%
Adams (1965)	10.31%	Adams (1965)	10.31%
Osgood (1962)	10.31%	Osgood (1962)	10.31%
Fouraker (1963)	9.28%	Fouraker (1963)	9.28%
Stevens (1963)	9.28%	Stevens (1963)	9.28%
Druckman (1967)	8.25%	Druckman (1968)	8.25%
Vidmar (1971)	8.25%	Druckman (1967)	8.25%
Bartos (1965)	8.25%	Vidmar (1971)	8.25%
Osgood (1959)	8.25%	Bartos (1965)	8.25%
Sawyer (1965)	8.25%	Kelley (1966)	8.25%
Druckman (1968)	7.73%	Osgood (1959)	8.25%
Brown (1968)	7.22%	Sawyer (1965)	8.25%
Hermann (1968)	7.22%	Brown (1968)	7.22%
Kelley (1966)	7.22%	Hermann (1968)	7.22%
Luce (1957)	7.22%	Luce (1957)	7.22%
Gallo (1966)	6.19%	Thibaut (1965)	6.19%
Pilisuk (1968)	6.19%	Lamm (1970)	6.19%

Appendix A-3: Top 30 Publications (Not) Accounted for Self-References

Source: Author's own illustration.

Period 2			
Accounted for Self-References		Not Accounted for Self-References	
Publication	**CV**	**Publication**	**CV**
Pruitt (1981)	32.63%	Pruitt (1981)	33.16%
Raiffa (1982)	31.56%	Raiffa (1982)	31.56%
Fisher (1991)	29.44%	Fisher (1991)	29.44%
Rubin (1975)	28.12%	Rubin (1975)	28.38%
Walton (1965)	28.12%	Walton (1965)	28.12%
Lax (1986)	21.35%	Lax (1986)	21.49%
Pruitt (1975)	16.05%	Pruitt (1975)	16.45%
Lewicki (1997)	15.12%	Lewicki (1997)	15.38%
Rubin (1994)	14.59%	Rubin (1994)	14.59%
Schelling (1980)	12.73%	Neale (1991)	14.06%
Neale (1991)	12.60%	Thompson (1990)	12.73%
Thompson (1990)	12.07%	Schelling (1980)	12.73%
Bazerman (1985)	11.80%	Bazerman (1985)	12.47%
Thompson (1990)	11.01%	Thompson (1990)	11.94%
Siegel (1960)	10.88%	Siegel (1960)	10.88%
Neale (1987)	9.68%	Neale (1987)	10.34%
Bazerman (1983)	9.68%	Bazerman (1983)	10.34%
Deutsch (1973)	9.02%	Deutsch (1973)	9.02%
Kimmel (1980)	8.75%	Kimmel (1980)	8.75%
Kelley (1966)	8.75%	Kelley (1966)	8.75%
Nash (1950)	7.96%	Neale (1986)	8.49%
Thibaut (1959)	7.96%	Neale (1985)	8.22%
Neale (1986)	7.82%	Nash (1950)	7.96%
Pruitt (1993)	7.56%	Thibaut (1959)	7.96%
Neale (1985)	7.43%	Pruitt (1993)	7.69%
Pruitt (1983)	7.43%	Pruitt (1983)	7.43%
Kahneman (1979)	7.16%	Kahneman (1979)	7.16%
Hofstede (1984)	6.90%	Hofstede (1984)	6.90%
Dwyer (1981)	6.50%	Dwyer (1981)	6.63%
Loewenstein (1989)	6.37%	Bazerman (1987)	6.63%

Appendix A-3: Continued

Source: Author's own illustration.

Period 3			
Accounted for Self-References		Not Accounted for Self-References	
Publication	**CV**	**Publication**	**CV**
Fisher (1997)	22.84%	Fisher (1997)	22.84%
Raiffa (1982)	20.07%	Raiffa (1982)	20.07%
Pruitt (1981)	16.94%	Pruitt (1981)	16.94%
Lax (1986)	15.38%	Lax (1986)	15.47%
Thompson (2008)	14.09%	Thompson (2008)	15.10%
Pruitt (1993)	14.00%	Lewicki (2007)	14.00%
Lewicki (2007)	13.90%	Pruitt (1993)	14.00%
Walton (1965)	12.15%	Walton (1965)	12.15%
Baron (1986)	10.31%	Baron (1986)	10.31%
Thompson (1990)	10.04%	Thompson (1990)	10.13%
Pruitt (2004)	9.94%	Pruitt (2004)	9.94%
Rubin (1975)	9.76%	Rubin (1975)	9.76%
Bazerman (2000)	9.21%	Bazerman (2000)	9.21%
Thompson (1990)	8.29%	Thompson (1990)	8.84%
Neale (1991)	8.29%	De Dreu (2000)	8.66%
Bazerman (1992)	8.01%	Neale (1991)	8.29%
Pruitt (1975)	7.92%	Bazerman (1992)	8.10%
De Dreu (2000)	7.18%	Pruitt (1975)	7.92%
Carnevale (1986)	6.45%	Carnevale (1986)	6.63%
Rubinstein (1982)	6.45%	Bazerman (1985)	6.45%
Schelling (1980)	6.45%	Rubinstein (1982)	6.45%
Bazerman (1985)	6.35%	Schelling (1980)	6.45%
Moore (1999)	5.89%	Weingart (1993)	6.26%
Weingart (1993)	5.89%	Moore (1999)	5.89%
Rosenschein (1994)	5.71%	De Dreu (1995)	5.89%
Allred (1997)	5.43%	Rosenschein (1994)	5.71%
Thompson (1991)	5.43%	Allred (1997)	5.52%
Carnevale (1992)	5.34%	Thompson (1991)	5.52%
Bazerman (1983)	5.34%	Carnevale (1992)	5.52%
Faratin (2002)	4.79%	Bazerman (1983)	5.52%

Appendix A-3: Continued

Source: Author's own illustration.

Period 4			
Accounted for Self-References		Not Accounted for Self-References	
Publication	CV	Publication	CV
Fisher (2011)	20.37%	Fisher (2011)	20.37%
Raiffa (1982)	17.22%	Raiffa (1982)	17.22%
Lewicki (2015)	14.54%	Lewicki (2015)	14.63%
Walton (1965)	14.26%	Walton (1965)	14.26%
Pruitt (1993)	13.89%	Galinsky (2001)	13.89%
Pruitt (1981)	13.70%	Pruitt (1993)	13.89%
Thompson (2015)	13.06%	Pruitt (1981)	13.70%
Galinsky (2001)	12.96%	Van Kleef (2004)	13.15%
Van Kleef (2004)	11.85%	Thompson (2015)	13.15%
Lax (1986)	10.93%	De Dreu (2000)	11.67%
De Dreu (2000)	10.83%	Lax (1986)	10.93%
Thompson (1990)	10.46%	Thompson (1990)	10.56%
Curhan (2006)	10.28%	Curhan (2006)	10.56%
Bazerman (2000)	10.00%	Bazerman (2000)	10.00%
Thompson (1990)	9.26%	Thompson (1990)	9.26%
Pruitt (1975)	9.07%	Pruitt (1975)	9.07%
Rubin (1975)	9.07%	Rubin (1975)	9.07%
Baron (1986)	8.70%	Sinaceur (2006)	8.89%
Pruitt (2004)	8.70%	Baron (1986)	8.70%
Sinaceur (2006)	8.52%	Pruitt (2004)	8.70%
Kopelman (2006)	7.78%	Van Kleef (2004)	8.15%
Hofstede (2001)	7.78%	Kopelman (2006)	7.96%
Thompson (1991)	7.69%	Thompson (1991)	7.78%
Preacher (2008)	7.59%	Hofstede (2001)	7.78%
Thompson (2010)	7.22%	Preacher (2008)	7.59%
Magee (2007)	7.13%	Kray (2001)	7.22%
Adair (2005)	7.13%	Weingart (1990)	7.22%
Weingart (1990)	7.04%	Thompson (2010)	7.22%
Van Kleef (2004)	6.94%	Magee (2007)	7.22%
Faratin (1998)	6.85%	Adair (2005)	7.22%

Appendix A-3: Continued

Source: Author's own illustration.

Author(s)	Year	Title	Journal/Publisher
Baron and Kenny	1986	The Moderator-Mediator Variable Distinction in Social Psychological Research	Journal of Personality and Social Psychology
De Dreu	2000	Influence of Social Motives on Integrative Negotiation: A Meta-Analytic Review and Test of Two Theories	Journal of Personality and Social Psychology
Deutsch	1960	The Effect of Motivational Orientation upon Trust and Suspicion	Human Relations
Deutsch and Krauss	1962	Studies of Interpersonal Bargaining	Journal of Conflict Resolution
Fisher and Ury	1981	Getting to YES: Negotiating Agreement without Giving In	Penguin Group
Galinsky and Mussweiler	2001	First Offers as Anchors: The Role of Perspective-Taking and Negotiator Focus	Journal of Personality and Social Psychology
Homans	1974	Social Behavior: Its Elementary Forms	Harcourt Brace Jovanovich
Kelley, Beckman and Fischer	1967	Negotiating the Division of a Reward under Incomplete Information	Journal of Experimental Social Psychology
Lax and Sebenius	1986	The Manager as Negotiator	The Free Press
Lewicki, Saunders and Minton	1997	Essentials of Negotiation	McGraw-Hill
Liebert, Smith and Hill	1968	The Effects of Information and Magnitude of Initial Offer on Interpersonal Negotiation	Journal of Experimental Social Psychology
Neale and Bazerman	1991	Cognition and Rationality in Negotiation	The Free Press
Pruitt and Drews	1969	The Effect of Time Pressure, Time Elapsed, and the Opponent's Concession Rate on Behavior in Negotiation	Journal of Experimental Social Psychology
Pruitt and Johnson	1970	Mediation As an Aid To Face Saving in Negotiation	Journal of Personality and Social Psychology
Pruitt and Lewis	1975	Development of Integrative Solutions in Bilateral Negotiation	Journal of Personality and Social Psychology
Pruitt	1981	Negotiation Behavior	Academic Press
Pruitt and Carnevale	1993	Negotiation in Social Conflict	McGraw-Hill
Raiffa	1982	The Art and Science of Negotiation	Harvard University Press
Rubin and Brown	1975	The Social Psychology of Bargaining and Negotiation	Academic Press
Rubin, Pruitt and Kim	1994	Social Conflict: Escalation, Stalemate and Settlement	McGraw-Hill
Schelling	1963	The Strategy of Conflict	Harvard University Press
Siegel and Fouraker	1960	Bargaining and Group Decision Making: Experiments in Bilateral Monopoly	McGraw-Hill
Thibaut and Kelley	1959	The Social Psychology of Groups	John Wiley & Sons
Thompson	1990	Negotiation Behavior and Outcomes: Empirical Evidence and Theoretical Issues	Psychological Bulletin
Thompson	2008	The Mind and Heart of the Negotiator	Pearson
Van Kleef, De Dreu and Manstead	2004	The Interpersonal Effects of Anger and Happiness in Negotiations	Journal of Personality and Social Psychology
Walton and McKersie	1965	A Behavioral Theory of Labor Negotiations	McGraw-Hill

Appendix A-4: Bibliographic Data of Most-Cited Publications of All Time Periods

Source: Author's own illustration.

Publications	Top 200		Top 300**		Top 400	
Co-Citations	Top 1.25%	Top 2%	Top 1.25%	Top 2%	Top 1.25%	Top 2%
Period 1	3 (0)*	3 (0)	**3 (0)**	3 (0)	3 (0)	3 (0)
Period 2	5 (1)	4 (0)	**7 (2)**	7 (2)	11 (5)	11 (5)
Period 3	8 (5)	8 (5)	**11 (7)**	14 (7)	14 (8)	17 (7)
Period 4	5 (2)	9 (3)	**8 (2)**	11 (4)	10 (3)	13 (4)

*The numbers without brackets represent the quantity of clusters in the respective period resulting from the combination of included publications and co-citations.
The numbers in brackets represent those clusters that consist out of four or less publications.
**The values highlighted in grey represent the numbers for the chosen approach in Study I.

Appendix A-5: Selection of Key Figures for the Co-Citation Analysis

Source: Author's own illustration.

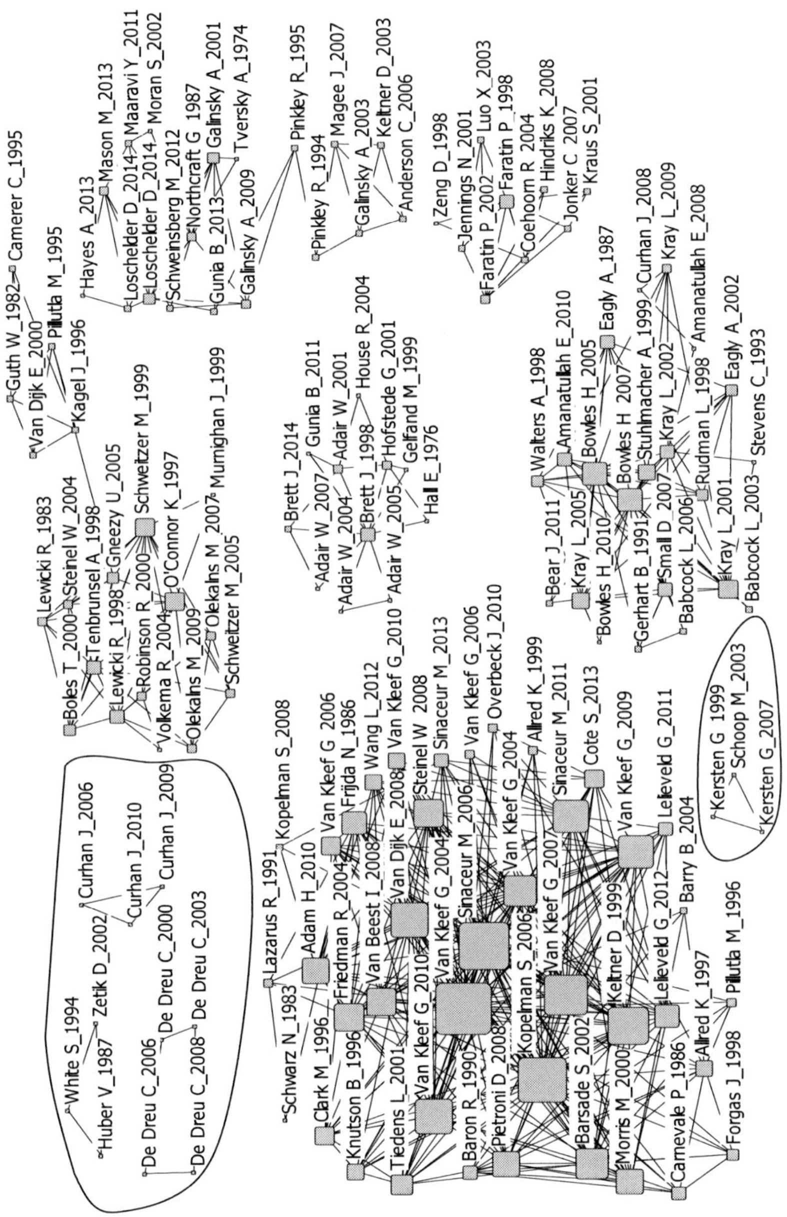

Appendix A-6: Network of Period 4 Based on Top 300 Publications and Top 2% of Co-Citation Relationships

Source: Author's own illustration.

A hierarchical model is set up that accounts for the effect sizes nested in studies by specifying a within-study level (effect sizes) as well as a between study level (studies). The observed effect sizes ES_{ms} represent the sum of their true effect sizes δ_{ms} and sampling errors ε_{ms} (here and in the following Bijmolt and Pieters 2001; Mazei et al. 2015) (see Formula A.1), whereby for the sampling error a normal distribution is assumed with zero mean and variance ϖ^2 ($\varepsilon_{ms} \sim N(0, \varpi^2)$). The sampling error of the respective observed effect size is known, as it can be calculated from the sample sizes of both groups and the observed effect size (see Formula 5.3 in chapter 5).

$$ES_{ms} = \delta_{ms} + \varepsilon_{ms} \tag{A.1}$$

m = consecutive number of effect sizes per study
s = consecutive number of studies
ES_{ms} = observed effect size m in study s
δ_{ms} = true effect size m in study s
ε_{ms} = sampling error of effect size m in study s

At the within-study level, the true effect sizes are decomposed in their single components: The true effect sizes can be described as the sum of the level two intercept (the intercept of the study level), moderators that impact the true effect sizes at the individual effect sizes' level and an error term h_{ms} (here and in the following Mazei et al. 2015) (see Formula A.2). The error term h_{ms} is assumed to be normally distributed ($h_{ms} \sim N(0, \sigma_{WS}^2)$).

$$\delta_{ms} = \mu_{0s} + \sum_{g} \rho_g * V_{gms} + h_{ms} \tag{A.2}$$

m = consecutive number of effect sizes per study
s = consecutive number of studies
δ_{ms} = true effect size m in study s
μ_{0s} = intercept of studies
g = consecutive number of moderators at the within-study level
ρ_g = coefficient of moderator g
V_{gms} = moderator g of effect size m in study s
h_{ms} = error term of true effect size m in study s at the within-study level

For the between-level model, the study specific intercept can be further decomposed (here and in the following Mazei et al. 2015). According to Formula A.3, it consists out of the next level's intercept (level three), moderators and an error component u_s. Also u_s is assumed to be normally distributed ($u_s \sim N(0, \sigma_{BS}^2)$). The moderators at the between-study level differ from the ones of the within-study level in such way that the former are deducible from the studies (e.g., the publication status of the studies), whereby the latter are only inherent in the individual effect sizes (e.g., the authenticity of the anger expression). In the present case, the control variables can be seen as study level moderators, while the moderators of the EASI model represent within-study level moderators.

$$\mu_{0s} = \zeta_0 + \sum_l \lambda_l * W_{ls} + u_s \tag{A.3}$$

s = consecutive number of studies
μ_{0s} = intercept of studies
ζ_0= level three intercept
l = consecutive number of moderators at the between-study level
λ_l = coefficient of moderator l
W_{ls} = moderator l of study s
u_s= error term of study level

Bringing together Formula A.1, A.2 and A.3 leads to the overall multilevel model, which accounts for the within-study as well as between-study level (see Formula A.4). At first, the model is applied to the dataset of EOs and SOs without moderators (see Appendix A-8). Ensuing, the moderators are added to the model accounting for EOs (see Appendix A-9).

$$ES_{ms} = \zeta_0 + \sum_l \lambda_l * W_{ls} + u_s + \sum_g \rho_g * V_{gms} + h_{ms} + \varepsilon_{ms} \tag{A.4}$$

m = consecutive number of effect sizes per study
s = consecutive number of studies
ES_{ms} = observed effect size m in study s
ζ_0 = level 3 intercept
l = consecutive number of moderators at the between-study level
λ_l = coefficient of moderator l
W_{ls} = moderator l of study s
u_s= error term of study level
g = consecutive number of moderators at the within-study level
ρ_g = coefficient of moderator g
V_{gms} = moderator g of effect size m in study s
h_{ms} = error term of true effect size m in study s at the within-study level
ε_{ms} = sampling error of effect size m in study s

Appendix A-7: Excursus to Multi-Level Models

Economic Outcomes

$\sigma_{BS}{}^2$	0.0129
$\sigma_{WS}{}^2$	0.2095
τ^2	0.2224
ICC	0.0582

Test for Heterogeneity

Q(df=163) = 724.0995, p-value < 0.0001

Model Results:

Estimate	s.e.	z-value	p-value	ci.lb	ci.ub	
0.2979	0.0438	6.8011	<0.0001	0.2121	0.3838	***

Subjective Outcomes

Model Results:

Estimate	s.e.	z-value	p-value	ci.lb	ci.ub	
-0.8336	0.0628	-13.2728	<0.0001	0.9567	-0.7105	***

Significance codes: 0.001(***) , 0.01(**) , 0.05(*) , 0.1(.)

Legend: df=degrees of freedom, s.e.=standard error, ci.lb=confidence interval lower boud, ci.up=confidence interval upper bound

Appendix A-8: Multilevel Model on the Main Effect of Anger Expression on Economic and Subjective Outcomes
Source: Author's own illustration.

τ^2	0.1628
R^2	26.80%

Test for Heterogeneity:

Q(df=147) = 486.2525, p-value < 0.0001

Model Results:

	Estimate	s.e.	z-value	p-value	ci.lb	ci.ub	
Intercept	-0.7460	0.2962	-2.5183	0.0118	-1.3265	-0.1654	*
Comparison Group	0.1455	0.0820	1.7749	0.0759	-0.0152	0.3063	°
Power							
Low	0.5961	0.1348	4.4234	<0.0001	0.3320	0.8602	***
Medium	0.3781	0.1344	2.8125	0.0049	0.1146	0.6416	**
Time Pressure	0.0492	0.1001	0.4918	0.6228	-0.1469	0.2453	
Competitiveness							
Competitive	0.2757	0.1603	1.7203	0.0854	-0.0384	0.5899	°
Balanced	0.2403	0.1931	1.2448	0.2132	-0.1381	0.6187	
Medium							
Visual	-0.2347	0.1743	-1.3464	0.1782	-0.5763	0.1069	
Vocal	-0.0126	0.1949	-0.0646	0.9485	-0.3945	0.3694	
Text	-0.3878	0.1535	-2.5271	0.0115	-0.6886	-0.0870	*
Authenticity	0.4037	0.1238	3.2604	0.0011	0.1610	0.6463	**
Direction	0.3914	0.1137	3.4427	0.0006	0.1686	0.6142	***
Negotiation Type							
Job	0.1537	0.1976	0.7779	0.4366	-0.2335	0.5409	
Game	-0.2554	0.1647	-1.5513	0.1208	-0.5781	0.0673	
Others	0.1446	0.1534	0.9430	0.3457	-0.1560	0.4452	
Publication Status	0.0540	0.1385	0.3901	0.6964	-0.2174	0.3255	
Author	0.0431	0.1111	0.3881	0.6979	-0.1746	0.2608	

Significance codes: 0.001(***) , 0.01(**) , 0.05(*) , 0.1(°)

Legend: df=degrees of freedom, s.e.=standard error, ci.lb=confidence interval lower boud, ci.up=confidence interval upper bound

Appendix A-9: Multilevel Meta-Regression on Economic Outcomes

Source: Author's own illustration.

TITLE ("negotiat*" OR "bargain*" OR "conflict*") AND TITLE ("emotion*" OR "anger" OR "angry" OR "happy" OR "happiness") AND (EXCLUDE (SUBJAREA, "MEDI") OR EXCLUDE (SUBJAREA, "BIOC") OR EXCLUDE (SUBJAREA, "ENGI") OR EXCLUDE (SUBJAREA, "AGRI") OR EXCLUDE (SUBJAREA, "MULT") OR EXCLUDE (SUBJAREA, "MATH") OR EXCLUDE (SUBJAREA, "HEAL") OR EXCLUDE (SUBJAREA, "ENVI") OR EXCLUDE (SUBJAREA, "PHAR") OR EXCLUDE (SUBJAREA, "EART") OR EXCLUDE (SUBJAREA, "ENER") OR EXCLUDE (SUBJAREA, "IMMU") OR EXCLUDE (SUBJAREA, "VETE") OR EXCLUDE (SUBJAREA, "CENG") OR EXCLUDE (SUBJAREA, "MATE") OR EXCLUDE (SUBJAREA, "NEUR") OR EXCLUDE (SUBJAREA, "NURS")) AND (EXCLUDE (LANGUAGE, "Spanish") OR EXCLUDE (LANGUAGE, "French") OR EXCLUDE (LANGUAGE, "Russian") OR EXCLUDE (LANGUAGE, "Czech") OR EXCLUDE (LANGUAGE, "Hungarian") OR EXCLUDE (LANGUAGE, "Italian") OR EXCLUDE (LANGUAGE, "Korean") OR EXCLUDE (LANGUAGE, "Lithuanian") OR EXCLUDE (LANGUAGE, "Turkish")) AND (EXCLUDE (EXACTKEYWORD, "Family Conflict") OR EXCLUDE (EXACTKEYWORD, "Work-family Conflict") OR EXCLUDE (EXACTKEYWORD, "Child Parent Relation") OR EXCLUDE (EXACTKEYWORD , "Parent-Child Relations") OR EXCLUDE (EXACTKEYWORD, "Marital Conflict") OR EXCLUDE (EXACTKEYWORD, "Middle Aged") OR EXCLUDE (EXACTKEYWORD, "Marriage") OR EXCLUDE (EXACTKEYWORD, "Burnout") OR EXCLUDE (EXACTKEYWORD, "Preschool Child") OR EXCLUDE (EXACTKEYWORD, "Child Behavior") OR EXCLUDE (EXACTKEYWORD, "Violence") OR EXCLUDE (EXACTKEYWORD, "Child, Preschool") OR EXCLUDE (EXACTKEYWORD, "Depression") OR EXCLUDE (EXACTKEYWORD, "Family") OR EXCLUDE (EXACTKEYWORD , "Parenting") OR EXCLUDE (EXACTKEYWORD, "Parents") OR EXCLUDE (EXACTKEYWORD, "Mother") OR EXCLUDE (EXACTKEYWORD, "Parent") OR EXCLUDE (EXACTKEYWORD, "War") OR EXCLUDE (EXACTKEYWORD, "Child Psychology") OR EXCLUDE (EXACTKEYWORD, "Childhood") OR EXCLUDE (EXACTKEYWORD, "Couples") OR EXCLUDE (EXACTKEYWORD, "Israel") OR EXCLUDE (EXACTKEYWORD, "Mental Health") OR EXCLUDE (EXACTKEYWORD, "Child Adjustment") OR EXCLUDE (EXACTKEYWORD, "Children") OR EXCLUDE (EXACTKEYWORD, "Father") OR EXCLUDE (EXACTKEYWORD, "Law") OR EXCLUDE (EXACTKEYWORD, "Adult") OR EXCLUDE (EXACTKEYWORD, "Young Adult")) AND (EXCLUDE (EXACTKEYWORD , "Emotional Intelligence") OR EXCLUDE (EXACTKEYWORD, "Emotion Regulation") OR EXCLUDE (EXACTKEYWORD, "Child"))

Appendix A-10: Search String in SCOPUS for the Meta-Analysis
Source: Author's own illustration.

TI(negotiat* OR bargain* OR conflict*) AND TI(emotion* OR happy OR happiness OR anger OR angry) AND Restrict With Subject (social psychology, social behaviour, self regulation, relationship quality, motivation, guilt, facial expressions, experiences (events), attention, adaptation, satisfaction, sadness, prosocial behaviour, negotiation, goals, face perception , externalization, cognitive control, choice behaviour, pride, dyads, behaviour, aggressive behaviour, interpersonal relationships, internalization, strategies, happiness, emotional responses, human sex differences, fear, emotional states, consequence, conflict resolution, interpersonal interaction, test construction, anger, conflict) AND Restrict With Classification (communication systems, professional personnel attitudes & characteristics, personality psychology, human experimental psychology, educational administration & personnel, psychophysiology, mediation & conflict resolution, management & management training, social perception & cognition classroom dynamics & student adjustment & attitudes organizational behaviour, general psychology, cognitive processes, social processes & social issues, personnel attitudes & job satisfaction, motivation & emotion, developmental psychology, industrial & organizational psychology, social psychology, personality traits & processes, psychosocial & personality development, group & interpersonal processes) AND Restrict With Methodology (experimental replication, prospective study, followup study, longitudinal study, quantitative study, empirical study) AND Restrict With Language (English) AND Restrict With SubjectMajor (organizational behaviour, interpersonal communication, decision making, bargaining, personality traits, happiness, cognitive processes, interpersonal relationships, intergroup dynamics, emotionality (personality), emotional control, interpersonal interaction, emotional adjustment, emotional responses, emotional states, anger, conflict resolution, negotiation, emotions, conflict)

Appendix A-11: Search String in EBSCO for the Meta-Analysis
Source: Author's own illustration.

Economic Outcomes							
Neutrality	τ^2=0.1856	I^2=78.87%					
Happiness	τ^2=0.2775	I^2=81.96%					
Test for Heterogeneity							
Neutrality	Q(df=94) = 396.6338, p-value < 0.0001						
Happiness	Q(df=68) = 310.2584, p-value < 0.0001						
Model Results:							
	Estimate	s.e.	z-value	p-value	ci.lb	ci.ub	
Happiness	0.3635	0.0723	5.0306	<0.0001	0.2219	0.5051	***
Neutrality	0.2520	0.0517	4.8790	<0.0001	0.1508	0.3533	***

Subjective Outcomes							
Model Results:							
	Estimate	s.e.	z-value	p-value	ci.lb	ci.ub	
Happiness	-0.9739	0.0814	-11.9646	<0.0001	-1.1334	-0.8143	***
Neutrality	-0.7639	0.0621	-12.2993	<0.0001	-0.8856	-0.6421	***

Significance codes: 0.001(***) , 0.01(**) , 0.05(*) , 0.1(.)

Legend: df=degrees of freedom, s.e.=standard error, ci.lb=confidence interval lower boud, ci.up=confidence interval upper bound

Appendix A-12: Main Effect of Anger Expression on Economic and Subjective Outcomes Separated for Emotional Neutrality and Happiness

Source: Author's own illustration.

τ^2	0.1429
I^2	72.73%
R^2	23.03%

Test for Heterogeneity:

Q(df=79) = 264.9743, p-value < 0.0001

Model Results:

	Estimate	s.e.	z-value	p-value	ci.lb	ci.ub	
Intercept	-0.5173	0.3583	-1.4437	0.1488	-1.2196	0.1850	
Power							
Low	0.5903	0.1973	2.9913	0.0028	0.2035	0.9770	**
Medium	0.2896	0.1602	1.8083	0.0706	-0.0243	0.6036	°
Time Pressure	0.0959	0.1335	0.7179	0.4728	-0.1658	0.3576	
Competitiveness							
Competitive	0.1951	0.2057	0.9483	0.3430	-0.2081	0.5982	
Balanced	0.1243	0.2565	0.4845	0.6280	-0.3784	0.6269	
Medium							
Visual	-0.3876	0.1915	-2.0237	0.0430	-0.7631	-0.0122	*
Vocal	-0.2595	0.2178	-1.1918	0.2333	-0.6863	0.1673	
Text	-0.4708	0.2143	-2.1969	0.0280	-0.8907	-0.0508	*
Authenticity	0.4114	0.1498	2.7466	0.0060	0.1178	0.7049	**
Direction	0.3202	0.1718	1.8644	0.0623	-0.0164	0.6569	°
Negotiation Type							
Job	0.2126	0.2486	0.8552	0.3924	-0.2747	0.6999	
Game	-0.2245	0.1791	-1.2537	0.2100	-0.5756	0.1265	
Others	0.2437	0.2261	1.0778	0.2811	-0.1994	0.6868	
Publication Status	0.1200	0.1470	0.8167	0.4141	-0.1680	0.4081	
Author	-0.0814	0.1192	-0.6832	0.4945	-0.3151	0.1522	

Significance codes: 0.001(***) , 0.01(**) , 0.05(*) , 0.1(°)

Legend: df=degrees of freedom, s.e.=standard error, ci.lb=confidence interval lower boud, ci.up=confidence interval upper bound

Appendix A-13: Meta-Regression on Economic Outcomes for Emotional Neutrality

Source: Author's own illustration.

τ^2	0.1743
I^2	71.24%
R^2	37.20%

Test for Heterogeneity:

Q(df=53) = 177.4249, p-value < 0.0001

Model Results:

	Estimate	s.e.	z-value	p-value	ci.lb	ci.ub	
Intercept	-0.9049	0.5139	-1.7609	0.0783	-1.9122	0.1023	°
Power							
Low	0.3522	0.2050	1.7183	0.0857	-0.0495	0.7540	°
Medium	0.4201	0.2132	1.9702	0.0488	0.0022	0.8379	*
Time Pressure	-0.1319	0.1749	-0.7540	0.4508	-0.4747	0.2109	
Competitiveness							
Competitive	0.3703	0.2690	1.3768	0.1686	-0.1569	0.8974	
Balanced	0.3172	0.3560	0.8910	0.3730	-0.3806	1.0149	
Medium							
Visual	0.2249	0.3124	0.7199	0.4716	-0.3874	0.8372	
Vocal	0.3817	0.2988	1.2775	0.2014	-0.2039	0.9673	
Text	-0.1508	0.2672	-0.5644	0.5725	-0.6744	0.3729	
Authenticity	0.4419	0.2290	1.9296	0.0537	-0.0069	0.8907	°
Direction	0.5394	0.1714	3.1460	0.0017	0.2033	0.8754	**
Negotiation Type							
Job	0.3488	0.3162	1.1031	0.2700	-0.2709	0.9684	
Game	-0.3133	0.2783	-1.1257	0.2603	-0.8587	0.2322	
Others	0.0475	0.2191	0.2169	0.8283	-0.3819	0.4770	
Publication Status	-0.1679	0.2314	-0.7255	0.4681	-0.6216	0.2857	
Author	0.2158	0.2089	1.0333	0.3015	-0.1569	0.8974	

Significance codes: 0.001(***) , 0.01(**) , 0.05(*) , 0.1(°)

Legend: df=degrees of freedom, s.e.=standard error, ci.lb=confidence interval lower boud, ci.up=confidence interval upper bound

Appendix A-14: Meta-Regression on Economic Outcomes for Happiness

Source: Author's own illustration.

Author(s)	Year	Title	Journal/Publisher
Adam and Brett	2015	Context Matters: The Social Effects of Anger in Cooperative, Balanced, and Competitive Negotiation Situations	Journal of Experimental Social Psychology
Adam and Brett	2018	Everything in Moderation: The Social Effects of Anger Depend on its Perceived Intensity	Journal of Experimental Social Psychology
Adam and Shirako	2013	Not all Anger is Created Equal: The Impact of the Expresser's Culture on the Social Effects of Anger in Negotiations	Journal of Applied Psychology
Adam, Shirako and Maddux	2010	Cultural Variance in the Interpersonal Effects of Anger in Negotiations.	Psychological Science
Belkin, Kurtzberg and Naquin	2013	Signaling Dominance in Online Negotiations: The Role of Affective Tone	Negotiation and Conflict Management Research
Belkin and Rothman	2017	Do I Trust You? Depends on What You Feel: Interpersonal Effects of Emotions on Initial Trust at Zero-Acquaintance	Negotiation and Conflict Management Research
Bertram	2018	The Social Effects of Anger in International Negotiations	Dissertation
Bhattacharjee and Moreno	2017	The Role of Informal Controls and a Bargaining Opponent's Emotions on Transfer Pricing Judgements	Contemporary Accounting Research
Butt	2003	The Role of Emotions in Dyadic Negotiation	Dissertation
Butt and Choi	2006	The Effects of Cognitive Appraisal and Emotion on Social Motive and Negotiation Behavior: The Critical Role of Agency of Negotiator Emotion	Human Performance
Campagna, Mislin, and Bottom	2019	Motivated by Guilt and Low Felt Trust: The Impact of Negotiators' Anger Expressions on the Implementation of Negotiated Agreements	Journal of Behavioral Decision Making
Campagna, Mislin, Kong and Bottom	2016	Strategic Consequences of Emotional Misrepresentation in Negotiations: The Blowback Effect	Journal of Applied Psychology
Cheshin, Rafaeli and Bos	2011	Anger and Happiness in Virtual Teams: Emotional Influences of Text and Behavior on Others' Affect in the Absence of Non-verbal Cues	Organizational Behavior and Human Decision Processes
Côté, Hideg and van Kleef	2013	The Consequences of Faking Anger in Negotiations	Journal of Experimental Social Psychology
De Melo, Carnevale and Gratch	2011	The Effect of Expression of Anger and Happiness in Computer Agents on Negotiations with Humans	10th International Conference on Autonomous Agents and Multiagent Systems
De Melo, Carnevale and Gratch	2012	The Effect of Virtual Agents' Emotion Displays and Appraisals on People's Decision Making in Negotiation	International Conference on Intelligent Virtual Agents
De Melo, Gratch and Carnevale	2014	Humans vs. Computers: Impact of Emotion Expressions on People's Decision Making	IEEE Transactions on Affective Computing
Dehghani, Carnevale, and Gratch	2014	Interpersonal Effects of Expressed Anger and Sorrow in Morally Charged Negotiation	Judgment and Decision Making
Hareli, Akron and Hess	2013	The Effect of the Negotiator's Social Power as a Function of the Counterpart's Emotional Reactions in a Computer Mediated Negotiation	Europe's Journal of Psychology

Appendix A-15: Bibliographic Data of the Articles Included in the Meta-Analysis

Source: Author's own illustration.

Author(s)	Year	Title	Journal/Publisher
Harinck and van Kleef	2012	Be Hard on the Interests and Soft on the Values: Conflict Issue Moderates the Effects of Anger in Negotiations	British Journal of Social Psychology
Hideg and van Kleef	2017	When Expressions of Fake Emotions Elicit Negative Reactions: The role of Observers' Dialectical Thinking	Journal of Organizational Behavior
Hillebrandt and Barclay	2017	Comparing Integral and Incidental Emotions: Testing Insights From Emotions as Social Information Theory and Attribution Theory	Journal of Applied Psychology
Hunsaker	2018	Reciprocal Anger, Threat, or Tracking: Using Intensity of Anger to Disentangle Three Common Responses to Angry Negotiators in Distributive and Integrative Negotiations	Dissertation
Jäger, Loschelder and Friese	2016	Using Self-Regulation to Overcome the Detrimental Effects of Anger in Negotiations	Journal of Economic Psychology
Jang and Bottom	2020	The Disutility of Anger Expression in Negotiations with Strategic Risk and Reciprocal Interdependence	Working Paper
Khooshabeh, De Melo and Volkman	2013	Negotiation Strategies with Incongruent Facial Expressions of Emotion Cause Cardiovascular Threat	Proceedings of the Annual Meeting of the Cognitive Science Society
Kooshabeh et al.	2016	Neurophysiological Effects of Negotiation Framing	Cognitive Science
Kopelman, Rosette and Thompson	2006	The Three Faces of Eve: Strategic Displays of Positive, Negative, and Neutral Emotions in Negotiations	Organizational Behavior and Human Decision Processes
Kung and Chao	2019	The Impact of Mixed Emotions on Creativity in Negotiation: An Interpersonal Perspective	Frontiers in Psychology
Lelieveld et al.	2013	Behavioral and Neural Reactions to Emotions of Others in the Distribution of Resources	Social Neuroscience
Lelieveld et al.	2012	Why Anger and Disappointment Affect Other's Bargaining Behavior Differently: The Moderating Role of Power and the Mediating Role of Reciprocal and Complementary Emotions	Personality and Social Psychology Bulletin
Li	2005	Strategic Emotion in Negotiation: The Use and Consequences of Positive and Negative Emotions	Dissertation
Overbeck, Neale and Govan	2010	I Feel, Therefore You Act: Intrapersonal and Interpersonal Effects of Emotion on Negotiation as a Function of Social Power	Organizational Behavior and Human Decision Processes
Pietroni et al.	2009	When Happiness Pays in Negotiation - The Interpersonal Effects of 'Exit Option': Directed Emotions	Mind & Society
Pietroni et al.	2008	Emotions as Strategic Information: Effects of Other's Emotional Expressions on Fixed-pie Perception, Demands, and Integrative Behavior in Negotiation	Journal of Experimental Social Psychology
Rees et al.	2019	Anger as a Trigger for Information Search in Integrative Negotiations	Journal of Applied Psychology
Rothman	2011	Steering Sheep: How Expressed Emotional Ambivalence Elicits Dominance in Interdependent Decision Making Contexts	Organizational Behavior and Human Decision Processes
Rothman and Northcraft	2015	Unlocking Integrative Potential: Expressed Emotional Ambivalence and Negotiation Outcomes	Organizational Behavior and Human Decision Processes

Appendix A-15: Continued

Source: Author's own illustration.

Author(s)	Year	Title	Journal/Publisher
Sinaceur et al.	2013	The Advantages of Being Unpredictable: How Emotional Inconsistency Extracts Concessions in Negotiation	Journal of Experimental Social Psychology
Sinaceur et al.	2011	Hot or Cold: Is Communicating Anger or Threats More Effective in Negotiation?	Journal of Applied Psychology
Sinaceur and Tiedens	2006	Get Mad and Get More than Even: When and Why Anger Expression is Effective in Negotiations	Journal of Experimental Social Psychology
Steinel, van Kleef and Harinck	2008	Are You Talking to Me?! Separating the People from the Problem When Expressing Emotions in Negotiation	Journal of Experimental Social Psychology
Tamir and Ford	2012	When Feeling Bad Is Expected to Be Good: Emotion Regulation and Outcome Expectancies in Social Conflicts	Emotion
Tng and Au	2014	Strategic Display of Anger and Happiness in Negotiation: The Moderating Role of Perceived Authenticity	Negotiation Journal
Van Beest, van Kleef and van Dijk	2008	Get Angry, Get Out: The Interpersonal Effects of Anger Communication in Multiparty Negotiation	Journal of Experimental Social Psychology
Van Dijk et al.	2008	A Social Functional Approach to Emotions in Bargaining: When Communicating Anger Pays and When it Backfires	Journal of Personality and Social Psychology
Van Kleef and Côté	2007	Expressing Anger in Conflict: When it Helps and When it Hurts	Journal of Applied Psychology
Van Kleef and De Dreu	2010	Longer-term Consequences of Anger Expression in Negotiation: Retaliation or Spillover?	Journal of Experimental Social Psychology
Van Kleef, De Dreu and Manstead	2004	The Interpersonal Effects of Anger and Happiness in Negotiations	Journal of Personality and Social Psychology
Van Kleef, De Dreu and Manstead	2004	The Interpersonal Effects of Emotions in Negotiations: A Motivated Information Processing Approach	Journal of Personality and Social Psychology
Van Kleef et al.	2006	Power and Emotion in Negotiation: Power Moderates the Interpersonal Effects of Anger and Happiness on Concession Making	European Journal of Social Psychology
Van Kleef and van Lange	2008	What Other's Disappointment May Do to Selfish People: Emotion and Social Value Orientation in a Negotiation Context	Personality and Social Psychology Bulletin
Wang, Northcraft and van Kleef	2012	Beyond Negotiated Outcomes: The Hidden Costs of Anger Expression in Dyadic Negotiation	Organizational Behavior and Human Decision Processes
Wang, Han and Su	2017	Social Contagion or Strategic Choice? The Interpersonal Effects of Emotions During Chinese Negotiations	Chinese Management Studies
Zhang and Bottom	2019	Blowback: Retaliating against Anger Expressions in Group Decision Making and the Role of Gender	Working Paper

Appendix A-15: Continued

Source: Author's own illustration.

Study	ES	Study	Anger			Comparison			Happy/ Neutral	Medium	Context	Authen-ticity	Direction	Power	Time Pressure
			n	mean	sd	n	mean	sd							
Adam and Brett (2015)	1	1	41	1,930.41	857.97	43	1,455.81	1,057.86	neutral	text	balanced	genuine	behavior	medium	no
Adam and Brett (2015)	2	1	53	1,595.28	944.84	41	1,855.37	1,021.66	neutral	text	coop	genuine	behavior	medium	no
Adam and Brett (2015)	3	1	44	1,332.97	1,092.33	41	1,523.78	722.41	neutral	text	comp	genuine	behavior	medium	no
Adam and Brett (2015)	4	2	30	197.17	80.74	32	150.47	77.81	neutral	text	balanced	genuine	behavior	medium	no
Adam and Brett (2015)	5	2	31	173.06	72.80	34	208.38	87.12	neutral	text	coop	genuine	behavior	medium	no
Adam and Brett (2015)	6	2	31	145.16	87.39	34	157.21	64.51	neutral	text	comp	genuine	behavior	medium	no
Adam and Brett (2018)	7	3	43	233.93	101.06	43	162.31	146.27	neutral	text	comp	genuine	behavior	medium	no
Adam and Brett (2018)	8	3	43	221.74	123.43	43	162.31	146.27	neutral	text	comp	genuine	behavior	medium	no
Adam and Brett (2018)	9	3	43	169.35	113.90	43	162.31	146.27	neutral	text	comp	genuine	behavior	medium	no
Adam and Shirako (2013)	10	4	29	7.129.00	6,003.66	28	6,706.90	5,957.46	neutral	audio	comp	genuine	behavior	medium	no
Adam and Shirako (2013)	11	4	29	11,726.92	5,120.43	28	6,884.62	4,693.20	neutral	audio	comp	genuine	behavior	medium	no
Adam and Shirako (2013)	12	5	35	51,91	7,10	36	51,76	7,14	neutral	FTF	balanced	fake	person	medium	yes
Adam and Shirako (2013)	13	5	35	55.18	5.74	36	48.76	8.34	neutral	FTF	balanced	fake	person	medium	yes
Adam and Shirako (2013)	14	5	35	54.80	5.18	35	47.53	8.15	neutral	FTF	balanced	fake	person	medium	yes
Adam and Shirako (2013)	15	5	35	51.74	7.85	35	52.54	6.93	neutral	FTF	balanced	fake	person	medium	yes
Adam, Shirako and Maddux (2010)	16	6	32	5.63	0.42	32	5.05	0.38	neutral	FTF	comp	genuine	person	medium	no
Adam, Shirako and Maddux (2010)	17	6	34	4.70	0.45	34	5.32	0.33	neutral	FTF	comp	genuine	person	medium	no
Adam, Shirako and Maddux (2010)	18	7	34	388.00	37.00	34	348.00	30.00	neutral	text	comp	genuine	behavior	medium	no
Adam, Shirako and Maddux (2010)	19	7	44	330.00	30.00	44	375.00	26.00	neutral	text	comp	genuine	behavior	medium	no

Legend: ES = effect size number; n = group size; sd = standard deviation; comp = competitive; coop = cooperative

Appendix A-16: Characteristics of the Effect Sizes Measuring Economic Outcomes

Source: Author's own illustration.

Study	ES	Study	Anger			Comparison			Happy/ Neutral	Medium	Context	Authen-ticity	Direction	Power	Time Pressure
			n	mean	sd	n	mean	sd							
Belkin, Kurtzberg and Naquin (2013)	20	8	19	r = 0.20		20			happy	text	balanced	genuine	person	medium	no
Belkin and Rothman (2017)	21	9	44	38.58	13.65	45	38.49	12.47	neutral	visual	comp	genuine	person	high	no
Belkin and Rothman (2017)	22	9	44	38.58	13.65	52	38.90	10.38	happy	visual	comp	genuine	person	high	no
Bertram (2018)	23	10	76	F-value = 0.15		83			neutral	text	comp	genuine	behavior	medium	no
Bertram (2018)	24	11	67	F-value = 0.01		59			neutral	text	comp	genuine	behavior	medium	no
Bhattacharjee and Moreno (2017)	25	12	30	55.63	6.90	33	60.50	6.89	happy	FTF	comp	genuine	behavior	high	no
Bhattacharjee and Moreno (2017)	26	12	33	58.42	6.89	27	55.22	7.07	happy	FTF	coop	genuine	behavior	high	no
Butt (2003)	27	13	104	r = 0.09		104			happy	FTF	balanced	genuine	person	medium	no
Butt and Choi (2006)	28	14	80	t-value = 0.12		80			neutral	FTF	balanced	genuine	person	medium	yes
Campagna et al. (2016)	29	15	33	7.42	1.19	32	7.23	1.33	neutral	text	balanced	fake	behavior	low	yes
Campagna et al. (2016)	30	15	33	7.42	1.19	22	7.63	1.39	happy	text	balanced	fake	behavior	low	yes
Campagna et al. (2016)	31	16	25	4.36	0.98	23	4.12	0.83	happy	text	balanced	fake	behavior	low	yes
Campagna, Mislin and Bottom (2019)	32	17	56	461,166.7	30,118.1	52	469,582.2	32,606.9	happy	text	comp	fake	behavior	medium	yes
Campagna, Mislin and Bottom (2019)	33	18	24	340.71	137.61	25	292.82	134.89	neutral	text	comp	fake	behavior	high	no
Campagna, Mislin and Bottom (2019)	34	18	24	340.71	137.61	12	267.31	97.41	happy	text	comp	fake	behavior	high	no
Campagna, Mislin and Bottom (2019)	35	19	23	4.94	3.00	25	4.42	2.35	neutral	text	balanced	fake	behavior	low	yes
Campagna. Mislin and Bottom (2019)	36	19	23	4.94	3.00	25	4.70	3.76	happy	text	balanced	fake	behavior	low	yes

Legend: ES = effect size number; n = group size; sd = standard deviation; comp = competitive; coop = cooperative; r = correlation

Appendix A-16: Continued

Source: Author's own illustration.

Study	ES	Study	Anger			Comparison			Happy/ Neutral	Medium	Context	Authen-ticity	Direction	Power	Time Pressure
			n	mean	sd	n	mean	sd							
Cheshin, Rafaeli and Bos (2011)	37	20	147	r = -0.12		100			neutral	text	balanced	genuine	behavior	medium	no
Côté, Hideg and van Kleef (2013)	38	21	65	3,109.23	173.86	65	3,087.68	187.31	neutral	FTF	comp	fake	person	medium	yes
Côté, Hideg and van Kleef (2013)	39	21	65	2,943.08	196.63	65	2,873.83	189.17	neutral	FTF	comp	fake	person	medium	yes
Côté, Hideg and van Kleef (2013)	40	22	45	3,187.19	181.94	45	3,093.62	209.20	neutral	visual	comp	fake	person	medium	no
Côté, Hideg and van Kleef (2013)	41	22	45	3,006.52	188.15	45	3,093.62	209.20	neutral	visual	comp	genuine	person	medium	no
De Melo, Carnevale and Gratch (2011)	42	23	18	179.72	164.55	20	110.75	115.72	neutral	text	comp	genuine	behavior	medium	yes
De Melo, Carnevale and Gratch (2011)	43	23	23	127.83	141.94	25	90.80	103.14	neutral	visual	comp	genuine	behavior	medium	yes
De Melo, Carnevale and Gratch (2011)	44	23	18	179.72	164.55	24	43.75	128.09	happy	text	comp	genuine	behavior	medium	yes
De Melo, Carnevale and Gratch (2011)	45	23	23	127.83	141.94	16	55.00	105.66	happy	visual	comp	genuine	behavior	medium	yes
De Melo, Carnevale and Gratch (2012)	46	24	41	137.56	125.21	41	96.95	109.15	neutral	visual	comp	genuine	behavior	medium	yes
De Melo, Carnevale and Gratch (2012)	47	24	41	137.56	125.21	41	52.20	143.68	happy	visual	comp	genuine	behavior	medium	yes
De Melo, Gratch and Carnevale (2014)	48	25	18	166.75	160.19	18	157.25	127.38	neutral	visual	comp	genuine	behavior	medium	yes
De Melo, Gratch and Carnevale (2014)	49	25	18	286.67	218.55	18	101.50	111.57	neutral	visual	comp	genuine	behavior	medium	yes
Dehgani, Carnevale and Gratch (2014)	50	26	10	0.53	0.32	10	0.13	0.23	neutral	visual	comp	genuine	person	medium	yes
Dehgani, Carnevale and Gratch (2014)	51	26	6	0.17	0.36	6	0.24	0.23	neutral	visual	comp	genuine	person	medium	yes

Legend: ES = effect size number; n = group size; sd = standard deviation; comp = competitive; coop = cooperative; r = correlation

Appendix A-16: Continued

Source: Author's own illustration.

Study	ES	Study	Anger			Comparison			Happy/ Neutral	Medium	Context	Authen-ticity	Direction	Power	Time Pressure
			n	mean	sd	n	mean	sd							
Hareli et al. (2013)	52	27	83	518.77	133.98	83	565.06	140.16	happy	text	comp	genuine	behavior	medium	no
Hareli et al. (2013)	53	27	83	518.77	133.98	83	495.18	136.96	neutral	text	comp	genuine	behavior	medium	no
Hideg and van Kleef (2017)	54	28	52	3,093.75	404.97	52	2,956.25	404.97	neutral	visual	comp	fake	behavior	medium	no
Hideg and van Kleef (2017)	55	28	52	2,973.00	251.98	52	3,031.25	251.98	neutral	visual	comp	fake	behavior	medium	no
Hideg and van Kleef (2017)	56	28	52	2,925.00	359.98	52	2,956.25	404.97	neutral	visual	comp	genuine	behavior	medium	no
Hideg and van Kleef (2017)	57	28	52	3,006.25	251.98	52	3,031.25	251.98	neutral	visual	comp	genuine	behavior	medium	no
Hillebrandt and Barclay (2017)	58	29	36	94,914.29	5,710.52	36	81,571.43	9,217.27	happy	audio	comp	genuine	behavior	medium	no
Hillebrandt and Barclay (2017)	59	29	36	90,689.19	3,795.74	36	89,857.14	5,621.09	happy	audio	comp	genuine	person	medium	no
Hillebrandt and Barclay (2017)	60	29	36	92,027.03	4,159.90	36	88,513.51	6,858.17	happy	audio	comp	genuine	person	medium	no
Hillebrandt and Barclay (2017)	61	29	36	94,914.29	5,710.52	36	90,394.74	6,193.02	neutral	audio	comp	genuine	behavior	medium	no
Hillebrandt and Barclay (2017)	62	29	36	90,689.19	3,795.74	36	90,394.74	6,193.02	neutral	audio	comp	genuine	person	medium	no
Hillebrandt and Barclay (2017)	63	29	36	92,027.03	4,159.90	36	90,394.74	6,193.02	neutral	audio	comp	genuine	person	medium	no
Hillebrandt and Barclay (2017)	64	30	40	93,675.68	4,521.88	40	85,054.05	7,680.95	happy	audio	comp	genuine	behavior	medium	no
Hillebrandt and Barclay (2017)	65	30	40	88,864.89	6,033.07	40	89,552.63	5,406.19	happy	audio	comp	genuine	person	medium	no
Hillebrandt and Barclay (2017)	66	30	40	90,268.29	7,123.99	40	90,377.78	3,961.69	happy	audio	comp	genuine	person	medium	no
Hillebrandt and Barclay (2017)	67	30	40	93,675.68	4,521.88	40	89,952.38	6,176.08	neutral	audio	comp	genuine	behavior	medium	no

Legend: ES = effect size number; n = group size; sd = standard deviation; comp = competitive; coop = cooperative

Appendix A-16: Continued

Source: Author's own illustration.

Study	ES	Study	Anger			Comparison			Happy/ Neutral	Medium	Context	Authen-ticity	Direction	Power	Time Pressure
			n	mean	sd	n	mean	sd							
Hillebrandt and Barclay (2017)	68	30	40	88,864.89	6,033.07	40	89,952.38	6,176.08	neutral	audio	comp	genuine	person	medium	no
Hillebrandt and Barclay (2017)	69	30	40	90,268.29	7,123.99	40	89,952.38	6,176.08	neutral	audio	comp	genuine	person	medium	no
Hillebrandt and Barclay (2017)	70	31	157	r = 0.09		157			happy	text	comp	genuine	person	medium	yes
Hunsaker (2018)	71	32	71	202.73	115.44	71	143.80	137.74	neutral	text	comp	genuine	behavior	medium	no
Hunsaker (2018)	72	32	71	221.96	131.79	71	143.80	137.74	neutral	text	comp	genuine	behavior	medium	no
Hunsaker (2018)	73	32	71	173.65	129.85	71	143.80	137.74	neutral	text	comp	genuine	behavior	medium	no
Jäger, Loschelder and Friese (2016)	74	33	23	t-value = 2.42		23			neutral	text	comp	genuine	behavior	medium	no
Jang and Bottom (2020)	75	34	66	r = -0.04		66			neutral	FTF	balanced	fake	person	medium	no
Khooshabeh, De Melo and Volkman (2013)	76	35	20	113.30	155.00	20	142.50	193.00	happy	text	comp	genuine	person	medium	no
Khooshabeh, De Melo and Volkman (2013)	77	35	20	150.20	223.00	20	129.40	217.00	happy	text	coop	genuine	person	medium	no
Khooshabeh et al. (2016)	78	36	25	1,300.00	1,450.00	25	245.00	1,545.00	neutral	text	balanced	genuine	behavior	medium	no
Kopelman, Rosette and Thompson (2006)	79	37	95	14,618.00	3,618.00	91	16,624.00	1,924.00	neutral	visual	comp	fake	person	low	no
Kopelman, Rosette and Thompson (2006)	80	37	95	14,618.00	3,618.00	95	16,136.00	2,754.00	happy	visual	comp	fake	person	low	no
Kopelman, Rosette and Thompson (2006)	81	38	29	r = -0.31		23			neutral	visual	comp	fake	person	high	no
Kopelman, Rosette and Thompson (2006)	82	38	29	r = -0.42		29			happy	visual	comp	fake	person	high	no
Kung and Chao (2019)	83	39	35	8,270.00	1,569.53	37	8,068.92	1,690.76	happy	FTF	balanced	genuine	behavior	medium	yes
Lelieveld et al. (2012)	84	40	18	36.26	22.43	18	50.80	20.01	neutral	text	comp	genuine	behavior	high	no

Legend: ES = effect size number; n = group size; sd = standard deviation; comp = competitive; coop = cooperative

Appendix A-16: Continued

Source: Author's own illustration.

Study	ES	Study	Anger			Comparison			Happy/ Neutral	Medium	Context	Authen- ticity	Direction	Power	Time Pressure
			n	mean	sd	n	mean	sd							
Lelieveld et al. (2012)	85	40	18	58.05	6.64	18	51.11	12.19	neutral	text	comp	genuine	behavior	low	no
Lelieveld et al. (2012)	86	41	20	39.05	22.90	20	50.48	15.85	neutral	text	comp	genuine	behavior	high	no
Lelieveld et al. (2012)	87	41	20	58.85	10.02	20	50.50	9.45	neutral	text	comp	genuine	behavior	low	no
Lelieveld et al. (2013)	88	42	9	r = -0.25		9			happy	text	comp	genuine	person	high	no
Li (2005)	89	43	35	6,277.42	1,113.48	33	4,818.75	1,691.79	happy	FTF	balanced	fake	person	medium	yes
Li (2005)	90	43	35	6,277.42	1,113.48	35	4,557.05	1,608.12	neutral	FTF	balanced	fake	person	medium	yes
Overbeck, Neale and Govan (2010)	91	44	21	r = -0.06		41			happy	FTF	balanced	genuine	person	low	yes
Overbeck, Neale and Govan (2010)	92	44	21	r = 0.17		41			happy	FTF	balanced	genuine	person	high	yes
Pietroni et al. (2008)	93	45	25	6.57	1.04	25	7.20	1.33	happy	text	balanced	genuine	behavior	medium	yes
Pietroni et al. (2008)	94	45	25	7.34	0.99	25	8.10	0.75	happy	text	balanced	genuine	behavior	medium	yes
Pietroni et al. (2008)	95	46	16	6.59	1.18	16	7.32	0.75	happy	text	balanced	genuine	behavior	high	yes
Pietroni et al. (2009)	96	47	24	407.00	69.68	24	596.00	122.40	happy	FTF	comp	genuine	behavior	high	no
Rees et al. (2019)	97	48	109	9,447.06	3,363.35	115	11,037.39	3,744.83	happy	text	coop	genuine	behavior	medium	no
Rees et al. (2019)	98	49	149	10,351.01	3,686.55	150	11,587.33	3,919.41	happy	visual	coop	genuine	behavior	medium	no
Rothman (2011)	99	50	12	4.83	0.58	9	4.86	0.38	neutral	visual	comp	genuine	person	high	no
Rothman (2011)	100	50	12	4.83	0.58	12	5.17	0.58	happy	visual	comp	genuine	person	high	no
Rothman and Northcraft (2015)	101	51	18	0.49	0.11	22	0.49	0.10	neutral	visual	coop	genuine	person	low	no
Rothman and Northcraft (2015)	102	51	18	0.49	0.11	20	0.48	0.10	happy	visual	coop	genuine	person	low	no
Rothman and Northcraft (2015)	103	52	42	0.51	0.06	44	0.52	0.09	neutral	visual	coop	genuine	person	medium	no

Legend: ES = effect size number; n = group size; sd = standard deviation; comp = competitive; coop = cooperative; r = correlation

Appendix A-16: Continued

Source: Author's own illustration.

Study	ES	Study	Anger			Comparison			Happy/ Neutral	Medium	Context	Authen-ticity	Direction	Power	Time Pressure
			n	mean	sd	n	mean	sd							
Rothman and Northcraft (2015)	104	52	42	0.51	0.06	42	0.54	0.07	happy	visual	coop	genuine	person	medium	no
Sinaceur et al. (2013)	105	53	38	14.50	1.00	38	12.50	1.10	happy	text	comp	genuine	behavior	medium	no
Sinaceur et al. (2013)	106	53	38	15.50	0.90	38	14.10	0.80	happy	text	comp	genuine	behavior	medium	no
Sinaceur et al. (2011)	107	54	19	15.02	2.70	19	13.14	2.89	neutral	text	comp	genuine	behavior	medium	yes
Sinaceur and Tiedens (2006)	108	55	38	1.04	2.03	38	-0.29	1.96	neutral	FTF	comp	genuine	person	low	no
Sinaceur and Tiedens (2006)	109	55	38	-0.33	2.28	38	-0.40	1.79	neutral	FTF	comp	genuine	person	high	no
Sinaceur and Tiedens (2006)	110	56	20	65.77	8.81	21	55.68	8.10	neutral	FTF	balanced	genuine	person	low	no
Sinaceur and Tiedens (2006)	111	56	12	t-value = 0.29		8			neutral	FTF	balanced	genuine	person	high	no
Steinel, van Kleef and Harinck (2008)	112	57	20	499,00	91.00	20	571.00	76.00	happy	text	comp	genuine	behavior	medium	yes
Steinel, van Kleef and Harinck (2008)	113	57	20	582.00	61.00	20	528.00	90.00	happy	text	comp	genuine	person	medium	yes
Tamir and Ford (2012)	114	58	23	5.07	0.61	23	4.54	0.60	neutral	audio	comp	genuine	person	high	no
Tamir and Ford (2012)	115	58	23	5.07	0.61	23	4.67	0.61	happy	audio	comp	genuine	person	high	no
Tng and Au (2014)	116	59	24	123.33	66.75	24	75.21	38.32	happy	text	comp	genuine	behavior	medium	no
Tng and Au (2014)	117	59	24	46.46	55.63	24	75.21	46.61	happy	text	comp	fake	behavior	medium	no
Tng and Au (2014)	118	59	24	34.79	25.90	24	48.54	30.52	happy	text	comp	fake	behavior	medium	no
Van Beest, van Kleef and van Dijk (2008)	119	60	40	6.53	12.72	40	7.50	13.93	neutral	text	comp	genuine	behavior	medium	no
Van Beest, van Kleef and van Dijk (2008)	120	60	40	12.18	16.90	40	26.10	13.58	neutral	text	comp	genuine	behavior	medium	no

Legend: ES = effect size number; n = group size; sd = standard deviation; comp = competitive; coop = cooperative

Appendix A-16: Continued

Source: Author's own illustration.

Study	ES	Study	Anger			Comparison			Happy/ Neutral	Medium	Context	Authen-ticity	Direction	Power	Time Pressure
			n	mean	sd	n	mean	sd							
Van Beest, van Kleef and van Dijk (2008)	121	61	27	33.12	3.87	27	30.47	4.08	neutral	text	comp	genuine	behavior	medium	no
Van Beest, van Kleef and van Dijk (2008)	122	61	27	2.52	4.79	27	1.66	4.08	neutral	text	comp	genuine	behavior	medium	no
Van Dijk et al. (2008)	123	62	57	57.65	8.13	57	53.67	9.76	happy	text	comp	genuine	behavior	high	no
Van Dijk et al. (2008)	124	63	35	62.04	8.95	17	56.50	6.27	happy	text	comp	genuine	behavior	high	no
Van Dijk et al. (2008)	125	63	43	49.96	9.61	9	57.58	12.56	happy	text	comp	genuine	behavior	high	no
Van Dijk et al. (2008)	126	64	26	32.56	16.31	26	45.21	9.03	happy	text	comp	genuine	behavior	high	no
Van Dijk et al. (2008)	127	64	26	47.52	7.11	26	47.67	4.89	happy	text	comp	genuine	behavior	low	no
Van Kleef, De Dreu and Manstead (2004a)	128	65	39	489.96	87.90	39	540.44	97.10	neutral	text	comp	genuine	behavior	medium	yes
Van Kleef, De Dreu and Manstead (2004a)	129	65	39	489.96	87.90	39	552.90	75.18	happy	text	comp	genuine	behavior	medium	yes
Van Kleef, De Dreu and Manstead (2004a)	130	66	44	633.53	111.80	44	681.98	92.26	happy	text	comp	genuine	behavior	medium	yes
Van Kleef, De Dreu and Manstead (2004a)	131	67	17	464.78	68.27	17	540.56	95.65	happy	text	comp	genuine	behavior	medium	yes
Van Kleef, De Dreu and Manstead (2004b)	132	68	18	513.00	86.00	18	573.00	72.00	neutral	text	comp	genuine	behavior	medium	yes
Van Kleef, De Dreu and Manstead (2004b)	133	68	18	513.00	86.00	18	593.00	71.00	happy	text	comp	genuine	behavior	medium	yes
Van Kleef, De Dreu and Manstead (2004b)	134	69	22	222.00	83.00	22	141.00	86.00	happy	text	comp	genuine	behavior	medium	no
Van Kleef, De Dreu and Manstead (2004b)	135	69	22	191.00	116.00	22	194.00	114.00	happy	text	comp	genuine	behavior	medium	yes
Van Kleef, De Dreu and Manstead (2004b)	136	70	20	95,136.00	5,383.00	20	88,663.00	3,504.00	happy	audio	comp	genuine	behavior	low	no
Van Kleef, De Dreu and Manstead (2004b)	137	70	20	91,000.00	2,052.00	20	89,250.00	2,447.00	happy	audio	comp	genuine	behavior	high	no

Legend: ES = effect size number; n = group size; sd = standard deviation; comp = competitive; coop = cooperative

Appendix A-16: Continued

Source: Author's own illustration.

Study	ES	Study	Anger			Comparison			Happy/ Neutral	Medium	Context	Authen-ticity	Direction	Power	Time Pressure
			n	mean	sd	n	mean	sd							
Van Kleef et al. (2006)	138	71	20	276.00	73.00	20	201.00	103.00	happy	text	comp	genuine	behavior	low	yes
Van Kleef et al. (2006)	139	71	20	159.00	58.00	20	154.00	59.00	happy	text	comp	genuine	behavior	high	yes
Van Kleef et al. (2006)	140	72	14	17.01	6.12	14	10.88	8.96	happy	text	comp	genuine	behavior	low	no
Van Kleef et al. (2006)	141	72	14	8.07	4.44	14	9.94	5.82	happy	text	comp	genuine	behavior	high	no
Van Kleef et al. (2006)	142	73	16	99,844.00	8,390.00	16	89,063.00	2,720.00	happy	audio	comp	genuine	behavior	low	no
Van Kleef et al. (2006)	143	73	16	90,882.00	2,643.00	16	87,500.00	6,831.00	happy	audio	comp	genuine	behavior	high	no
Van Kleef et al. (2006)	144	74	23	4.03	1.00	23	3.36	1.20	happy	text	comp	genuine	behavior	low	no
Van Kleef et al. (2006)	145	74	23	2.70	1.35	23	3.03	1.23	happy	text	comp	genuine	behavior	high	no
Van Kleef et al. (2006)	146	75	15	13.99	11.05	15	0.80	2.18	happy	text	comp	genuine	behavior	low	no
Van Kleef et al. (2006)	147	75	15	6.85	6.26	15	2.84	6.18	happy	text	comp	genuine	behavior	high	no
Van Kleef and Côté (2007)	148	76	44	15,479.00	4,057.00	44	20,761.00	4,799.00	neutral	audio	comp	genuine	behavior	low	no
Van Kleef and Côté (2007)	149	76	44	17,083.00	3,850.00	44	20,761.00	4,799.00	neutral	audio	comp	genuine	behavior	low	no
Van Kleef and Côté (2007)	150	76	44	21,430.00	7,204.00	44	23,522.00	5,714.00	neutral	audio	comp	genuine	behavior	high	no
Van Kleef and Côté (2007)	151	76	44	26,407.00	6,143.00	44	23,522.00	5,714.00	neutral	audio	comp	genuine	behavior	high	no
Van Kleef and Côté (2007)	152	77	21	F-value = 0.89		21			neutral	text	comp	genuine	behavior	high	yes
Van Kleef and Côté (2007)	153	77	21	F-value = 2.66		21			neutral	text	comp	genuine	behavior	high	yes
Van Kleef and van Lange (2008)	154	78	18	551.00	50.00	18	564.00	76.00	neutral	text	comp	genuine	behavior	medium	yes
Van Kleef and van Lange (2008)	155	78	18	512.00	82.00	18	534.00	127.00	neutral	text	coop	genuine	behavior	medium	yes

Legend: ES = effect size number; n = group size; sd = standard deviation; comp = competitive; coop = cooperative

Appendix A-16: Continued

Source: Author's own illustration.

Study	ES	Study	Anger			Comparison			Happy/ Neutral	Medium	Context	Authen-ticity	Direction	Power	Time Pressure
			n	mean	sd	n	mean	sd							
Van Kleef and De Dreu (2010)	156	79	18	32.11	4.06	18	38.29	7.29	neutral	text	comp	genuine	behavior	medium	no
Van Kleef and De Dreu (2010)	157	79	18	36.15	4.34	18	36.18	4.32	neutral	text	comp	genuine	behavior	medium	no
Van Kleef and De Dreu (2010)	158	80	14	33.09	8.51	14	36.81	4.31	neutral	text	coop	genuine	behavior	medium	no
Van Kleef and De Dreu (2010)	159	80	18	32.35	7.29	18	35.13	5.86	neutral	text	comp	genuine	behavior	medium	no
Van Kleef and De Dreu (2010)	160	80	14	29.36	6.33	14	36.81	4.31	neutral	text	coop	genuine	behavior	medium	no
Van Kleef and De Dreu (2010)	161	80	18	38.21	5.99	18	35.13	5.86	neutral	text	comp	genuine	behavior	medium	no
Wang, Han and Su (2017)	162	81	71	197.90	107.39	71	153.33	117.49	happy	FTF	balanced	genuine	person	medium	no
Wang, Northcraft and van Kleef (2012)	163	82	30	344.17	96.24	31	374.35	105.78	neutral	FTF	comp	genuine	person	medium	no
Wang, Northcraft and van Kleef (2012)	164	83	24	289.00	94.49	24	394.20	117.75	neutral	FTF	comp	genuine	person	low	no
Wang, Northcraft and van Kleef (2012)	165	83	24	507.60	131.22	24	504.00	141.67	neutral	FTF	comp	genuine	person	high	no
Zhang and Bottom (2019)	166	84	217	69.76	28.96	231	64.71	24.35	neutral	text	balanced	genuine	person	medium	no
Zhang and Bottom (2019)	167	85	336	63.14	29.77	343	62.23	29.15	neutral	text	balanced	genuine	person	medium	no
Zhang and Bottom (2019)	168	86	72	74.94	23.44	42	73.57	22.66	neutral	FTF	balanced	genuine	person	medium	no

Legend: ES = effect size number; n = group size; sd = standard deviation; comp = competitive; coop = cooperative

Appendix A-16: Continued

Source: Author's own illustration.

Study	ES	Study	Anger			Comparison			Happy/ Neutral
			n	mean	sd	n	mean	sd	
Adam and Brett (2018)	1	1	43	2.47	1.02	43	2.94	0.85	neutral
Adam and Brett (2018)	2	1	43	2.38	0.92	43	2.94	0.85	neutral
Adam and Brett (2018)	3	1	43	1.98	0.72	43	2.94	0.85	neutral
Belkin and Rothman (2017)	4	2	83	2.79	1.42	115	4.18	1.29	neutral
Belkin and Rothman (2017)	5	2	83	2.79	1.42	114	5.22	1.40	happy
Belkin and Rothman (2017)	6	3	44	3.30	1.47	45	4.71	1.33	neutral
Belkin and Rothman (2017)	7	3	44	3.30	1.47	52	5.45	1.01	happy
Belkin and Rothman (2017)	8	4	48	2.48	0.91	42	3.38	1.03	neutral
Belkin and Rothman (2017)	9	4	48	2.48	0.91	45	4.73	1.29	happy
Belkin and Rothman (2017)	10	4	48	2.42	1.04	42	3.83	1.22	neutral
Belkin and Rothman (2017)	11	4	48	2.42	1.04	45	5.44	1.11	happy
Bhattacharjee and Moreno (2017)	12	5	30	7.46	1.71	8	1.41	33	happy
Bhattacharjee and Moreno (2017)	13	5	33	7.12	2.36	8	1.68	27	happy
Butt (2003)	14	6	104	r = -0.28		104			neutral
Butt and Choi (2006)	15	7	80	r = -0.25		80			neutral
Campagna et al. (2016)	16	8	47	1.93	1.10	45	3.85	1.56	neutral
Campagna et al. (2016)	17	8	47	2.56	1.16	45	4.47	1.15	neutral
Campagna et al. (2016)	18	8	47	1.93	1.10	48	4.06	1.56	happy
Campagna et al. (2016)	19	8	47	2.56	1.16	48	4.40	1.52	happy
Campagna et al. (2016)	20	8	47	r = 0.38		45			neutral

Legend: ES = effect size number; n = group size; sd = standard deviation; r = correlation

Appendix A-17: Characteristics of the Effect Sizes Measuring Subjective Outcomes

Source: Author's own illustration.

Study	ES	Study	Anger			Comparison			Happy/ Neutral
			n	mean	sd	n	mean	sd	
Campagna et al. (2016)	21	8	47	r = 0.40		48			happy
Campagna et al. (2016)	22	9	33	2.37	0.80	32	2.89	0.94	neutral
Campagna et al. (2016)	23	9	33	2.77	0.80	32	3.45	0.77	neutral
Campagna et al. (2016)	24	9	33	2.37	0.80	22	3.34	0.62	happy
Campagna et al. (2016)	25	9	33	2.77	0.80	22	3.72	0.43	happy
Campagna et al. (2016)	26	9	33	2.18	1.92	32	3.28	1.72	neutral
Campagna et al. (2016)	27	9	33	2.18	1.92	22	3.43	1.66	happy
Campagna et al. (2016)	28	10	25	2.33	0.86	23	3.13	0.88	neutral
Campagna et al. (2016)	29	10	25	2.72	0.80	23	3.54	0.68	neutral
Campagna et al. (2016)	30	10	25	2.33	0.86	23	3.48	0.70	happy
Campagna et al. (2016)	31	10	25	2.72	0.80	23	3.78	0.52	happy
Campagna et al. (2016)	32	11	20	2.35	1.13	32	3.61	1.62	neutral
Campagna et al. (2016)	33	11	20	3.30	1.19	32	4.28	1.29	neutral
Campagna et al. (2016)	34	11	20	2.35	1.13	29	3.93	1.37	happy
Campagna et al. (2016)	35	11	20	3.30	1.19	29	4.56	1.19	happy
Campagna et al. (2016)	36	11	20	2.02	2.12	32	2.66	1.81	neutral
Campagna et al. (2016)	37	11	20	2.02	2.12	29	2.35	1.68	happy
Campagna, Mislin and Bottom (2019)	38	12	45	2.71	0.93	45	3.35	0.99	happy
Campagna, Mislin and Bottom (2019)	39	13	56	3.28	0.95	52	4.72	1.42	happy
Campagna, Mislin and Bottom (2019)	40	14	24	2.82	1.12	25	3.70	0.95	neutral

Legend: ES = effect size number; n = group size; sd = standard deviation; r = correlation

Appendix A-17: Continued

Source: Author's own illustration.

Study	ES	Study	Anger			Comparison			Happy/ Neutral
			n	mean	sd	n	mean	sd	
Campagna, Mislin and Bottom (2019)	41	14	24	2.82	1.12	12	4.62	1.17	happy
Campagna, Mislin and Bottom (2019)	42	15	23	2,26	1.09	18	2.88	0.79	neutral
Campagna, Mislin and Bottom (2019)	43	15	23	2.26	1.09	18	2.98	0.79	happy
Campagna, Mislin and Bottom (2019)	44	15	23	r = -0.25		25			neutral
Campagna, Mislin and Bottom (2019)	45	15	23	r = -0.25		18			happy
Côté, Hideg and van Kleef (2013)	46	16	65	4.29	0.96	65	5.26	0.88	neutral
Côté, Hideg and van Kleef (2013)	47	16	65	3.47	0.99	65	4.26	1.02	neutral
Côté, Hideg and van Kleef (2013)	48	17	46	3.59	1.22	47	4.98	1.06	neutral
Côté, Hideg and van Kleef (2013)	49	17	47	4.44	1.07	47	4.98	1.06	neutral
De Melo, Carnevale and Gratch (2012)	50	18	41	4.32	1.66	41	4.80	1.29	neutral
De Melo, Carnevale and Gratch (2012)	51	18	41	4.32	1.66	41	5.63	1.32	happy
De Melo, Gratch and Carnevale (2014)	52	19	18	2.54	1.05	18	2.80	1.00	neutral
De Melo, Gratch and Carnevale (2014)	53	19	18	3.81	1.62	18	2.56	1.21	neutral
Harinck and van Kleef (2012)	54	20	24	4.38	0.77	24	3.32	1.17	neutral
Harinck and van Kleef (2012)	55	20	24	3.58	1.07	24	3.63	1.31	neutral
Hideg and van Kleef (2017)	56	21	52	3.40	2.88	52	4.50	2.88	neutral
Hideg and van Kleef (2017)	57	21	52	3.85	1.22	52	4.20	1.22	neutral
Hideg and van Kleef (2017)	58	21	52	3.90	2.74	52	4.50	2.88	neutral
Hideg and van Kleef (2017)	59	21	52	3.80	1.22	52	4.20	1.22	neutral
Hillebrandt and Barclay (2017)	60	22	40	3.32	0.78	40	2.23	0.87	neutral

Legend: ES = effect size number; n = group size; sd = standard deviation; r = correlation

Appendix A-17: Continued

Source: Author's own illustration.

Study	ES	Study	Anger			Comparison			Happy/ Neutral
			n	mean	sd	n	mean	sd	
Hillebrandt and Barclay (2017)	61	22	40	2.19	0.87	40	2.23	0.87	neutral
Hillebrandt and Barclay (2017)	62	22	40	2.81	0.93	40	2.23	0.87	neutral
Hunsaker (2018)	63	23	141	r = -0.49		141			neutral
Hunsaker (2018)	64	24	53	r = -0.59		53			neutral
Jang and Bottom (2020)	65	25	66	r = -0.45		66			neutral
Jang and Bottom (2020)	66	25	66	r = -0.42		66			neutral
Jang and Bottom (2020)	67	25	66	r = -0.39		66			neutral
Kopelman, Rosette and Thompson (2006)	68	26	35	r = -0.03		35			neutral
Kopelman, Rosette and Thompson (2006)	69	26	35	r = -0.27		35			happy
Kopelman, Rosette and Thompson (2006)	70	27	29	3.03	1.50	23	3.74	1.37	neutral
Kopelman, Rosette and Thompson (2006)	71	27	29	3.03	1.50	29	4.11	1.32	happy
Li (2005)	72	28	35	5.40	2.07	33	6.61	1.63	happy
Li (2005)	73	28	35	6.46	1.95	33	7.33	1.47	happy
Pietroni et al. (2008)	74	29	25	2.92	1.75	25	4.79	1.61	happy
Pietroni et al. (2008)	75	30	16	4.25	1.65	16	5.50	1.10	happy
Pietroni et al. (2009)	76	31	24	2.67	1.63	24	4.13	1.87	happy
Sinaceur et al. (2011)	77	32	19	5.19	1.68	19	1.65	1.29	neutral
Tamir and Ford (2012)	78	32	23	4.52	0.76	23	4.52	0.75	neutral
Tamir and Ford (2012)	79	32	23	4.52	0.76	23	4.89	0.77	happy
Tng and Au (2014)	80	33	24	3.46	0.59	24	3.90	0.59	happy

Legend: ES = effect size number; n = group size; sd = standard deviation; r = correlation

Appendix A-17: Continued

Source: Author's own illustration.

Study	ES	Study	Anger			Comparison			Happy/ Neutral
			n	mean	sd	n	mean	sd	
Tng and Au (2014)	81	33	24	4.69	0.78	24	3.96	0.59	happy
Tng and Au (2014)	82	33	24	4.38	0.47	24	3.92	0.50	happy
Van Beest, van Kleef and van Dijk (2008)	83	34	40	3.01	1.11	40	4.23	0.88	neutral
Van Beest, van Kleef and van Dijk (2008)	84	34	40	3.16	1.11	40	4.68	0.92	neutral
Van Beest, van Kleef and van Dijk (2008)	85	35	27	3.01	1.12	27	4.79	0.73	neutral
Van Beest, van Kleef and van Dijk (2008)	86	35	27	3.27	1.35	27	4.47	0.62	neutral
Van Kleef and Côté (2007)	87	36	42	3.52	1.51	42	3.07	1.82	neutral
Van Kleef and Côté (2007)	88	36	42	4.53	1.76	42	2.65	1.32	neutral
Van Kleef and De Dreu (2010)	89	37	18	3.92	1.29	18	5.48	1.06	neutral
Van Kleef and De Dreu (2010)	90	38	32	3.61	1.25	32	4.58	1.11	neutral
Van Kleef and De Dreu (2010)	91	38	32	4.43	0.87	32	4.58	1.11	neutral
Van Kleef and De Dreu (2010)	92	38	32	3.51	1.41	32	4.28	1.85	neutral
Van Kleef and De Dreu (2010)	93	38	32	4.42	1.92	32	4.28	1.85	neutral
Van Kleef, De Dreu and Manstead (2004b)	94	39	44	3.65	1.51	44	4.43	1.24	happy
Van Kleef, De Dreu and Manstead (2004b)	95	39	44	3.03	1.50	44	4.77	1.25	happy
Wang, Northcraft and van Kleef (2012)	96	40	30	4.35	1.81	31	2.10	1.07	neutral
Wang, Northcraft and van Kleef (2012)	97	41	24	2.87	1.37	24	2.13	0.92	neutral
Wang, Northcraft and van Kleef (2012)	98	41	24	3.30	1.53	24	2.22	1.00	neutral
Wang, Han and Su (2017)	99	42	71	4.60	0.75	71	5.03	0.84	happy

Legend: ES = effect size number; n = group size; sd = standard deviation

Appendix A-17: Continued

Source: Author's own illustration.

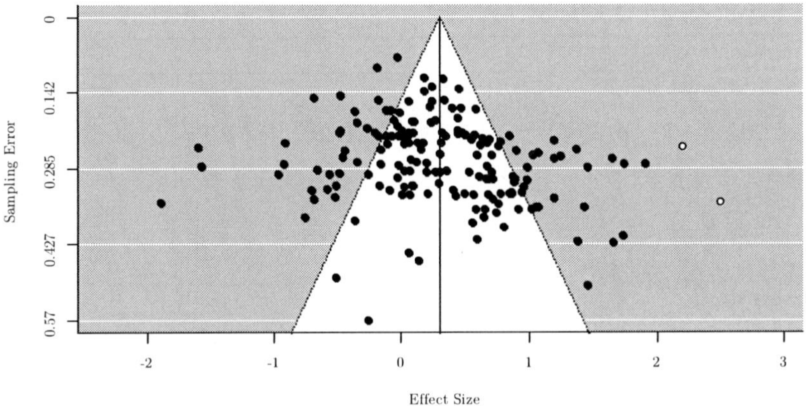

Appendix A-18: Funnel Plot of Economic Outcomes Corrected with Estimator R_0

Source: Author's own illustration.

Estimator R_0, right side						
Number of Missing Studies: 2						
Model Results:						
Estimate	s.e.	z-value	p-value	ci.lb	ci.ub	
0.3022	0.0498	6.0696	<0.0001	0.2046	0.3998	***

Estimator L_0, left side						
Number of Missing Studies: 30						
Model Results:						
Estimate	s.e.	z-value	p-value	ci.lb	ci.ub	
0.0945	0.0521	1.8137	0.0697	-0.0076	0.1967	°

Significance codes: 0.001(***) , 0.01(**) , 0.05(*) , 0.1(.)

Legend: s.e.=standard error, ci.lb=confidence interval lower boud, ci.up=confidence interval upper bound

Appendix A-19: Results of the Trim and Fill Method
Source: Author's own illustration.

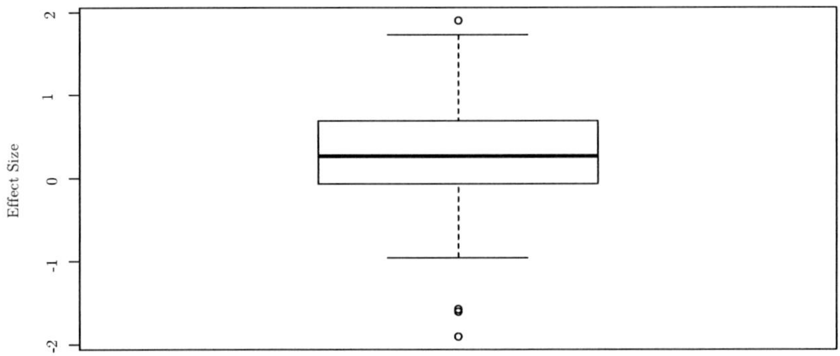

Appendix A-20: Boxplot for Economic Outcomes

Source: Author's own illustration.

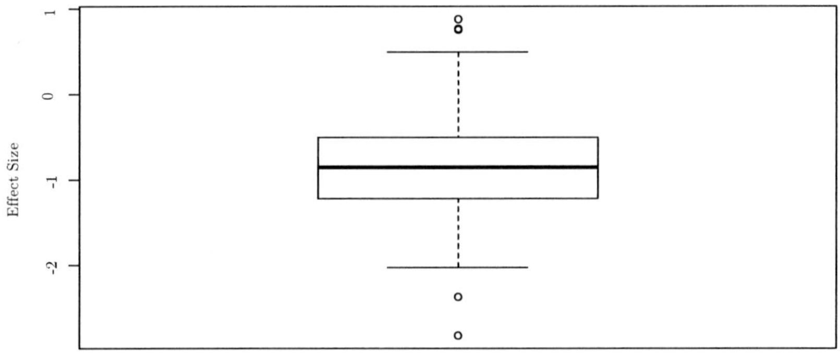

Appendix A-21: Boxplot for Subjective Outcomes

Source: Author's own illustration.

References

Adabi, S., A. Movaghar, A. M. Rahmani, and H. Beigy (2013), Market-based grid resource allocation using new negotiation model, in: *Journal of Network and Computer Applications*, 36 (1), 543–65.

Adair, W., J. Brett, A. Lempereur, T. Okumura, P. Shikhirev, C. Tinsley, and A. Lytle (2004), Culture and Negotiation Strategy, in: *Negotiation Journal*, 20 (1), 87–111.

Adair, W. L. (2003), Integrative Sequences and Negotiation Outcome in Same- and Mixed-Culture Negotiations, in: *International Journal of Conflict Management*, 14 (3/4), 273–96.

———**, K. J. Behfar, M. Olekalns, and D. L. Shapiro (2016)**, Celebrating the Work of Jeanne M. Brett: Building Bridges and Making Connections, in: *Negotiation and Conflict Management Research*, 9 (4), 292–308.

——— **and J. M. Brett (2005)**, The Negotiation Dance: Time, Culture, and Behavioral Sequences in Negotiation, in: *Organization Science*, 16 (1), 33–51.

———**, T. Okumura, and J. M. Brett (2001)**, Negotiation Behavior When Cultures Collide: The United States and Japan, in: *Journal of Applied Psychology*, 86 (3), 371–85.

———**, L. Weingart, and J. M. Brett (2007)**, The Timing and Function of Offers in U.S. and Japanese Negotiations, in: *Journal of Applied Psychology*, 92 (4), 1056–68.

Adam, H. and J. M. Brett (2015), Context matters: The social effects of anger in cooperative, balanced, and competitive negotiation situations, in: *Journal of Experimental Social Psychology*, 61, 44–58.

——— **and** ——— **(2018)**, Everything in moderation: The social effects of anger depend on its perceived intensity, in: *Journal of Experimental Social Psychology*, 76, 12–18.

——— **and A. Shirako (2013)**, Not All Anger Is Created Equal: The Impact of the Expresser's Culture on the Social Effects of Anger in Negotiations, in: *Journal of Applied Psychology*, 98 (5), 785–98.

———, ———, **and W. W. Maddux (2010)**, Cultural Variance in the Interpersonal Effects of Anger in Negotiations, in: *Psychological Science*, 21 (6), 882–89.

Adam, M. T. P., T. Teubner, and H. Gimpel (2018), No Rage Against the Machine: How Computer Agents Mitigate Human Emotional Processes in Electronic Negotiations, in: *Group Decision and Negotiation*, 27 (4), 543–71.

Adams, J. S. (1965), Inequity in Social Exchange, in: *Advances in Experimental Social Psychology*, 2, 267–99.

Adler, N. J., R. Brahm, and J. L. Graham (1992), Strategy Implementation: A Comparison of Face-to-Face Negotiations in the People's Republic of China and the United States, in: *Strategic Management Journal*, 13 (6), 449–66.

——— **and J. L. Graham (1989)**, Cross-Cultural Interaction: The International Comparison Fallacy?, in: *Journal of International Business Studies*, 20 (3), 515–37.

Agndal, H. (2007), Current Trends in Business Negotiation Research - An Overview of Articles published 1996-2005, working paper, Stockholm School of Economics.

———, **L. J. Åge, and J. Eklinder-Frick (2017)**, Two decades of business negotiation research: an overview and suggestions for future studies, in: *Journal of Business and Industrial Marketing*, 32 (4), 487–504.

Aguinis, H., D. R. Dalton, F. A. Bosco, C. A. Pierce, and C. M. Dalton (2011), Meta-Analytic Choices and Judgment Calls: Implications for Theory Building and Testing, Obtained Effect Sizes, and Scholarly Impact, in: *Journal of Management*, 37 (1), 5–38.

Ahlgren, P., B. Jarneving, and R. Rousseau (2003), Requirements for a Cocitation Similarity Measure, with Special Reference to Pearson's Correlation Coefficient, in: *Journal of the American Society for Information Science and Technology*, 54 (6), 550–60.

Allred, K. G. (1999), Anger and Retaliation: Toward an Understanding of Impassioned Conflict in Organizations, in *Research on Negotiation in Organizations*, R. J. Lewicki, R. J. Bies, and B. H. Sheppard, eds., Bingley: Emerald, 27–58.

———, **J. S. Mallozzi, F. Matsui, and C. P. Raia (1997)**, The Influence of Anger and Compassion on Negotiation Performance, in: *Organizational Behavior and Human Decision Processes*, 70 (3), 175–87.

Altman, D. G. and J. M. Bland (2005), Standard deviations and standard errors, in: *British Medical Journal*, 331, 903.

Amanatullah, E. T. and M. W. Morris (2010), Negotiating Gender Roles: Gender Differences in Assertive Negotiating Are Mediated by Women's Fear of Backlash and Attenuated When Negotiating on Behalf of Others, in: *Journal of Personality and Social Psychology*, 98 (2), 256–67.

———, ———, and J. R. Curhan (2008), Negotiators Who Give Too Much: Unmitigated Communion, Relational Anxieties, and Economic Costs in Distributive and Integrative Bargaining, in: *Journal of Personality and Social Psychology*, 95 (3), 723–38.

Ames, D. R. (2008), Assertiveness Expectancies: How Hard People Push Depends on the Consequences They Predict, in: *Journal of Personality and Social Psychology*, 95 (6), 1541–57.

Anderson, C. and J. L. Berdahl (2002), The Experience of Power: Examining the Effects of Power on Approach and Inhibition Tendencies, in: *Journal of Personality and Social Psychology*, 83 (6), 1362–77.

——— and A. D. Galinsky (2006), Power, optimism, and risk-taking, in: *European Journal of Social Psychology*, 36 (4), 511–36.

Angelmar, R. and L. W. Stern (1978), Development of a Content Analytic System for Analysis of Bargaining Communication in Marketing, in: *Journal of Marketing Research*, 15 (1), 93–102.

Aparicio, G., T. Iturralde, and A. Maseda (2019), Conceptual structure and perspectives on entrepreneurship education research: A bibliometric review, in: *European Research on Management and Business Economics*, 25 (3), 105–13.

Babcock, L., M. J. Gelfand, D. Small, and H. Stayn (2006), Gender Differences in the Propensity to Initiate Negotiation, in *Social Psychology and Economics*, D. De Cremer, M. Zeelenberg, and J. K. Murnighan, eds., Mahwah: Erlbaum, 239–62.

——— and S. Laschever (2009), *Women don't ask: Negotiation and the gender divide*, Princeton: Princeton University Press.

Bacharach, S. B. and E. J. Lawler (1981), *Bargaining: Power, Tactics, and Outcomes*, San Francisco: Jossey-Bass.

Backhaus, K. (2015), Every Future Needs a Past: A Personal View of the History of JBBM, in: *Journal of Business-to-Business Marketing*, 22 (1–2), 37–46.

———, **J. van Doorn, and R. Wilken (2008)**, The Impact of Team Characteristics on the Course and Outcome of Intergroup Price Negotiations, in: *Journal of Business-to-Business Marketing*, 15 (4), 365–96.

———, **B. Erichson, W. Plinke, and R. Weiber (2016)**, *Multivariate Analysemethoden*, 14th ed., Heidelberg: Springer Gabler.

———, **M. Hummelsberger, and R. Wilken (2005)**, Erfolgsdeterminanten in der Verhandlungsanalyse - eine systematisierende und kritische Übersicht, working paper, Förderkreis für Industriegütermarketing e.V.

——— **and U. König (2019)**, How Collaboration and Digitization Transform Large Project Business, in *Collaboration in the Digital Age - How Technology Enables Individuals, Teams and Businesses*, K. Riemer, S. Schellhammer, and M. Meinert, eds., Wiesbaden: Springer, 207–26.

———, **K. Lügger, and M. Koch (2011)**, The structure and evolution of business-to-business marketing: A citation and co-citation analysis, in: *Industrial Marketing Management*, 40 (6), 940–51.

——— **and M. A. Pesch (2018)**, Verhandlungen – Spiegeln die Lehrbücher den Stand der Forschung wider?, in: *Die Unternehmung*, 72 (1), 3–26.

——— **and M. Voeth (2014)**, *Industriegütermarketing*, 10th ed., München: Vahlen.

Bacon, N. and P. Blyton (2002), Militant and moderate trade union orientations: What are the effects on workplace trade unionism, union-management relations and employee gains?, in: *International Journal of Human Resource Management*, 13 (2), 302–19.

Bagozzi, R. P. (1978), Marketing as Exchange: A Theory of Transactions in the Marketplace, in: *American Behavioral Scientist*, 21 (4), 535–56.

Balliet, D., N. P. Li, S. J. Macfarlan, and M. van Vugt (2011), Sex Differences in Cooperation: A Meta-Analytic Review of Social Dilemmas, in: *Psychological Bulletin*, 137 (6), 881–909.

Baltes, B. B., M. W. Dickson, M. P. Sherman, C. C. Bauer, and J. S. LaGanke (2002), Computer-Mediated Communication and Group Decision Making: A Meta-Analysis, in: *Organizational Behavior and Human Decision Processes*, 87 (1), 156–79.

Balzert, H., M. Schröder, and C. Schäfer (2011), *Wissenschaftliches Arbeiten - Ethik, Inhalt & Form wiss. Arbeiten, Handwerkszeug, Quellen, Projektmanagement, Präsentation*, Herdecke: W3L.

Bar-Zion, B. (2004), Understanding barriers to peace: Reflecting on israeli-palestinian economic negotiations, in: *Negotiation Journal*, 20 (3), 383–400.

Baron, R. A. (1990), Environmentally Induced Positive Affect: Its Impact on Self-Efficacy, Task Performance, Negotiation, and Conflict, in: *Journal of Applied Social Psychology*, 20 (5), 368–84.

Baron, R. M. and D. A. Kenny (1986), The Moderator-Mediator Variable Distinction in Social Psychological Research: Conceptual, Strategic, and Statistical Considerations, in: *Journal of Personality and Social Psychology*, 51 (6), 1173–82.

Barrett, L. F. and J. A. Russell (1998), Independence and Bipolarity in the Structure of Current Affect, in: *Journal of Personality and Social Psychology*, 74 (4), 967–84.

Barry, B. (2008), Negotiator affect: the state of the art (and the science), in: *Group Decision and Negotiation*, 17 (1), 97–105.

——— **and R. A. Friedman (1998)**, Bargainer Characteristics in Distributive and Integrative Negotiation, in: *Journal of Personality and Social Psychology*, 74 (2), 345–59.

———**, I. S. Fulmer, and G. A. van Kleef (2004)**, I laughed, I cried, I settled: The Role of Emotion in Negotiation, in *The Handbook of Negotiation and Culture*, M. Gelfand and J. M. Brett, eds., Stanford: California University Press, 71–94.

——— **and R. L. Oliver (1996)**, Affect in Dyadic Negotiation: A Model and Propositions, in: *Organizational Behavior and Human Decision Processes*, 67 (2), 127–43.

Barsade, S. G. (2002), The Ripple Effect: Emotional Contagion and Its Influence on Group Behavior, in: *Administrative Science Quarterly*, 47 (4), 644–75.

——— **and D. E. Gibson (2007)**, Why Does Affect Matter in Organizations?, in: *Academy of Management Perspectives*, 21 (1), 36–59.

Bartos, O. J. (1965), Concession-Making in Experimental Negotiations, in *Sociological Theories in Progress*, J. Berger, M. Zelditch, and B. Anderson, eds., Boston: Houghton-Mifflin, 3–28.

——— **(1967)**, How predictable are negotiations?, in: *Journal of Conflict Resolution*, 11 (4), 481–96.

Bass, B. M. (1966), Effects on the subsequent performance of negotiators of studying issues or planning strategies alone or in groups, in: *Psychological Monographs: General and Applied*, 80 (6), 1–31.

Bateman, T. S. (1980), Contingent Concession Strategies in Dyadic Bargaining, in: *Organizational Behavior and Human Performance*, 26 (2), 212–21.

Batistič, S., M. Černe, and B. Vogel (2017), Just how multi-level is leadership research? A document co-citation analysis 1980–2013 on leadership constructs and outcomes, in: *The Leadership Quarterly*, 28 (1), 86–103.

Bazerman, M. H. and J. S. Carroll (1987), Negotiator Cognition, in *Research in Organizational Behavior*, B. Staw and L. L. Cummings, eds., Greenwich: JAI Press, 247–88.

——— **and D. Chugh (2006)**, Bounded Awareness: Focusing Failures in Negotiation, in *Negotiation Theory and Research*, L. L. Thompson, ed., New York: Psychology Press, 7–26.

———**, J. R. Curhan, and D. A. Moore (2001)**, The Death and Rebirth of the Social Psychology of Negotiation, in *Blackwell Handbook of Social Psychology: Interpersonal Processes*, G. J. O. Fletcher and M. S. Clark, eds., Oxford: Blackwell Publishing, 196–229.

———**,** ———**,** ———**, and K. L. Valley (2000)**, Negotiation, in: *Annual Review of Psychology*, 51, 279–314.

———**, T. Magliozzi, and M. A. Neale (1985)**, Integrative Bargaining in a Competitive Market, in: *Organizational Behavior and Human Decision Processes*, 35 (3), 294–313.

——— **and M. A. Neale (1983)**, Heuristics in Negotiation: Limitations to Effective Dispute Resolution, in *Negotiating in Organizations*, M. H. Bazerman and R. J. Lewicki, eds., Beverly Hills: SAGE Publications, 51–67.

——— **and** ——— **(1991)**, Negotiator Rationality and Negotiator Cognition: The Interactive Roles of Prescriptive and Descriptive Research, in *Negotiation Analysis*, H. P. Young, ed., Ann Arbor: The University of Michigan Press, 109–29.

——— **and** ——— **(1992)**, *Negotiating Rationally*, New York: Free Press.

Bear, J. (2011), "Passing the Buck": Incongruence between Gender Role and Topic Leads to Avoidance of Negotiation, in: *Negotiation and Conflict Management Research*, 4 (1), 47–72.

Beersma, B. and C. K. W. De Dreu (2002), Integrative and Distributive Negotiation in Small Groups: Effects of Task Structure, Decision Rule, and Social Motive, in: *Organizational Behavior and Human Decision Processes*, 87 (2), 227–52.

van Beest, I., E. van Dijk, C. K. W. De Dreu, and H. A. M. Wilke (2005), Do-no-harm in coalition formation: Why losses inhibit exclusion and promote fairness cognitions, in: *Journal of Experimental Social Psychology*, 41 (6), 609–17.

———**, G. A. van Kleef, and E. van Dijk (2008)**, Get angry, get out: The interpersonal effects of anger communication in multiparty negotiation, in: *Journal of Experimental Social Psychology*, 44 (4), 993–1002.

———**, H. Wilke, and E. van Dijk (2003)**, The Excluded Player in Coalition Formation, in: *Personality and Social Psychology Bulletin*, 29 (2), 237–47.

Begg, C. B. and M. Mazumdar (1994), Operating Characteristics of a Rank Correlation Test for Publication Bias, in: *Biometrics*, 50 (4), 1088–1101.

Beggan, J. K., D. M. Messick, and S. T. Allison (1988), Social Values and Egocentric Bias: Two Tests of the Might Over Morality Hypothesis, in: *Journal of Personality and Social Psychology*, 55 (4), 606–11.

Belkin, L. Y., T. R. Kurtzberg, and C. E. Naquin (2013), Signaling Dominance in Online Negotiations: The Role of Affective Tone, in: *Negotiation and Conflict Management Research*, 6 (4), 285–304.

——— **and N. B. Rothman (2017)**, Do I Trust You? Depends on What You Feel: Interpersonal Effects of Emotions on Initial Trust at Zero-Acquaintance, in: *Negotiation and Conflict Management Research*, 10 (1), 3–27.

Belter, C. W. (2016), Citation Analysis as a Literature Search Method for Systematic Reviews, in: *Journal of the Association for Information Science and Technology*, 67 (11), 2766–77.

Ben-Yoav, O. and D. G. Pruitt (1984), Accountability to Constituents: A Two-Edged Sword, in: *Organizational Behavior and Human Performance*, 34 (3), 283–95.

Benton, A. A. and D. Druckman (1974), Constituent's Bargaining Orientation and Intergroup Negotiations, in: *Journal of Applied Social Psychology*, 4 (2), 141–50.

———**, H. H. Kelley, and B. Liebling (1972)**, Effects of Extremity of Offers and Consession Rate on the Outcomes of Bargaining, in: *Journal of Personality and Social Psychology*, 24 (1), 73–83.

Bertram, S. (2018), The Social Effects of Anger in International Negotiations: The Role of Self-Regulation and Bargaining Power, doctoral dissertation, Victoria University of Wellington.

Bhattacharjee, S. and K. K. Moreno (2017), The Role of Informal Controls and a Bargaining Opponent's Emotions on Transfer Pricing Judgments, in: *Contemporary Accounting Research*, 34 (1), 427–54.

Bichler, M., G. Kersten, and S. Strecker (2003), Towards a Structured Design of Electronic Negotiations, in: *Group Decision and Negotiation*, 12 (4), 311–35.

Bijmolt, T. H. A., H. J. van Heerde, and R. G. M. Pieters (2005), New Empirical Generalizations on the Determinants of Price Elasticity, in: *Journal of Marketing Research*, 42 (2), 141–56.

——— **and R. G. M. Pieters (2001)**, Meta-Analysis in Marketing when Studies Contain Multiple Measurements, in: *Marketing Letters*, 12 (2), 157–69.

Blake, R. R. and J. S. Mouton (1961), Loyalty of Representatives to Ingroup Positions During Intergroup Competition, in: *Sociometry*, 24 (2), 177–83.

Blount, S. (1995), When Social Outcomes Aren't Fair: The Effect of Causal Attributions on Preferences, in: *Organizational Behavior and Human Decision Processes*, 63 (2), 131–44.

Boles, T. L., R. T. A. Croson, and J. K. Murnighan (2000), Deception and Retribution in Repeated Ultimatum Bargaining, in: *Organizational Behavior and Human Decision Processes*, 83 (2), 235–59.

Borah, L. A. (1963), The Effects of Threat in Bargaining: Critical and Experimental Analysis, in: *Journal of Abnormal and Social Psychology*, 66 (1), 37–44.

Borenstein, M. (2005), Software for Publication Bias, in *Publication Bias in Meta-Analysis: Prevention, Assessment and Adjustments*, H. R. Rothstein, A. J. Sutton, and M. Borenstein, eds., Chichester: John Wiley & Sons, 193–220.

———**, L. V. Hedges, J. P. T. Higgins, and H. R. Rothstein (2009)**, *Introduction to Meta-Analysis*, Chichester: John Wiley & Sons.

Borgatti, S. P. (2005), Centrality and network flow, in: *Social Networks*, 27 (1), 55–71.

———**, M. G. Everett, and L. C. Fremman (2002)**, *UCINET for Windows: Software for Social Network Analysis*, Cambridge: Harvard University Press.

———**,** ———**, and J. C. Johnson (2018)**, *Analyzing Social Networks*, 2nd ed., London: SAGE Publications.

———**, A. Mehra, D. J. Brass, and G. Labianca (2009)**, Network Analysis in the Social Sciences, in: *Science*, 323 (5916), 892–96.

Börner, K., C. Chen, and K. W. Boyack (2005), Visualizing Knowledge Domains, in: *Annual Review of Information Science and Technology*, 37 (1), 179–255.

Bottom, W. P. and A. Studt (1993), Framing Effects and the Distributive Aspect of Integrative Bargaining, in: *Organizational Behavior and Human Decision Processes*, 56 (3), 459–74.

Bowles, H. R., L. Babcock, and L. Lai (2007), Social Incentives for Gender Differences in the Propensity to Initiate Negotiations Sometimes It Does Hurt to Ask, in: *Organizational Behavior and Human Decision Processes*, 103 (1), 84–103.

———, ———, and K. L. McGinn (2005), Constraints and Triggers: Situational Mechanics of Gender in Negotiation, in: *Journal of Personality and Social Psychology*, 89 (6), 951–65.

——— and F. Flynn (2010), Gender and persistence in negotiation: A dyadic perspective, in: *Academy of Management Journal*, 53 (4), 769–87.

Boyack, K. W., R. Klavans, and K. Börner (2005), Mapping the backbone of science, in: *Scientometrics*, 64 (3), 351–74.

Brett, J. M. (2000), Culture and Negotiation, in: *International Journal of Psychology*, 35 (2), 97–104.

——— (2014), *Negotiating Globally: How to Negotiate Deals, Resolve Disputes, and Make Decisions Across Cultural Boundaries*, 3rd ed., San Francisco: Jossey-Bass.

———, W. Adair, A. Lempereur, T. Okumura, P. Shikhirev, C. Tinsley, and A. Lytle (1998), Culture and Joint Gains in Negotiation, in: *Negotiation Journal*, 14 (1), 61–86.

——— and T. Okumura (1998), Inter- and Intracultural Negotiation: U.S. and Japanese Negotiators, in: *Academy of Management Journal*, 41 (5), 495–510.

———, M. Olekalns, R. Friedman, N. Goates, C. Anderson, and C. C. Lisco (2007), Stick and Stones: Language, Face, and Online Dispute Resolution, in: *Academy of Management Journal*, 50 (1), 85–99.

———, D. L. Shapiro, and A. L. Lytle (1998), Breaking the Bonds of Reciprocity in Negotiations, in: *Academy of Management Journal*, 41 (4), 410–24.

——— and L. L. Thompson (2016), Negotiation, in: *Organizational Behavior and Human Decision Processes*, 136, 68–79.

vom Brocke, J., A. Simons, K. Riemer, B. Niehaves, R. Plattfaut, and A. Cleven (2015), Standing on the Shoulders of Giants: Challenges and Recommendations of Literature Search in Information Systems Research, in: *Communications of the Association for Information Systems*, 37, 205–24.

Brooks, A. W. and M. E. Schweitzer (2011), Can Nervous Nelly negotiate? How anxiety causes negotiators to make low first offers, exit early, and earn less profit, in: *Organizational Behavior and Human Decision Processes*, 115 (1), 43–54.

Brown, B. R. (1968), The Effects of Need to Maintain Face on Interpersonal Bargaining, in: *Journal of Experimental Social Psychology*, 4 (1), 107–22.

Brown, G. and M. Baer (2011), Location in Negotiation: Is There a Home Field Advantage?, in: *Organizational Behavior and Human Decision Processes*, 114 (2), 190–200.

Brown, R. (1965), *Social Psychology*, New York: Free Press.

Bucy, E. P. (2000), Emotional and Evaluative Consequences of Inappropriate Leader Displays, in: *Communication Research*, 27 (2), 194–226.

Buelens, M., M. van De Woestyne, S. Mestdagh, and D. Bouckenooghe (2008), Methodological Issues in Negotiation Research: A State-of-the-Art-Review, in: *Group Decision and Negotiation*, 17 (4), 321–45.

Butler, J. K. (1999), Trust Expectations, Information Sharing, Climate of Trust, and Negotiation Effectiveness and Efficiency, in: *Group & Organization Management*, 24 (2), 217–38.

Butt, A. N. (2003), The Role of Emotions in Dyadic Negotiation: An Empirical Study, doctoral dissertation, Faculty of Management, McGill University Montreal.

——— **and J. N. Choi (2006)**, The Effects of Cognitive Appraisal and Emotion on Social Motive and Negotiation Behavior: The Critical Role of Agency of Negotiator Emotion, in: *Human Performance*, 19 (4), 305–25.

Cacioppo, J. T. and R. E. Petty (1982), The Need for Cognition, in: *Journal of Personality and Social Psychology*, 42 (1), 116–31.

Calero Medina, C. M. and T. N. van Leeuwen (2012), Seed Journal Citation Network Maps: A Method Based on Network Theory, in: *Journal of the American Society for Information Science and Technology*, 63 (6), 1226–34.

Camerer, C. and R. H. Thaler (1995), Anomalies: Ultimatums, Dictators and Manners, in: *Journal of Economic Perspectives*, 9 (2), 209–19.

Campagna, R. L., A. A. Mislin, and W. P. Bottom (2019), Motivated by guilt and low felt trust: The impact of negotiators' anger expressions on the implementation of negotiated agreements, in: *Journal of Behavioral Decision Making*, 32 (4), 450–70.

———, ———, D. T. Kong, and W. P. Bottom (2016), Strategic Consequences of Emotional Misrepresentation in Negotiation: The Blowback Effect, in: *Journal of Applied Psychology*, 101 (5), 605–24.

Campbell, N. C. G., J. L. Graham, A. Jolibert, and H. G. Meissner (1988), Marketing Negotiations in France, Germany, the United Kingdom, and the United States, in: *Journal of Marketing*, 52 (2), 49–62.

Cao, M., X. Luo, X. Luo, and X. Dai (2015), Automated negotiation for e-commerce decision making: A goal deliberated agent architecture for multi-strategy selection, in: *Decision Support Systems*, 73, 1–14.

Caputo, A. (2013a), A literature review of cognitive biases in negotiation processes, in: *International Journal of Conflict Management*, 24 (4), 374–98.

——— (2013b), The third who joins a negotiation: A systematic review of the literature, in: *International Journal of Business and Globalisation*, 10 (3), 256–77.

Carnevale, P. J. and A. M. Isen (1986), The Influence of Positive Affect and Visual Access on the Discovery of Integrative Solutions in Bilateral Negotiation, in: *Organizational Behavior and Human Decision Processes*, 37 (1), 1–13.

——— and E. J. Lawler (1986), Time Pressure and the Development of Integrative Agreements in Bilateral Negotiations, in: *Journal of Conflict Resolution*, 30 (4), 636–59.

——— and D. G. Pruitt (1992), Negotiation and Mediation, in: *Annual Review of Psychology*, 43 (1), 531–82.

Chabowski, B. R., S. Samiee, and G. T. M. Hult (2013), A bibliometric analysis of the global branding literature and a research agenda, in: *Journal of International Business Studies*, 44 (6), 622–34.

Chandrakar, R. (2006), Digital object identifier system : an overview, in: *The Electronic Library*, 24 (4), 445–52.

Chang, Y.-W., M.-H. Huang, and C.-W. Lin (2015), Evolution of research subjects in library and information science based on keyword, bibliographical coupling, and co-citation analyses, in: *Scientometrics*, 105 (3), 2071–87.

Chen, I.-S. and F.-T. Tseng (2016), The relevance of communication media in conflict contexts and their effectiveness: A negotiation experiment, in: *Computers in Human Behavior*, 59, 134–41.

Chertkoff, J. M. and M. Conley (1967), Opening Offer and Frequency of Concession as Bargaining Strategies, in: *Journal of Personality and Social Psychology*, 7 (2), 181–85.

Cheshin, A., A. Rafaeli, and N. Bos (2011), Anger and happiness in virtual teams: Emotional influences of text and behavior on others' affect in the absence of non-verbal cues, in: *Organizational Behavior and Human Decision Processes*, 116 (1), 2–16.

Chow, S. L. (1987), Meta-Analysis of Pragmatic and Theoretical Research: A Critique, in: *The Journal of Psychology*, 121 (3), 259–71.

Citera, M., R. Beauregard, and T. Mitsuya (2005), An experimental study of credibility in E-negotiations, in: *Psychology and Marketing*, 22 (2), 163–79.

Clandinin, J. and M. L. Hamilton (2011), Standing on the shoulders of giants, in: *Teaching and Teacher Education*, 27 (1), 1–2.

Clarivate Analytics (2019a), Web of Science Search, (accessed March 10, 2019), [available at https://apps.webofknowledge.com/WOS_AdvancedSearch_input.do?SID=C4nYL7sXYduD8NrjVbN&product=WOS&search_mode=AdvancedSearch].

——— **(2019b)**, Web of Science, (accessed March 7, 2019), [available at https://clarivate.com/webofsciencegroup/solutions/web-of-science/].

——— **(2019c)**, Research Areas, (accessed March 20, 2019), [available at https://images.webofknowledge.com/images/help/WOS/hp_research_areas_easca.html].

——— **(2019d)**, Web of Science Core Collection Help, (accessed March 20, 2019), [available at https://images.webofknowledge.com/images/help/WOS/hp_subject_category_terms_tasca.html].

Clark, M. S., S. P. Pataki, and V. H. Carver (1996), Some Thoughts and Findings on Self-Presentation of Emotions in Relationships, in *Knowledge Structures in Close Relationships: A Social Psychological Approach*, G. J. O. Fletcher and J. Fitness, eds., Mahwah: Erlbaum, 247–274.

——— **and C. Taraban (1991)**, Reactions to and Willingness to Express Emotion in Communal and Exchange Relationships, in: *Journal of Experimental Social Psychology*, 27 (4), 324–36.

Clopton, S. W. (1984), Seller and Buying Firm Factors Affecting Industrial Buyers' Negotiation Behavior and Outcomes, in: *Journal of Marketing Research*, 21 (1), 39–53.

Clore, G. (1994), Why emotions are felt, in *The nature of emotion*, P. Ekman and R. J. Davidson, eds., New York: Cambridge University Press, 103–11.

Cobo, M. J., A. G. López-Herrera, E. Herrera-Viedma, and F. Herrera (2011), Science Mapping Software Tools: Review, Analysis, and Cooperative Study Among Tools, in: *Journal of the American Society for Information Science and Technology*, 62 (7), 1382–1402.

———**, M. A. Martínez, M. Gutiérrez-Salcedo, H. Fujita, and E. Herrera-Viedma (2015)**, 25 years at Knowledge-Based Systems: A bibliometric analysis, in: *Knowledge-Based Systems*, 80, 3–13.

Cochran, W. G. (1954), The Combination of Estimates from Different Experiments, in: *Biometrics*, 10 (1), 101–29.

Coehoorn, R. M. and N. R. Jennings (2004), Learning an Opponent's Preferences to Make Effective Multi-Issue Negotiation Trade-Offs, in: *Proceedings of the 6th International Conference on Electronic Commerce*, 59–68.

Cohen, J. (1960), A Coefficient of Agreement for Nominal Scales, in: *Educational and Psychological Measurement*, 20 (1), 37–46.

——— **(1992)**, A Power Primer, in: *Psychological Bulletin*, 112 (1), 155–59.

———**, P. Cohen, S. G. West, and L. S. Aiken (2003)**, *Applied multiple regression/correlation analysis for the behavioral sciences*, 3rd ed., Mahwah: Lawrence Erlbaum Associates.

Combs, J. G., D. J. Ketchen, T. R. Crook, and P. L. Roth (2011), Assessing Cumulative Evidence within "Macro" Research: Why Meta-Analysis Should be Preferred Over Vote Counting, in: *Journal of Management Studies*, 48 (1), 178–97.

Cooper, H. and L. V. Hedges (2009), Research Synthesis as a Scientific Process, in *The Handbook of Research Synthesis and Meta-Analysis*, H. Cooper, L. V. Hedges, and J. C. Valentine, eds., New York: Russell Sage Foundation, 3–18.

Cortina, J. M. (2003), Apples and Oranges (and Pears, Oh My!): The Search for Moderators in Meta-Analysis, in: *Organizational Research Methods*, 6 (4), 415–39.

——— **and R. G. Folger (1998)**, When Is It Acceptable to Accept a Null Hypothesis: no Way, Jose?, in: *Organizational Research Methods*, 1 (3), 334–50.

Côté, S., I. Hideg, and G. A. van Kleef (2013), The consequences of faking anger in negotiations, in: *Journal of Experimental Social Psychology*, 49 (3), 453–63.

Cox, A. D., D. Cox, and R. D. Anderson (2005), Reassessing the pleasures of store shopping, in: *Journal of Business Research*, 58, 250–59.

Crane, D. (1972), *Invisible Colleges: Diffusion of Knowledge in Scientific Communication*, Chicago: University of Chicago Press.

Creswell, J. W. (2012), *Educational Research: Planning, Conducting, and Evaluating Quantitative and Qualitative Research*, 4th ed., Boston: Pearson.

Cronin, B. (2001), Bibliometrics and beyond: some thoughts on web-based citation analysis, in: *Journal of Information Science*, 27 (1), 1–7.

Culnan, M. J. (1986), The Intellectual Development of Management Information Systems, 1972-1982: A Co- Citation Analysis, in: *Management Science*, 32 (2), 156–72.

Cummings, L. L. and D. L. Harnett (1969), Bargaining Behaviour in a Symmetric Bargaining Triad: The Impact of Risk-Taking Propensity, Information, Communication and Terminal Bid, in: *The Review of Economic Studies*, 36 (4), 485–501.

Curhan, J. R., H. A. Elfenbein, and N. Eisenkraft (2010), The Objective Value of Subjective Value: A Multi-Round Negotiation Study, in: *Journal of Applied Social Psychology*, 40 (3), 690–709.

———, ———, and G. J. Kilduff (2009), Getting Off on the Right Foot: Subjective Value Versus Economic Value in Predicting Longitudinal Job Outcomes From Job Offer Negotiations, in: *Journal of Applied Psychology*, 94 (2), 524–34.

———, ———, and H. Xu (2006), What Do People Value When They Negotiate? Mapping the Domain of Subjective Value in Negotiation, in: *Journal of Personality and Social Psychology*, 91 (3), 493–512.

———, M. A. Neale, L. Ross, and J. Rosencranz-Engelmann (2008), Relational accommodation in negotiation: Effects of egalitarianism and gender on economic efficiency and relational capital, in: *Organizational Behavior and Human Decision Processes*, 107 (2), 192–205.

Daft, R. L. and R. H. Lengel (1986), Organizational Information Requirements, Media Richness and Structural Design, in: *Mangement Science*, 32 (5), 554–71.

Dalton, D. R. and C. M. Dalton (2005), Strategic Management Studies are a Special Case for Meta-Analysis, in: *Research Methodology in Strategy and Management*, 2 (2), 31–63.

Dehghani, M., P. J. Carnevale, and J. Gratch (2014), Interpersonal effects of expressed anger and sorrow in morally charged negotiation, in: *Judgment and Decision Making*, 9 (2), 104–13.

Denson, N. and M. H. Seltzer (2011), Meta-Analysis in Higher Education: An Illustrative Example Using Hierarchical Linear Modeling, in: *Research in Higher Education*, 52 (3), 215–44.

Deutsch, M. (1958), Trust and Suspicion, in: *Journal of Conflict Resolution*, 2 (4), 265–79.

——— **(1960a)**, The Effect of Motivational Orientation upon Trust and Suspicion, in: *Human Relations*, 13 (2), 123–39.

——— **(1960b)**, Trust, Trustworthiness, and the F-Scale, in: *Journal of Abnormal and Social Psychology*, 61 (1), 138–40.

——— **(1973)**, *The resolution of conflict: Constructive and destructive processes*, New Haven: Yale University Press.

——— **and R. M. Krauss (1960)**, The Effect of Threat upon Interpersonal Bargaining, in: *Journal of Abnormal and Social Psychology*, 61 (2), 181–89.

——— **and ——— (1962)**, Studies of interpersonal bargaining, in: *Journal of Conflict Resolution*, 6 (1), 52–76.

van Dijk, E., I. van Beest, G. A. van Kleef, and G.-J. Lelieveld (2018), Communication of anger versus disappointment in bargaining and the moderating role of power, in: *Journal of Behavioral Decision Making*, 31 (5), 632–43.

———**, G. A. van Kleef, W. Steinel, and I. van Beest (2008)**, A social functional approach to emotions in bargaining: When communicating anger pays and when it backfires, in: *Journal of Personality and Social Psychology*, 94 (4), 600–614.

——— **and R. Vermunt (2000)**, Strategy and Fairness in Social Decision Making: Sometimes It Pays to Be Powerless, in: *Journal of Experimental Social Psychology*, 36 (1), 1–25.

Dimotakis, N., D. E. Conlon, and R. Ilies (2012), The Mind and Heart (Literally) of the Negotiator: Personality and Contextual Determinants of Experiential Reactions and Economic Outcomes in Negotiation, in: *Journal of Applied Psychology*, 97 (1), 183–93.

Ding, Y., G. G. Chowdhury, and S. Foo (1999), Mapping the intellectual structure of information retrieval studies: An author co-citation analysis, 1987-1997, in: *Journal of Information Science*, 25 (1), 67–78.

———, ———, and ——— (2000), Journal as markers of intellectual space: Journal co-citation analysis of information retrieval area, 1987-1997, in: *Scientometrics*, 47 (1), 55–73.

De Dreu, C. K. W. (2003), Time Pressure and Closing of the Mind in Negotiation, in: *Organizational Behavior and Human Decision Processes*, 91 (2), 280–95.

———, B. Beersma, K. Stroebe, and M. C. Euwema (2006), Motivated Information Processing, Strategic Choice, and the Quality of Negotiated Agreement, in: *Journal of Personality and Social Psychology*, 90 (6), 927–43.

——— and T. L. Boles (1998), Share and Share Alike or Winner Take All?: The Influence of Social Value Orientation upon Choice and Recall of Negotiation Heuristics, in: *Organizational Behavior and Human Decision Processes*, 76 (3), 253–76.

——— and P. J. Carnevale (2003), Motivational Bases of Information Processing and Strategy in Conflict and Negotiation, in *Advances in Experimental Social Psychology*, M. P. Zanna, ed., San Diego: Elsevier Academic Press, 235–91.

———, E. Giebels, and E. van de Vliert (1998), Social Motives and Trust in Integrative Negotiation: The Disruptive Effects of Punitive Capability, in: *Journal of Applied Psychology*, 83 (3), 408–22.

——— and G. A. van Kleef (2004), The influence of power on the information search, impression formation, and demands in negotiation, in: *Journal of Experimental Social Psychology*, 40 (3), 303–19.

———, S. L. Koole, and F. L. Oldersma (1999), On the Seizing and Freezing of Negotiator Inferences: Need for Cognitive Closure Moderates the Use of Heuristics in Negotiation, in: *Personality and Social Psychology Bulletin*, 25 (3), 348–62.

———, ———, and W. Steinel (2000), Unfixing the Fixed Pie: A Motivated Information-Processing Approach to Integrative Negotiation, in: *Journal of Personality and Social Psychology*, 79 (6), 975–87.

——— **and P. A. M. van Lange (1995)**, The Impact of Social Value Orientations on Negotiator Cognition and Behavior, in: *Personality and Social Psychology Bulletin*, 21 (11), 1178–88.

——— **and C. McCusker (1997)**, Gain-Loss Frames and Cooperation in Two-Person Social Dilemmas: A Transformational Analysis, in: *Journal of Personality and Social Psychology*, 72 (5), 1093–1106.

———, **B. A. Nijstad, and D. van Knippenberg (2008)**, Motivated Information Processing in Group Judgment and Decision Making, in: *Personality and Social Psychology Review*, 12 (1), 22–49.

———, **L. R. Weingart, and S. Kwon (2000)**, Influence of Social Motives on Integrative Negotiation: A Meta-Analytic Review and Test of Two Theories, in: *Journal of Personality and Social Psychology*, 78 (5), 889–905.

Druckman, D. (1967), Dogmatism, Prenegotiation Experience, and Simulated Group Representation as Determinants of Dyadic Behavior in a Bargaining Situation, in: *Journal of Personality and Social Psychology*, 6 (3), 279–90.

——— **(1968)**, Prenegotiation Experience and Dyadic Conflict Resolution in a Bargaining Situation, in: *Journal of Experimental Social Psychology*, 4 (4), 367–83.

——— **and M. Olekalns (2008)**, Emotions in negotiation, in: *Group Decision and Negotiation*, 17 (1), 1–11.

——— **and V. Robinson (1998)**, From Research to Application: Utilizing Research Findings in Negotiation Training Programs, in: *International Negotiation*, 3 (1), 7–38.

———, **D. Solomon, and K. Zechmeister (1972)**, Effects of Representational Role Obligations on the Process of Children's Distribution of Resources, in: *Sociometry*, 35 (3), 387–410.

Dupont, C. (1996), Negotiation as Coalition Building, in: *Interational Negotiation*, 1, 47–64.

——— **(1999)**, A Model of the Negotiation Process with Different Strategies, in *International Business Negotiations*, P. Ghauri and J.-C. Usunier, eds., Oxford: Pergamon, 39–68.

Duval, S. and R. Tweedie (2000a), A Nonparametric "Trim and Fill" Method of Accounting for Publication Bias in Meta-Analysis, in: *Journal of the American Statistical Association*, 95 (449), 89–98.

——— and ——— (**2000b**), Trim and Fill: A Simple Funnel-Plot-Based Method of Testing and Adjusting for Publication Bias in Meta-Analysis, in: *Biometrics*, 56 (2), 455–63.

Dwyer, F. R. and O. C. Walker (**1981**), Bargaining in an Asymmetrical Power Structure, in: *Journal of Marketing*, 45 (1), 104–15.

Eagly, A. H. (**1987**), *Sex differences in Social Behavior: A Social-Role Interpretation*, Hillsdale: Erlbaum.

——— **and S. Chaiken** (**1993**), *The psychology of attitudes*, New York: Harcourt Brace Jovanovich.

——— **and S. J. Karau** (**2002**), Role Congruity Theory of Prejudice Toward Female Leaders, in: *Psychological Review*, 109 (3), 573–98.

van Eck, N. J. and L. Waltman (**2014**), Visualizing bibliometric networks, in *Measuring scholarly impact: Methods and practice*, Y. Ding, R. Rousseau, and D. Wolfram, eds., Cham: Springer, 285–320.

Eden, D. (**2002**), From the Editors: Replication, Meta-Analysis, Scientific Progress, and AMJ's Publication Policy, in: *Academy of Management Journal*, 45 (5), 841–46.

Egger, M., G. D. Smith, M. Schneider, and C. Minder (**1997**), Bias in meta-analysis detected by a simple, graphical test, in: *British Medical Journal*, 315 (7109), 629–634.

———**, T. Zellweger-Zähner, M. Schneider, C. Junker, C. Lengeler, and G. Antes** (**1997**), Language bias in randomised controlled trials published in English and German, in: *The Lancet*, 350 (9074), 326–29.

Eisend, M. (**2014**), *Metaanalyse*, München: Rainer Hampp Verlag.

Ekman, P. (**1984**), Expression and the Nature of Emotion, in *Approaches to Emotion*, K. R. Scherer and P. Ekman, eds., Hillsdale: Lawrence Erlbaum Associates, 319–43.

——— (**2004**), *Gefühle lesen: Wie Sie Emotionen erkennen und richtig interpretieren*, Heidelberg: Spektrum.

Elfenbein, H. A. (**2015**), Individual Differences in Negotiation: A Nearly Abandoned Pursuit Revived, in: *Current Directions in Psychological Science*, 24 (2), 131–36.

———**, M. D. Foo, J. White, H. H. Tan, and V. C. Aik** (**2007**), Reading your counterpart: The benefit of emotion recognition accuracy for effectiveness in negotiation, in: *Journal of Nonverbal Behavior*, 31 (4), 205–23.

Eliason, S. R. (1993), *Maximum Likelihood Estimation - Logic and Practice*, Newbury Park: SAGE Publications.

Ellis, P. D. (2010), *The Essential Guide to Effect Sizes: Statistical Power, Meta-Analysis, and the Interpretation of Research Results*, Cambridge: Cambridge University Press.

Elsevier (2019), Historical depth, (accessed March 20, 2019), [available at https://www.elsevier.com/solutions/scopus/how-scopus-works/content].

Enzle, M. E., M. D. Harvey, and E. F. Wright (1992), Implicit Role Obligations Versus Social Responsibility in Constituency Representation, in: *Journal of Personality and Social Psychology*, 62 (2), 238–45.

Esser, J. K. and S. S. Komorita (1975), Reciprocity and Concession Making in Bargaining, in: *Journal of Personality and Social Psychology*, 31 (5), 864–72.

Evans, F. B. (1963), Selling as a Dyadic Relationship – A New Approach, in: *American Behavioral Scientist*, 6 (9), 76–79.

Faratin, P., C. Sierra, and N. R. Jennings (1998), Negotiation Decision Functions for Autonomous Agents, in: *Robotics and Autonomous Systems*, 24 (3–4), 159–82.

———, ———, and ——— (2002), Using similarity criteria to make issue trade-offs in automated negotiations, in: *Artificial Intelligence*, 142 (2), 205–37.

Farley, J. U., D. R. Lehmann, and L. H. Mann (1998), Designing the Next Study for Maximum Impact, in: *Journal of Marketing Research*, 35 (4), 496–501.

Faure, G. O., V. Le Dong, and M. F. Shakun (1990), Social-emotional aspects of negotiation, in: *European Journal of Operational Research*, 46 (2), 177–80.

——— and J. Z. Rubin (1993), *Culture and Negotiation*, Thousand Oaks: SAGE Publications.

Fernandez-Alles, M. and A. Ramos-Rodriguez (2009), Intellectual Structure of Human Resources Management Research: A Bibliometric Analysis of the Journal Human Resource Management, in: *Journal of the American Society for Information Science and Technology*, 60 (1), 161–75.

Filipowicz, A., S. Barsade, and S. Melwani (2011), Understanding Emotional Transitions: The Interpersonal Consequences of Changing Emotions in Negotiations, in: *Journal of Personality and Social Psychology*, 101 (3), 541–56.

Fischer, K. W., P. R. Shaver, and P. Carnochan (1990), How Emotions Develop and How they Organise Development, in: *Cognition and Emotion*, 4 (2), 81–127.

Fisher, G. (1980), *International Negotiation. A Cross-Cultural Perspective*, Yarmouth: Intercultural Press.

Fisher, R. and W. Ury (1981), *Getting to YES: Negotiating Agreement without Giving In*, New York: Penguin Group.

———, ———, and B. Patton (1991), *Getting to Yes: Negotiating Agreement without Giving In*, 2nd ed., New York: Penguin Group.

———, ———, and ——— (1997), *Getting to Yes: Negotiating Agreement without Giving In*, 2nd ed., New York: Random House Business.

———, ———, and ——— (2011), *Getting to YES: Negotiating Agreement without Giving In*, 3rd ed., New York: Random House Business.

Fiske, S. T. (1993), Controlling Other People: The Impact of Power on Stereotyping, in: *American Psychologist*, 48 (6), 621–28.

Fitzgerald, S. M. and P. D. Rumrill (2003), Meta-analysis as a tool for understanding existing research literature, in: *Work*, 21 (1), 97–103.

Forgas, J. P. (1992), Affect in Social Judgements and Decisions: A Multiprocess Model, in: *Advances in Experimental Social Psychology*, 25, 227–75.

——— (1995), Mood and Judgment: The Affect Infusion Model (AIM), in: *Psychological Bulletin*, 117 (1), 39–66.

——— (1998), On Feeling Good and Getting Your Way: Mood Effects on Negotiator Cognition and Bargaining Strategies, in: *Journal or Personality and Social Psychology*, 74 (3), 565–77.

Fouraker, L. E. and S. Siegel (1963), *Bargaining Behavior*, New York: McGraw-Hill.

Francis, J. N. P. (1991), When in Rome? The Effects of Cultural Adaptation on Intercultural Business Negotiations, in: *Journal of International Business Studies*, 22 (3), 403–28.

Freeman, L. C. (1979), Centrality in social networks: Conceptual clarification, in: *Social Networks*, 1, 215–39.

Frey, R. L. and J. S. Adams (1972), The Negotiator's Dilemma: Simultaneous In-Group and Out-Group Conflict, in: *Journal of Experimental Social Psychology*, 8 (4), 331–46.

Friedman, R., C. Anderson, J. M. Brett, M. Olekalns, N. Goates, and C. C. Lisco (2004), The Positive and Negative Effects of Anger on Dispute Resolution: Evidence From Electronically Mediated Disputes, in: *Journal of Applied Psychology*, 89 (2), 369–76.

Frijda, N. H. (1986), *The Emotions*, New York: Cambridge University Press.

——— **(2008)**, The Psychologists' Point of View, in *Handbook of Emotions*, M. Lewis, J. M. Haviland-Jones, and L. F. Barrett, eds., New York: The Guilford Press, 68–87.

——— **and B. Mesquita (1994)**, The Social Roles and Functions of Emotions, in *Emotion and culture: Empirical studies of mutual influence*, S. Kitayama and H. R. Markus, eds., Washington: American Psychological Association, 51–87.

Fulmer, I. S. and B. Barry (2004), The Smart Negotiator: Cognitive Ability and Emotional Intelligence in Negotiation, in: *International Journal of Conflict Management*, 15 (3), 245–72.

Galin, A., M. Gross, and G. Gosalker (2007), E-Negotiation versus Face-to-Face Negotiation What has Changed – if Anything?, in: *Computers in Human Behavior*, 23 (1), 787–97.

Galinsky, A. D., D. H. Gruenfeld, and J. C. Magee (2003), From Power to Action, in: *Journal of Personality and Social Psychology*, 85 (3), 453–66.

———**, G. Ku, and T. Mussweiler (2009)**, To Start Low or To Start High? The Case of Auctions Versus Negotiations, in: *Current Directions in Psychological Science*, 18 (6), 357–61.

——— **and T. Mussweiler (2001)**, First Offers as Anchors: The Role of Perpective-Taking and Negotiator Focus, in: *Journal of Personality and Social Psychology*, 81 (4), 657–69.

Gallo, P. S. (1966), Effects of Increased Incentives Upon the Use of Threat in Bargaining, in: *Journal of Personality and Social Psychology*, 4 (1), 14–20.

García-Lillo, F., E. Claver-Cortés, B. Marco-Lajara, and M. Úbeda-García (2017), Mapping the Intellectual Structure of Research on "Born Global" Firms and INVs: A Citation/Co-citation Analysis, in: *Management International Review*, 57 (4), 631–52.

Garfield, E. (1965), Can Citation Indexing be Automated?, in *Statistical Association Methods for Mechanized Documentation*, M. E. Stevens, V. E. Giuliano, and L. B. Heilprin, eds., Wahington D.C.: National Bureau of Standards, 189–92.

——— **(1979)**, *Citation Indexing: Its Theory and Application in Science, Technology, and Humanities*, Philadelphia: isi Press.

Gaspar, J. P., R. Methasani, and M. Schweitzer (2019), Fifty shades of deception: Characteristics and consequences of lying in negotiations, in: *Academy of Management Perspectives*, 33 (1), 62–81.

Geiger, I. (2014), Media Effects on the Formation of Negotiator Satisfaction: The Example of Face-to-Face and Text Based Electronically Mediated Negotiations, in: *Group Decision and Negotiation*, 23 (4), 735–63.

Gelfand, M. J. and S. Christakopoulou (1999), Culture and Negotiator Cognition: Judgment Accuracy and Negotiation Processes in Individualistic and Collectivistic Cultures, in: *Organizational Behavior and Human Decision Processes*, 79 (3), 248–69.

Gerber, A. S. and D. P. Green (2009), Field Experiments and Natural Experiments, in *The Oxford Handbook of Political Science*, R. E. Goodin, ed., Oxford: Oxford University Press, 1108–32.

Gerhart, B. and S. Rynes (1991), Determinants and Consequences of Salary Negotiations by Male and Female MBA Graduates, in: *Journal of Applied Psychology*, 76 (2), 256–62.

Geyskens, I., R. Krishnan, J. E. M. Steenkamp, and P. V. Cunha (2009), A Review and Evaluation of Meta-Analysis Practices in Management Research, in: *Journal of Management*, 35 (2), 393–419.

Girvan, M. and M. E. J. Newman (2002), Community structure in social and biological networks, in: *Proceedings of the National Academy of Sciences*, 99 (12), 7821–26.

Glänzel, W. and B. Thijs (2004), The influence of author self-citations on bibliometric macro indicators, in: *Scientometrics*, 59 (3), 281–310.

——— **and** ——— **(2012)**, Using "core documents" for detecting and labelling new emerging topics, in: *Scientometrics*, 91 (2), 399–416.

Gmür, M. (2003), Co-citation analysis and the search for invisible colleges: A methodological evaluation, in: *Scientometrics*, 57 (1), 27–57.

Gneezy, U. (2005), Deception: The Role of Consequences, in: *American Economic Review*, 95 (1), 384–94.

Goodwin, S. A., A. Gubin, S. T. Fiske, and V. Y. Yzerbyt (2000), Power Can Bias Impression Processes: Stereotyping Subordinates by Default and by Design, in: *Group Processes & Intergroup Relations*, 3 (3), 227–56.

Gouldner, A. W. (1960), The Norm of Reciprocity: A Preliminary Statement, in: *American Sociological Review*, 25 (2), 161–78.

Graham, J. L. (1985), Cross-Cultural Marketing Negotiations: A Laboratory Experiment, in: *Marketing Science*, 4 (2), 130–46.

——— **(1986)**, The Problem-Solving Approach to Negotiations in Industrial Marketing, in: *Journal of Business Research*, 14 (6), 549–66.

———, D. K. Kim, C.-Y. Lin, and M. Robinson (1988), Buyer-Seller Negotiations Around the Pacific Rim: Differences in Fundamental Exchange Processes, in: *Journal of Consumer Research*, 15 (1), 48–54.

———, A. T. Mintu, and W. Rodgers (1994), Explorations of Negotiation Behaviors in Ten Foreign Cultures Using a Model Developed in the United States, in: *Management Science*, 40 (1), 72–95.

Graham, S. M., J. Y. Huang, M. S. Clark, and V. S. Helgeson (2008), The Positives of Negative Emotions: Willingness to Express Negative Emotions Promotes Relationships, in: *Personality and Social Psychology Bulletin*, 34 (3), 394–406.

Greenhalgh, L. and D. I. Chapman (1995), Joint Decision Making: The Inseperability of Relationships and Negotiation, in *Negotiation as a Social Process*, R. M. Kramer and D. M. Messick, eds., Thousand Oaks: SAGE Publications, 166–85.

Greer, L. L., B. A. de Jong, M. E. Schouten, and J. E. Dannals (2018), Why and when hierarchy impacts team effectiveness: A meta-analytic integration, in: *Journal of Applied Psychology*, 103 (6), 591–613.

Gruder, C. L. and N. A. Rosen (1971), Effects of intragroup relations on intergroup bargaining, in: *International Journal of Group Tensions*, 1 (4), 301–17.

Gruenfeld, D. H. (1995), Status, Ideology, and Integrative Complexity on the United-States-Supreme-Court - Rethinking the Politics of Political Decision-Making, in: *Journal of Personality and Social Psychology*, 68 (1), 5–20.

Di Guardo, M. C. and K. R. Harrigan (2012), Mapping research on strategic alliances and innovation: A co-citation analysis, in: *Journal of Technology Transfer*, 37 (6), 789–811.

Gunia, B. C., J. M. Brett, and M. J. Gelfand (2016), The science of culture and negotiation, in: *Current Opinion in Psychology*, 8, 78–83.

———, ———, A. K. Nandkeolyar, and D. Kamdar (2011), Paying a Price: Culture, Trust, and Negotiation Consequences, in: *Journal of Applied Psychology*, 96 (4), 774–89.

———, R. I. Swaab, N. Sivanathan, and A. D. Galinsky (2013), The Remarkable Robustness of the First-Offer Effect: Across Culture, Power, and Issues, in: *Personality and Social Psychology Bulletin*, 39 (12), 1547–58.

Güth, W., R. Schmittberger, and B. Schwarze (1982), An Experimental Analysis of Ultimatum Bargaining, in: *Journal of Economic Behavior and Organization*, 3 (4), 367–88.

Hägler, M. (2016), Porsche-Prozess: Ex-Porsche-Chef Wiedeking freigesprochen, in: *Süddeutsche Zeitung*, (accessed November 22, 2019), [available at https://www.sueddeutsche.de/wirtschaft/porsche-prozess-ex-porsche-chef-wiedeking-frei-gesprochen-richter-kritisieren-staatsanwaelte-massiv-1.2913235].

Hair, J. F., W. C. Black, B. J. Babin, and R. E. Anderson (2014), *Multivariate Data Analysis*, 7th ed., Harlow: Pearson Education.

Halevy, N. (2008), Team Negotiation: Social, Epistemic, Economic, and Psychological Consequences of Subgroup Conflict, in: *Personality and Social Psychology Bulletin*, 34 (12), 1687–1702.

Hall, E. T. (1976), *Beyond Culture*, New York: Anchor.

——— and M. R. Hall (1990), *Understanding Cultural Differences*, Boston: Intercultural Press.

Hall, J. (1971), Decisions, Decisions, Decisions, in: *Psychology Today*, 5 (1), 51–54.

Hamilton, P., E. Lawson, C. Gaudet, J. Chisholm, J. Kaur, and S. Abercromby (2018), At the intersection of idealized youth and marginalized almost-adulthood: how girls negotiate young motherhood in London, Ontario, in: *Journal of Youth Studies*, 21 (9), 1182–97.

Hamner, W. C. (1974), Effects of Bargaining Strategy and Pressure to Reach Agreement in a Stalemated Negotiation, in: *Journal of Personality and Social Psychology*, 30 (4), 458–67.

Hanneman, R. A. and M. Riddle (2005), *Introduction to social network methods*, Riverside: University of California.

Hareli, S., S. David, S. Akron, and U. Hess (2013), The Effect of the Negotiator's Social Power as a Function of the Counterpart's Emotional Reactions in a Computer Mediated Negotiation, in: *Europe's Journal of Psychology*, 9 (4), 820–31.

Harinck, F. and G. A. van Kleef (2012), Be hard on the interests and soft on the values: Conflict issue moderates the effects of anger in negotiations, in: *British Journal of Social Psychology*, 51 (4), 741–52.

Harnett, D. L. and L. L. Cummings (1971), *Bargaining behavior in an asymmetric triad: The role of information, communication, and risk-taking propensity*, New York: Gordon & Breach.

———, ———, and W. C. Hamner (1973), Personality, Bargaining Style and Payoff in Bilateral Monopoly Bargaining Among European Managers, in: *Sociometry*, 36 (3), 325–45.

Harsanyi, J. C. (1962), Bargaining in ignorance of the opponent's utility function, in: *Journal of Conflict Resolution*, 6 (1), 29–38.

Hartmann, G. W. (1934), Personality traits associated with variations in happiness, in: *Journal of Abnormal and Social Psychology*, 29 (2), 202–12.

Hatfield, E., J. T. Cacioppo, and R. L. Rapson (1992), Primitive Emotional Contagion, in: *Review of Personality and Social Psychology*, 14, 151–77.

Hausken, K. (1997), Game-theoretic and Behavioral Negotiation Theory, in: *Group Decision and Negotiation*, 6 (6), 511–28.

Hawranek, D. (2009), The Downfall of a Corporate Upstart, in: *Spiegel Online*, (accessed November 22, 2019), [available at https://www.spiegel.de/international/business/the-porsche-story-the-downfall-of-a-corporate-upstart-a-637542-3.html].

Hayes, A. F. (2013), *Introduction to mediation, moderation, and conditional process analysis : A regression-based approach*, London: The Guilford Press.

Heaton, A. W. and A. W. Kruglanski (1991), Person Perception by Introverts and Extraverts Under Time Pressure: Effects of Need for Closure, in: *Personality and Social Psychology Bulletin*, 17 (2), 161–65.

Hedges, L. V. (1981), Distribution Theory for Glass's Estimator of Effect Size and Related Estimators, in: *Journal of Educational Statistics*, 6 (2), 107–28.

——— **(2009)**, Effect Sizes in Nested Designs, in *The Handbook of Research Synthesis and Meta-Analysis*, H. Cooper, L. V. Hedges, and J. C. Valentine, eds., New York: Russell Sage Foundation, 337–56.

Henchy, T. and D. C. Glass (1968), Evaluation Apprehension and the Social Facilitation of Dominant and Subordinate Responses, in: *Journal of Personality and Social Psychology*, 10 (4), 446–54.

Herbst, U. (2007), *Präferenzmessung in industriellen Verhandlungen*, Wiesbaden: Deutscher Universitäts-Verlag.

———, **M. Voeth, and C. Meister (2011)**, What do we know about buyer-seller negotiations in marketing research? A status quo analysis, in: *Industrial Marketing Management*, 40 (6), 967–78.

Hermann, M. G. and N. Kogan (1968), Negotiation in leader and delegate groups, in: *Journal of Conflict Resolution*, 12 (3), 332–44.

Hideg, I. and G. A. van Kleef (2017), When expressions of fake emotions elicit negative reactions: The role of observers' dialectical thinking, in: *Journal of Organizational Behavior*, 38 (8), 1196–1212.

Higgins, J. P. T., S. G. Thompson, J. J. Deeks, and D. G. Altman (2003), Measuring inconsistency in meta-analyses, in: *British Medical Journal*, 327 (7414), 557–60.

Hillebrandt, A. and L. J. Barclay (2017), Comparing Integral and Incidental Emotions: Testing Insights From Emotions as Social Information Theory and Attribution Theory, in: *Journal of Applied Psychology*, 102 (5), 732–52.

Hindriks, K. and D. Tykhonov (2008), Opponent Modelling in Automated Multi-Issue Negotiation Using Bayesian Learning, in: *Proceedings of the 7th International Conference on Autonomous Agents and Multiagent Systems*, 331–38.

Hine, M. J., S. A. Murphy, M. Weber, and G. Kersten (2009), The Role of Emotion and Language in Dyadic E-negotiations, in: *Group Decision and Negotiation*, 18 (3), 193–211.

Hofstede, G. (1984), *Culture's Consequences: International Differences in Work-Related Values*, Newbury Park: SAGE Publications.

——— **(2001)**, *Culture's Consequences: Comparing Values, Behaviors, Institutions and Organizations Across Nations*, Thousand Oaks: SAGE Publications.

Holsapple, C. W., H. Lai, and A. B. Whinston (1998), A Formal Basis for Negotiation Support System Research, in: *Group Decision and Negotiation*, 7 (3), 203–27.

Holt, J. L. and C. J. DeVore (2005), Culture, gender, organizational role, and styles of conflict resolution: A meta-analysis, in: *International Journal of Intercultural Relations*, 29 (2), 165–96.

Homans, G. C. (1974), *Social Behavior: Its Elementary Forms*, New York: Harcourt Brace Jovanovich.

House, R. J., P. J. Hanges, M. Javidan, P. W. Dorfman, and V. Gupta (2004), *Culture, Leadership, and Organizations*, Thousand Oaks: SAGE Publications.

Hox, J. J., M. Moerbeek, and R. van de Schoot (2017), *Multilevel Analysis - Techniques and Applications*, Abingdon-on-Thames: Routledge.

Huber, V. L. and M. A. Neale (1987), Effects of Self- and Competitor Goals on Performance in an Interdependent Bargaining Task, in: *Journal of Applied Psychology*, 72 (2), 197–203.

Hüffmeier, J., P. A. Freund, A. Zerres, K. Backhaus, and G. Hertel (2014), Being Tough or Being Nice? A Meta-Analysis on the Impact of Hard- and Softline Strategies in Distributive Negotiations, in: *Journal of Management*, 40 (3), 866–92.

———, A. Zerres, P. A. Freund, K. Backhaus, R. Trötschel, and G. Hertel (2019), Strong or Weak Synergy? Revising the Assumption of Team-Related Advantages in Integrative Negotiations, in: *Journal of Management*, 45 (7), 2721–50.

Hume, D. (2012), Emotions and Moods, in *Organizational Behavior*, S. P. Robbins and T. A. Judge, eds., 15th ed., Upper Saddle River: Prentice Hall, 258–97.

Hunsaker, D. A. (2017), Anger in Negotiations: A Review of Causes, Effects, and Unanswered Questions, in: *Negotiation and Conflict Management Research*, 10 (3), 220–41.

——— **(2018)**, Reciprocal Anger, Threat, Or Tracking: Using Intensity of Anger to Disentangle Three Common Responses to Angry Negotiators in Distributive and Integrative Negotiations, doctoral dissertation, David Eccles School of Business, University of Utah.

Hunter, J. E. and F. L. Schmidt (2004), *Methods of Meta-Analysis - Correcting Error and Bias in Research Findings*, 2nd ed., Thousand Oaks: SAGE Publications.

Iklé, F. C. and N. Leites (1962), Political negotiation as a process of modifying utilities, in: *Journal of Conflict Resolution*, 6 (1), 19–28.

Jäger, A., D. D. Loschelder, and M. Friese (2016), Using Self-Regulation to Overcome the Detrimental Effects of Anger in Negotiations, in: *Journal of Economic Psychology*, 58, 31–43.

Jang, D. and W. P. Bottom (2020), The Disutility of Anger Expression in Negotiations with Strategic Risk and Reciprocal Interdependence, working paper.

———, **H. A. Elfenbein, and W. P. Bottom (2018)**, More than a phase: Form and features of a general theory of negotiation, in: *Academy of Management Annals*, 12 (1), 318–56.

Janssens, A. C. J. W. and M. Gwinn (2015), Novel citation-based search method for scientific literature: application to meta-analyses, in: *BMC Medical Research Methodology*, 15 (84), 1–11.

———, ———, **J. E. Brockman, K. Powell, and M. Goodman (2019)**, Novel Citation-Based Search Method for Scientific Literature: A Validation Study, in: *PeerJ — the Journal of Life and Environmental Sciences*, 1–17.

Jennings, N. R., P. Faratin, A. R. Lomuscio, S. Parsons, M. Wooldridge, and C. Sierra (2001), Automated Negotiation: Prospects, Methods and Challenges, in: *Group Decision and Negotiation*, 10, 199–215.

Jeong, Y. K., M. Song, and Y. Ding (2014), Content-based author co-citation analysis, in: *Journal of Informetrics*, 8 (1), 197–211.

Jinha, A. E. (2010), Article 50 million: An estimate of the number of scholarly articles in existence, in: *Learned Publishing*, 23 (3), 258–63.

Johnson, N. A. and R. B. Cooper (2009), Media, affect, concession, and agreement in negotiation: IM versus telephone, in: *Decision Support Systems*, 46 (3), 673–84.

Jonker, C. M., V. Robu, and J. Treur (2007), An agent architecture for multi-attribute negotiation using incomplete preference information, in: *Autonomous Agents and Multi-Agent Systems*, 15 (2), 221–52.

Kagel, J. H., C. Kim, and D. Moser (1996), Fairness in Ultimatum Games with Asymmetric Information and Asymmetric Payoffs, in: *Games and Economic Behavior*, 13 (1), 100–110.

Kahneman, D. (1992), Reference Points, Anchors, Norms, and Mixed Feelings, in: *Organizational Behavior and Human Decision Processes*, 51 (2), 296–312.

——— **and A. Tversky (1979)**, Prospect Theory: An Analysis of Decision under Risk, in: *Econometrica*, 47 (2), 263–92.

Karagözoğlu, E. and Ü. B. Urhan (2017), The Effect of Stake Size in Experimental Bargaining and Distribution Games: A Survey, in: *Group Decision and Negotiation*, 26 (2), 285–325.

Kärki, R. (1996), Searching for bridges between disciplines: an author co-citation analysis on the research into scholarly communication, in: *Journal of Information Science*, 22 (5), 323–34.

Keith, B., K. Vitasek, K. Manrodt, and J. Kling (2016), *Strategic Sourcing in the New Economy - Harnessing the Potential of Sourcing Business Models for Modern Procurement*, New York: Palgrave Macmillan.

Kelley, H. H. (1965), Experimental Studies of Threats in Interpersonal Negotiations, in: *Journal of Conflict Resolution*, 9 (1), 79–105.

——— **(1966)**, A Classroom Study of the Dilemmas in Interpersonal Negotiations, in *Strategic Interaction and Conflict*, K. Archibald, ed., Berkeley: Institute of International Studies, 49–73.

———**, L. L. Beckman, and C. S. Fischer (1967)**, Negotiating the Division of a Reward under Incomplete Information, in: *Journal of Experimental Social Psychology*, 3 (4), 361–98.

——— **and J. W. Thibaut (1978)**, *Interpersonal Relations: A Theory of Interdependence*, New York: John Wiley & Sons.

Keltner, D., D. H. Gruenfeld, and C. Anderson (2003), Power, approach, and inhibition, in: *Psychological Review*, 110 (2), 265–84.

——— **and J. Haidt (1999)**, Social Functions of Emotions at Four Levels of Analysis, in: *Cognition and Emotion*, 13 (5), 505–21.

———**, K. Oatley, and J. M. Jenkins (2013)**, *Understanding Emotions*, 3rd ed., Danvers: John Wiley & Sons.

Kennedy, J. A. and L. J. Kray (2015), A pawn in someone else's game?: The cognitive, motivational, and paradigmatic barriers to women's excelling in negotiation, in: *Research in Organizational Behavior*, 35, 3–28.

Kersten, G. E. (1997), Support for Group Decisions and Negotiations: An Overview, in *Multicriteria Analysis*, J. Claco, ed., Berlin Heidelberg: Springer, 332–46.

——— **and H. Lai (2007)**, Negotiation Support and E-negotiation Systems: An Overview, in: *Group Decision and Negotiation*, 16 (6), 553–86.

——— **and G. Lo (2001)**, Negotiation Support Systems and Software Agents in E-Business Negotiations, in: *The First International Conference on Electronic Business*, 1–9.

———, **W. Michalowski, S. Szpakowicz, and Z. Koperczak (1991)**, Restructurable Representations of Negotiation, in: *Management Science*, 37 (10), 1269–90.

——— **and S. J. Noronha (1999)**, WWW-based Negotiation Support: Design, Implementation, and Use, in: *Decision Support Systems*, 25 (2), 135–54.

Khooshabeh, P., C. De Melo, B. Volkman, J. Gratch, J. Blascovich, and P. J. Carnevale (2013), Negotiation Strategies with Incongruent Facial Expressions of Emotion Cause Cardiovascular Threat, in: *Proceedings of the Annual Meeting of the Cognitive Science Society*, 770–75.

Kimmel, M. J., D. G. Pruitt, J. M. Magenau, and E. Konar-Goldband (1980), Effects of Trust, Aspiration, and Gender on Negotiation Tactics, in: *Journal of Personality and Social Psychology*, 38 (1), 9–22.

Klavans, R. and K. W. Boyack (2006), Identifying a Better Measure of Relatedness for Mapping Science, in: *Journal of the American Society for Information Science and Technology*, 57 (2), 251–63.

——— **and** ——— **(2017)**, Which Type of Citation Analysis Generates the Most Accurate Taxonomy of Scientific and Technical Knowledge?, in: *Journal of the Association for Information Science and Technology*, 68 (4), 984–98.

van Kleef, G. A. (2007), Emotion in Conflict and Negotiation: Introducing the Emotions as Social Information (EASI) Model, in *Proceedings of the International Association of Conflict Management*, 1–29.

——— **(2009)**, How Emotions Regulate Social Life - The Emotions as Social Information (EASI) Model, in: *Current Directions in Psychological Science*, 18 (3), 184–88.

——— **(2014)**, Understanding the positive and negative effects of emotional expressions in organizations: EASI does it, in: *Human Relations*, 67 (9), 1145–64.

———, **C. Anastasopoulou, and B. A. Nijstad (2010)**, Can expressions of anger enhance creativity? A test of the emotions as social information (EASI) model, in: *Journal of Experimental Social Psychology*, 46 (6), 1042–48.

——— **and S. Côté (2007)**, Expressing Anger in Conflict: When It Helps and When It Hurts, in: *Journal of Applied Psychology*, 92 (6), 1557–69.

——— **and** ——— **(2018)**, Emotional Dynamics in Conflict and Negotiation: Individual, Dyadic, and Group Processes, in: *Annual Review of Organizational Psychology and Organizational Behavior*, 5 (1), 437–64.

——— **and C. K. W. De Dreu** (**2010**), Longer-term consequences of anger expression in negotiation: Retaliation or spillover?, in: *Journal of Experimental Social Psychology*, 46 (5), 753–60.

———, ———, **and A. S. R. Manstead** (**2004a**), The Interpersonal Effects of Anger and Happiness in Negotiations, in: *Journal of Personality and Social Psychology*, 86 (1), 57–76.

———, ———, **and** ——— (**2004b**), The Interpersonal Effects of Emotions in Negotiations: A Motivated Information Processing Approach, in: *Journal of Personality and Social Psychology*, 87 (4), 510–28.

———, ———, **and** ——— (**2006**), Supplication and Appeasement in Conflict and Negotiation: The Interpersonal Effects of Disappointment, Worry, Guilt, and Regret, in: *Journal of Personality and Social Psychology*, 91 (1), 124–42.

———, ———, **and** ——— (**2010**), An Interpersonal Approach to Emotion in Social Decision Making: The Emotions as Social Information Model, in *Advances in Experimental Social Psychology*, M. P. Zanna, ed., Burlington: Academic Press, 45–96.

———, ———, **D. Pietroni, and A. S. R. Manstead** (**2006**), Power and emotion in negotiation: Power moderates the interpersonal effects of anger and happiness on concession making, in: *European Journal of Social Psychology*, 36 (4), 557–81.

———, **A. C. Homan, and A. Cheshin** (**2012**), Emotional influence at work: Take it EASI, in: *Organizational Psychology Review*, 2 (4), 311–39.

——— **and P. A. M. van Lange** (**2008**), What Other's Disappointment May Do to Selfish People: Emotion and Social Value Orientation in a Negotiation Context, in: *Personality and Social Psychology Bulletin*, 34 (8), 1084–95.

——— **and M. Sinaceur** (**2013**), The demise of the "rational" negotiator: emotional forces in conflict and negotiation, in *Handbook of Research on Negotiation*, M. Olekalns and W. L. Adair, eds., Cheltenham: Edward Elgar, 103–32.

Klimoski, R. J. (**1972**), The Effects of Intragroup Forces on Intergroup Conflict Resolution, in: *Organizational Behavior and Human Performance*, 8 (3), 363–83.

——— **and R. A. Ash** (**1974**), Accountability and Negotiator Behavior, in: *Organizational Behavior and Human Performance*, 11 (3), 409–25.

Knutson, B. (**1996**), Facial Expressions of Emotion Influence Interpersonal Trait Inferences, in: *Journal of Nonverbal Behavior*, 20 (3), 165–82.

Kogan, N., H. Lamm, and G. Trommsdorff (1972), Negotiation Constraints in the Risk-Taking Domain: Effects of Being Observed by Partners of Higher or Lower Status, in: *Journal of Personality and Social Psychology*, 23 (2), 143–56.

Komorita, S. S. and M. Barnes (1969), Effects of pressures to reach agreement in bargaining, in: *Journal of Personality and Social Psychology*, 13 (3), 245–52.

——— **and A. R. Brenner (1968)**, Bargaining and Concession Making Under Bilateral Monopoly, in: *Journal of Personality and Social Psychology*, 9 (1), 15–20.

——— **and J. K. Esser (1975)**, Frequency of Reciprocated Concessions in Bargaining, in: *Journal of Personality and Social Psychology*, 32 (4), 699–705.

Kong, D. T., K. T. Dirks, and D. L. Ferrin (2014), Interpersonal trust within negotations: meta-analytic evidence, critical contingencies, and directions for future research, in: *Academy of Management Journal*, 57 (5), 1235–55.

Kopelman, S. and A. S. Rosette (2008), Cultural variation in response to strategic emotions in negotiations, in: *Group Decision and Negotiation*, 17 (1), 65–77.

———, ———, and L. L. Thompson (2006), The three faces of Eve: Strategic displays of positive, negative, and neutral emotions in negotiations, in: *Organizational Behavior and Human Decision Processes*, 99 (1), 81–101.

Kramer, R. M., E. Newton, and P. L. Pommerenke (1993), Self-Enhancement Biases and Negotiator Judgment: Effects of Self-Esteem and Mood, in: *Organizational Behavior and Human Decision Processes*, 56 (1), 110–33.

Kraus, S. (2001), *Strategic Negotiation in Multiagent Environments*, Cambridge: MIT Press.

———, K. Sycara, and A. Evenchik (1998), Reaching agreements through argumentation: a logical model and implementation, in: *Artificial Intelligence*, 104 (1–2), 1–69.

Kray, L. J., A. D. Galinsky, and L. L. Thompson (2002), Reversing the Gender Gap in Negotiations: An Exploration of Stereotype Regeneration, in: *Organizational Behavior and Human Decision Processes*, 87 (2), 386–409.

——— **and M. J. Gelfand (2009)**, Relief Versus Regret: The Effect of Gender and Negotiating Norm Ambiguity on Reactions to Having One's First Offer Accepted, in: *Social Cognition*, 27 (3), 418–36.

——— **and L. L. Thompson (2005)**, Gender Stereotypes and Negotiation Performance: An Examination of Theory and Research, in: *Research in Organizational Behavior*, 26, 103–82.

———, ———, **and A. Galinsky (2001)**, Battle of the Sexes: Gender Stereotype Confirmation and Reactance in Negotiations, in: *Journal of Personality and Social Psychology*, 80 (6), 942–58.

Kristensen, H. and T. Gärling (1997), Determinants of Buyers' Aspiration and Reservation Price, in: *Journal of Economic Psychology*, 18 (5), 487–503.

Kruglanski, A. W. (1989), The Psychology of Being "Right": The Problem of Accuracy in Social Perception and Cognition, in: *Psychological Bulletin*, 106 (3), 395–409.

——— **and T. Freund (1983)**, The Freezing and Unfreezing of Lay-Inferences: Effects on Impressional Primacy, Ethnic Stereotyping, and Numerical Anchoring, in: *Journal of Experimental Social Psychology*, 19 (5), 448–68.

——— **and D. M. Webster (1996)**, Motivated Closing of the Mind: "Seizing" and "Freezing"," in: *Psychological Review*, 103 (2), 263–83.

Kubota, J. T., J. Li, E. Bar-David, M. R. Banaji, and E. A. Phelps (2013), The Price of Racial Bias: Intergroup Negotiations in the Ultimatum Game, in: *Psychological Science*, 24 (12), 2498–2504.

Kugler, K. G., J. A. M. Reif, T. Kaschner, and F. C. Brodbeck (2018), Gender differences in the initiation of negotiations: A meta-analysis, in: *Psychological Bulletin*, 144 (2), 198–222.

Kuhlman, D. M. and A. F. J. Marshello (1975), Individual Differences in Game Motivation as Moderators of Preprogrammed Strategy Effects in Prisoner's Dilemma, in: *Journal of Personality and Social Psychology*, 32 (5), 922–31.

——— **and D. L. Wimberley (1976)**, Expectations of Choice Behavior Held by Cooperators, Competitors, and Individualists Across Four Classes of Experimental Games, in: *Journal of Personality and Social Psychology*, 34 (1), 69–81.

Kung, F. Y. H. and M. M. Chao (2019), The Impact of Mixed Emotions on Creativity in Negotiation: An Interpersonal Perspective, in: *Frontiers in Psychology*, 9 (1), 1–15.

Kwon, K.-N., D. W. Schumann, and A. Fairhurst (2010), Consideration of an expected future deal in assessing the value of a present deal: Forward-looking bargain shopping, in: *Journal of Business Research*, 63 (1), 84–90.

Laengle, S., N. M. Modak, J. M. Merigo, and G. Zurita (2018), Twenty-Five Years of Group Decision and Negotiation: A Bibliometric Overview, in: *Group Decision and Negotiation*, 27 (4), 505–42.

Lamm, H. and N. Kogan (1970), Risk Taking in the Context of Intergroup Negotiation, in: *Journal of Experimental Social Psychology*, 6 (3), 351–63.

Landis, J. R. and G. G. Koch (1977), The Measurement of Observer Agreement for Categorical Data, in: *Biometrics*, 33 (1), 159–74.

van Lange, P. A. M., E. M. N. De Bruin, W. Otten, and J. A. Joireman (1997), Development of Prosocial, Individualistic, and Competitive Orientations: Theory and Preliminary Evidence, in: *Journal of Personality and Social Psychology*, 73 (4), 733–46.

——— **and D. M. Kuhlman (1994)**, Social Value Orientations and Impressions of Partner's Honesty and Intelligence : A Test of the Might Versus Morality Effect, in: *Journal of Personality and Social Psychology*, 67 (1), 126–41.

——— **and K. Visser (1999)**, Locomotion in Social Dilemmas: How People Adapt to Cooperative, Tit-for-Tat, and Noncooperative Partners, in: *Journal of Personality and Social Psychology*, 77 (4), 762–73.

Larsen, P. O. and M. von Ins (2010), The rate of growth in scientific publication and the decline in coverage provided by science citation index, in: *Scientometrics*, 84 (3), 575–603.

Lau, R. Y. K., O. Wong, Y. Li, and L. C. K. Ma (2008), Mining Trading Partners' Preferences for Efficient Multi-Issue Bargaining in E-Business, in: *Journal of Management Information Systems*, 25 (1), 79–103.

Lawler, E. J. and J. Yoon (1995), Structural Power and Emotional Processes in Negotiation: A Social Exchange Approach, in *Negotiation as a Social Process*, R. M. Kramer and D. M. Merrick, eds., Thousand Oaks: SAGE Publications, 143–65.

Lax, D. A. and J. K. Sebenius (1986), *The Manager as Negotiator*, New York: The Free Press.

Lazarus, R. (1991), *Emotion and Adaptation*, New York: Oxford University Press.

Lelieveld, G.-J., E. van Dijk, I. van Beest, and G. A. van Kleef (2012), Why Anger and Disappointment Affect Other's Bargaining Behavior Differently: The Moderating Role of Power and the Mediating Role of Reciprocal and Complementary Emotions, in: *Personality and Social Psychology Bulletin*, 38 (9), 1209–21.

———, ———, ———, **W. Steinel, and G. A. van Kleef (2011)**, Disappointed in You, Angry about Your Offer: Distinct Negative Emotions Induce Concessions via Different Mechanisms, in: *Journal of Experimental Social Psychology*, 47 (3), 635–41.

———, ———, **B. Güroğlu, I. van Beest, G. A. van Kleef, S. A. R. B. Rombouts, and E. A. Crone (2013)**, Behavioral and neural reactions to emotions of others in the distribution of resources, in: *Social Neuroscience*, 8 (1), 52–62.

Leung, X. Y., J. Sun, and B. Bai (2017), Bibliometrics of social media research: A co-citation and co-word analysis, in: *International Journal of Hospitality Management*, 66, 35–45.

Levi, D. (2001), *Group Dynamics for Teams*, Thousand Oaks: SAGE Publications.

Lewicki, R. J. (1983), Lying and Deception: A Behavioral Model, in *Negotiating in Organizations*, M. H. Bazerman and R. J. Lewicki, eds., Beverly Hills: SAGE Publications, 68–90.

———, **B. Barry, and D. M. Saunders (2007)**, *Essentials of Negotiation*, 4th ed., New York: McGraw-Hill.

———, ———, **and** ——— **(2015)**, *Essentials of Negotiation*, 6th edn., New York: McGraw-Hill.

——— **and A. Hiam (2006)**, *Mastering Business Negotiation: A Working Guide to Making Deals and Resolving Conflict*, San Francisco: Jossey-Bass.

———, ———, **and K. W. Olander (1998)**, *Verhandeln mit Strategie - Das große Handbuch der Verhandlungstechniken*, St. Gallen: Midas Management Verlag.

——— **and R. J. Robinson (1998)**, Ethical and Unethical Bargaining Tactics: An Empirical Study, in: *Journal of Business Ethics*, 17 (1994), 665–82.

———, **D. M. Saunders, and J. W. Minton (1997)**, *Essentials of Negotiation*, 2nd ed., Chicago: McGraw-Hill.

Lewis, S. A. and W. R. Fry (1977), Effects of Visual Access and Orientation on the Discovery of Integrative Bargaining Alternatives, in: *Organizational Behavior and Human Performance*, 20 (1), 75–92.

Leydesdorff, L. (2008), On the Normalization and Visualization of Author Co-Citation Data: Salton's Cosine versus the Jaccard Index, in: *Journal of the American Society for Information Science and Technology*, 59 (1), 77–85.

Li, S.-C. S. (2007), Computer-Mediated Communication and Group Decision Making: A Functional Perspective, in: *Small Group Research*, 38 (5), 593–614.

Li, S. (2005), Strategic Emotion in Negotiation: The Use and Consequences of Positive and Negative Emotions, doctoral dissertation, Northwestern University.

——— **and M. E. Roloff (2006)**, Strategie Emotion in Negotiation: Cognition, Emotion, and Culture, in *From Communication to Presence: Cognition, Emotions and Culture towards the Ultimate Communicative Experience*, G. Riva, M. T. Anguera, B. K. Wiederhold, and F. Mantovani, eds., Amsterdam: IOS Press, 166–85.

Liebert, R. M., W. P. Smith, J. H. Hill, and M. Keifer (1968), The Effects of Information and Magnitude of Initial Offer on Interpersonal Negotiation, in: *Journal of Experimental Social Psychology*, 4 (4), 431–41.

Lipsey, M. W. and D. B. Wilson (2001), *Practical Meta-Analysis*, Thousand Oaks: SAGE Publications.

Loewenstein, G. F., L. L. Thompson, and M. H. Bazerman (1989), Social Utility and Decision Making in Interpersonal Contexts, in: *Journal of Personality and Social Psychology*, 57 (3), 426–41.

Loewenstein, J., L. L. Thompson, and D. Gentner (1999), Analogical encoding facilitates knowledge transfer in negotiation, in: *Psychonomic Bulletin and Review*, 6 (4), 586–97.

Loschelder, D. D., J. Stuppi, and R. Trötschel (2014), "€14,875?!": Precision Boosts the Anchoring Potency of First Offers, in: *Social Psychological and Personality Science*, 5 (4), 491–99.

———, **R. I. Swaab, R. Trötschel, and A. D. Galinsky (2014)**, The First-Mover Disadvantage: The Folly of Revealing Compatible Preferences, in: *Psychological Science*, 25 (4), 954–62.

Luce, R. D. and H. Raiffa (1957), *Games and Decisions: Introduction and Critical Survey*, New York: John Wiley & Sons.

Luecke, R. (2003), *Negotiation*, Boston: Harvard Business School Press.

Lügger, K., I. Geiger, H. Neun, and K. Backhaus (2015), When East Meets West at the Bargaining Table: Adaptation, Behavior and Outcomes in Intra- and Intercultural German–Chinese Business Negotiations, in: *Journal of Business Economics*, 85 (1), 15–43.

Luo, X., N. R. Jennings, N. Shadbolt, H. Leung, and J. H. Lee (2003), A fuzzy constraint based model for bilateral, multi-issue negotiations in semi-competitive environments, in: *Artificial Intelligence*, 148 (1–2), 53–102.

Lutzker, D. R. (1960), Internationalism as a Predictor of Cooperative Behavior, in: *Journal of Conflict Resolution*, 4 (4), 426–30.

Lyubomirsky, S., L. King, and E. Diener (2005), The benefits of frequent positive affect: Does happiness lead to success?, in: *Psychological Bulletin*, 131 (6), 803–55.

Maaravi, Y., Y. Ganzach, and A. Pazy (2011), Negotiation as a Form of Persuasion: Arguments in First Offers, in: *Journal of Personality and Social Psychology*, 101 (2), 245–55.

Maeße, J., H. Pahl, and J. Sparsam (2017), *Die Innenwelt der Ökonomie - Wissen, Macht und Performativität in der Wirtschaftswissenschaft*, Wiesbaden: Springer.

Magee, J. C., A. D. Galinsky, and D. H. Gruenfeld (2007), Power, Propensity to Negotiate, and Moving First in Competitive Interactions, in: *Personality and Social Psychology Bulletin*, 33 (2), 200–212.

Malhotra, N. K. and D. F. Birks (2007), *Marketing Research: An Applied Approach*, 3rd ed., Harlow: Prentice Hall.

Mannix, E. A. (1993), The Influence of Power, Distribution Norms and Task Meeting Structure on Resource Allocation in Small Group Negotiation, in: *International Journal of Conflict Management*, 4 (1), 5–23.

Marín-Martínez, F. and J. Sánchez-Meca (2010), Weighting by Inverse Variance or by Sample Size in Random-Effects Meta-Analysis, in: *Educational and Psychological Measurement*, 70 (1), 56–73.

Marlowe, D. (1963), Psychological Needs and Cooperation: Competition in a Two-Person Game, in: *Psychological Reports*, 13 (2), 364.

Marquard, O. (2003), *Zukunft braucht Herkunft: Philosophische Essays*, Ditzingen: Reclam.

Marsilio, M., G. Cappellaro, and C. Cuccurullo (2011), The intellectual structure of research into PPPS: A bibliometric analysis, in: *Public Management Review*, 13 (6), 763–82.

Martinovski, B. (2010), Emotion in Negotiation, in *Handbook of Group Decision and Negotiation*, M. Kilgour and C. Eden, eds., Basel: Springer, 65–86.

Mason, M. F., A. J. Lee, E. A. Wiley, and D. R. Ames (2013), Precise offers are potent anchors: Conciliatory counteroffers and attributions of knowledge in negotiations, in: *Journal of Experimental Social Psychology*, 49 (4), 759–63.

Matsumoto, D. (1993), Ethnic Differences in Affect Intensity, Emotion Judgments, Display Rule Attitudes, and Self-Reported Emotional Expression in an American Sample, in: *Motivation and Emotion*, 17 (2), 107–23.

Mazei, J., J. Hüffmeier, P. A. Freund, A. F. Stuhlmacher, L. Bilke, and G. Hertel (2015), A Meta-Analysis on Gender Differences in Negotiation Outcomes and Their Moderators, in: *Psychological Bulletin*, 141 (1), 85–104.

McCain, K. W. (1986), Cocited Author Mapping as a Valid Representation of Intellectual Structure, in: *Journal of the American Society for Information Science*, 37 (3), 111–22.

——— **(1990)**, Mapping Authors in Intellectual Space: A Technical Overview, in: *Journal of the American Society for Information Science*, 41 (6), 433–43.

McClintock, C. G. (1977), Social Motivation in Settings of Outcome Interdependence, in *Negotiations: Social Psychological Perspective*, D. Druckman, ed., Beverly Hills: SAGE Publications, 49–77.

——— **and W. B. G. Liebrand (1988)**, Role of Interdependence Structure, Individual Value Orientation, and Another's Strategy in Social Decision Making: A Transformational Analysis, in: *Journal of Personality and Social Psychology*, 55 (3), 396–409.

———**, D. M. Messick, D. M. Kuhlman, and F. T. Campos (1973)**, Motivational Bases of Choice in Three-Choice Decomposed Games, in: *Journal of Experimental Social Psychology*, 9 (6), 572–90.

McGarry, D. D. (1955), *The Metalogicon of John of Salisbury - A Twelfth-Century Defense of the Verbal and Logical Arts of the Trivium*, Berkeley: University of California Press.

McGrath, J. E. (1966), A social psychological approach to the study of negotiation, in *Studies on Behaviour in Organizations: A Research Symposium*, R. Bowers, ed., Athens: University of Georgia Press, 101–34.

———**, J. L. Berdahl, and H. Arrow (1995)**, Traits, Expectations, Culture, and Clout: The Dynamics of Diversity in Work Groups, in *Diversity in Work Teams*, S. E. Jackson and M. N. Ruderman, eds., Washington: American Psychological Association, 17–45.

De Melo, C. M., P. J. Carnevale, and J. Gratch (2011), The Effect of Expression of Anger and Happiness in Computer Agents on Negotiations with Humans, in: *Proceedings of the 10th International Conference on Autonomous Agents and Multiagent Systems*, 937–44.

———, ———, and ——— (2012), The Effect of Virtual Agents' Emotion Displays and Appraisals on People's Decision Making in Negotiation, in *International Conference on Intelligent Virtual Agents*, Berlin: Springer, 53–66.

———, J. Gratch, and P. J. Carnevale (2014), Humans vs. Computers: Impact of Emotion Expressions on People's Decision Making, in: *IEEE Transactions on Affective Computing*, 6 (2), 127–36.

Messick, D. M. and C. G. McClintock (1968), Motivational Bases of Choice in Experimental Games, in: *Journal of Experimental Social Psychology*, 4 (1), 1–25.

Meyer, M., I. Lorscheid, and K. G. Troitzsch (2009), The development of social simulation as reflected in the first ten years of JASSS: A citation and co-citation analysis, in: *Journal of Artificial Societies and Social Simulation*, 12 (4), 12.

Montori, V. M., M. F. Swiontkowski, and D. Cook (2003), Methodologic Issues in Systematic Reviews and Meta-Analyses, in: *Clinical Orthopaedics and Related Research*, 413, 43–54.

Moore, D. A., T. R. Kurtzberg, L. L. Thompson, and M. W. Morris (1999), Long and Short Routes to Success in Electronically Mediated Negotiations: Group Affiliations and Good Vibrations, in: *Organizational Behavior and Human Decision Processes*, 77 (1), 22–43.

Mor, S. and A. Suppes (2011), The Role of Communication Media in Negotiations, in *Negotiation Excellence: Successful Deal Making*, M. Benoliel, ed., Singapore: World Scientific Publishing, 381–96.

Moran, S. and I. Ritov (2002), Initial Perceptions in Negotiations: Evaluation and Response to "Logrolling" Offers, in: *Journal of Behavioral Decision Making*, 15 (2), 101–24.

Morlacchi, P., I. F. Wilkinson, and L. C. Young (2005), Social Networks of Researchers in B2B Marketing: A Case Study of the IMP Group 1984 – 1999, in: *Journal of Business-to-Business Marketing*, 12 (1), 3–34.

Morris, M. W. and D. Keltner (2000), How Emotions Work: The Social Functions of Emotional Expression in Negotiations, in: *Research in Organizational Behavior*, 22, 1–50.

Mosterd, I. and C. G. Rutte (2000), Effects of Time Pressure and Accountability to Constituents on Negotiation, in: *International Journal of Conflict Management*, 11 (3), 227–47.

Mueller, B. (2019), What Is Brexit? A Simple Guide to Why It Matters and What Happens Next, in: *The New York Times*, (accessed November 21, 2019), [available at https://www.nytimes.com/interactive/2019/world/europe/what-is-brexit.html].

Mueller, J. S. and J. R. Curhan (2006), Emotional intelligence and counterpart mood induction in a negotiation, in: *International Journal of Conflict Management*, 17 (2), 110–28.

Mullen, B. and N. Miller (1991), Meta-Analysis, in *Research Methods in Social Relations*, C. M. Judd, E. R. Smith, and L. H. Kidder, eds., 6th edn., New York: Harcourt Brace Jovanovich College Publishers, 425–49.

Mullins, N. C., L. L. Hargens, P. K. Hecht, and E. L. Kick (1977), The Group Structure of Cocitation Clusters: A Comparative Study, in: *American Sociological Review* , 42 (4), 552–62.

Murnighan, J. K. (1978), Models of Coalition Behavior: Game Theoretic, Social Psychological, and Political Perspectives, in: *Psychological Bulletin*, 85 (5), 1130–53.

———**, L. Babcock, L. L. Thompson, and M. Pillutla (1999)**, The Information Dilemma in Negotiations: Effects of Experience, Incentives, and Integrative Potential, in: *International Journal of Conflict Management*, 10 (4), 313–39.

——— **and D. Brass (1991)**, Intraorganizational coalitions, in *The Handbook of Negotiation Research*, R. J. Lewicki, B. Sheppard, and M. Bazerman, eds., Greenwich: JAI Press, 283–306.

Myerson, R. B. and M. A. Satterthwaite (1983), Efficient Mechanisms for Bilateral Trading, in: *Journal of Economic Theory*, 29, 265–81.

Nadler, J., L. L. Thompson, and L. van Boven (2003), Learning Negotiation Skills: Four Models of Knowledge Creation and Transfer, in: *Management Science*, 49 (4), 529–40.

Naquin, C. E. and T. R. Kurtzberg (2009), Team Negotiation and Perceptions of Trustworthiness: The Whole Versus the Sum of the Parts, in: *Group Dynamics: Theory, Research, and Practice*, 13 (2), 133–50.

——— **and G. D. Paulson (2003)**, Online Bargaining and Interpersonal Trust, in: *Journal of Applied Psychology*, 88 (1), 113–20.

Narvaez-Berthelemot, N. and J. M. Russell (2001), World distribution of social science journals: A view from the periphery, in: *Scientometrics*, 51 (1), 223–39.

Nash, J. F. (1950), The Bargaining Problem, in: *Econometrica*, 18 (2), 155–62.

Neale, M. A. and M. H. Bazerman (1985), The Effects of Framing and Negotiator Overconfidence on Bargaining Behaviors and Outcomes, in: *The Academy of Management Journal*, 28 (1), 34–49.

——— and ——— (1991), *Cognition and Rationality in Negotiation*, New York: The Free Press.

——— and ——— (1992), Negotiating Rationally: The Power and Impact of the Negotiator's Frame, in: *Academy of Management Journal*, 6 (3), 42–51.

———, V. L. Huber, and G. B. Northcraft (1987), The Framing of Negotiations: Contextual versus Task Frames, in: *Organizational Behavior and Human Decision Processes*, 39 (2), 228–41.

——— and G. B. Northcraft (1986), Experts, Amateurs, and Refrigerators: Comparing Expert and Amateur Negotiators in a Novel Task, in: *Organizational Behavior and Human Decision Processes*, 38 (3), 305–17.

——— and ——— (1991), Behavioral Negotiation Theory - A Framework for Conceptualizing Dyadic Bargaining, in *Research in Organizational Behavior*, L. L. Cummings and B. M. Staw, eds., Greenwich: JAI Press, 147–90.

Nerur, S. P., A. A. Rasheed, and V. Natarajan (2008), The Intellectual Structure of the Strategic Management Field: An Author Co-Citation Analysis, in: *Strategic Management Journal*, 29 (3), 319–36.

von Neumann, J. and O. Morgenstern (1944), *Theory of Games and Economic Behavior*, New York: John Wiley & Sons.

Newman, M. E. J. and M. Girvan (2004), Finding and evaluating community structure in networks, in: *Physical Review*, 69 (2), 1–15.

Newton, I. (1676), Letter to Robert Hooke.

van den Noortgate, W. and P. Onghena (2003), Multilevel meta-analysis: A comparison with traditional meta-analytical procedures, in: *Educational and Psychological Measurement*, 63 (5), 765–90.

Northcraft, G. B. and M. A. Neale (1987), Experts, Amateurs, and Real Estate: An Anchoring - and - Adjustment Perspective on Property Pricing Decisions, in: *Organizational Behavior and Human Decision Processes*, 39 (1), 84–97.

O'Connor, K. M. (1997), Groups and Solos in Context: The Effects of Accountability on Team Negotiation, in: *Organizational Behavior and Human Decision Processes*, 72 (3), 384–407.

——— **and J. A. Arnold (2001)**, Distributive Spirals: Negotiation Impasses and the Moderating Role of Disputant Self-Efficacy, in: *Organizational Behavior and Human Decision Processes*, 84 (1), 148–76.

Olekalns, M. and D. Druckman (2014), With Feeling: How Emotions Shape Negotiation, in: *Negotiation Journal*, 30, 455–78.

———**, C. Robert, T. Probst, P. L. Smith, and P. J. Carnevale (2005)**, The impact of message frame on negotiators' impressions, emotions, and behaviors, in: *International Journal of Conflict Management*, 16 (4), 379–402.

——— **and P. L. Smith (2007)**, Loose with the Truth: Predicting Deception in Negotiation, in: *Journal of Business Ethics*, 76 (2), 225–38.

——— **and** ——— **(2009)**, Mutually Dependent: Power, Trust , Affect and the Use of Deception in Negotiation, in: *Journal of Business Ethics*, 85, 347–65.

———, ———**, and R. Kibby (1996)**, Social value orientations and negotiator outcomes, in: *European Journal of Social Psychology*, 26 (2), 299–313.

———, ———**, and T. Walsh (1996)**, The Process of Negotiating: Strategy and Timing as Predictors of Outcomes, in: *Organizational Behavior and Human Decision Processes*, 68 (1), 68–77.

Oliver, J. R. (1997), A Machine-Learning Approach to Automated Negotiation and Prospects for Electronic Commerce, in: *Journal of Management Information Systems*, 13 (3), 83–112.

Oliver, R. L., P. V. S. Balakrishnan, and B. Barry (1994), Outcome Satisfaction in Negotiation: A Test of Expectancy Disconfirmation, in: *Organizational Behavior and Human Decision Processes*, 60 (2), 252–75.

Oosterbeek, H., R. Sloof, and G. van de Kuilen (2004), Cultural differences in ultimatum game experiments: Evidence from a meta-analysis, in: *Experimental Economics*, 7 (2), 171–88.

Orr, D. and C. Guthrie (2006), Anchoring, Information, Expertise, and Negotiation: New Insights from Meta-Analysis, in: *Ohio State Journal on Dispute Resolution*, 21 (3), 597–628.

Ortony, A. and T. J. Turner (1990), What's Basic About Basic Emotions?, in: *Psychological Review*, 97 (3), 315–31.

Osgood, C. E. (1959), Suggestions for winning the real war with communism, in: *Journal of Conflict Resolution*, 3 (4), 295–325.

——— **(1962)**, *An alternative to war or surrender*, Urbana: University of Illinois Press.

Otte, E. and R. Rousseau (2002), Social network analysis: a powerful strategy, also for the information sciences, in: *Journal of Information Science*, 28 (6), 441–53.

Overbeck, J. R. and Y. K. Kim (2013), Power, status, and influence in negotiation, in *Handbook of Research on Negotiation*, M. Olekalns and W. L. Adair, eds., Cheltenham: Edward Elgar, 133–60.

———**, M. A. Neale, and C. L. Govan (2010)**, I feel, therefore you act: Intrapersonal and interpersonal effects of emotion on negotiation as a function of social power, in: *Organizational Behavior and Human Decision Processes*, 112 (2), 126–39.

Parkinson, B. (1996), Emotions are social, in: *British Journal of Psychology*, 87 (4), 663–83.

Parsons, S., C. Sierra, and N. Jennings (1998), Agents that Reason and Negotiate by Arguing, in: *Journal of Logic Computation*, 8 (3), 261–92.

Paskin, N. (2010), Digital Object Identifier (DOI®) System, in *Encyclopedia of Library and Information Sciences*, M. J. Bates and M. N. Maack, eds., London: Taylor & Francis, 1586–92.

Pennington, A. L. (1968), Customer-Salesman Bargaining Behavior in Retail Transactions, in: *Journal of Marketing Research*, 5 (3), 255–62.

Perrey, E., U. Schäffer, and S. D. Becker (2012), Struktur und Entwicklung der Rechnungslegungsforschung in deutschen wissenschaftlichen Zeitschriften, in: *Zeitschrift für Betriebswirtschaft*, 82 (11), 1157–91.

Persson, O., R. Danell, and J. Wiborg Schneider (2009), How to use Bibexcel for various types of bibliometric analysis - A Festschrift for Olle Persson at his 60th Birthday, in *Celebrating Scholarly Communication Studies*, F. Aström, R. Danell, B. Larsen, and J. Wiborg Schneider, eds., Leuven: International Society for Scientometrics and Informetrics, 9–24.

Pesic, M. (2016), *Emotionen in Verhandlungen*, Hamburg: Dr. Kovac.

Petty, R. E. and J. T. Cacioppo (1986), The elaboration likelihood model of persuasion, in *Advances in experimental social psychology*, L. Berkowitz, ed., New York: Academic Press, 123–205.

Phillips, E. and R. Cheston (1979), Conflict Resolution: What Works?, in: *California Management Review*, 21 (4), 76–83.

Pietroni, D., G. A. van Kleef, C. K. W. De Dreu, and S. Pagliaro (2008), Emotions as Strategic Information: Effects of Other's Emotional Expressions on Fixed-Pie Perception, Demands, and Integrative Behavior in Negotiation, in: *Journal of Experimental Social Psychology*, 44 (6), 1444–54.

———, ———, E. Rubaltelli, and R. Rumiati (2009), When happiness pays in negotiation: The interpersonal effects of "exit option": Directed emotions, in: *Mind and Society*, 8 (1), 77–92.

Pilisuk, M. and P. Skolnick (1968), Inducing Trust: A Test of the Osgood Proposal, in: *Journal of Personality and Social Psychology*, 8 (2), 121–33.

Pilkington, A. and J. Meredith (2009), The evolution of the intellectual structure of operations management—1980–2006: A citation/co-citation analysis, in: *Journal of Operations Management*, 27 (3), 185–202.

Pillutla, M. M. and J. K. Murnighan (1995), Being Fair or Appearing Fair: Strategic Behavior in Ultimatum Bargaining, in: *The Academy of Management Journal*, 38 (5), 1408–26.

——— and ——— (1996), Unfairness, Anger, and Spite: Emotional Rejections of Ultimatum Offers, in: *Organizational Behavior and Human Decision Processes*, 68 (3), 208–24.

Pinkley, R. L. (1995), Impact of Knowledge Regarding Alternatives to Settlement in Dyadic Negotiations: Whose Knowledge Counts?, in: *Journal of Applied Psychology*, 80 (3), 403–17.

———, M. A. Neale, and R. J. Bennett (1994), The Impact of Alternatives to Settlement in Dyadic Negotiation, in: *Organizational Behavior and Human Decision Processes*, 57 (1), 97–116.

Polke-Majewski, K. (2009), Wiedeking, ein stolzer Verlierer, in: *Zeit Online*, (accessed November 21, 2019), [available at https://www.zeit.de/online/2009/20/vw-porsche-wiedeking].

Polzer, J. T., E. A. Mannix, and M. A. Neale (1998), Interest Alignment and Coalitions in Multiparty Negotiation, in: *The Academy of Management Journal*, 41 (1), 42–54.

Porter, A. L. (1977), Citation Analysis: Queries and Caveats, in: *Social Studies of Science*, 7 (2), 257–67.

Preacher, K. J. and A. F. Hayes (2008), Asymptotic and resampling strategies for assessing and comparing indirect effects in multiple mediator models, in: *Behavior Research Methods*, 40 (3), 879–91.

Pritchard, A. (1969), Statistical bibliography or bibliometrics?, in: *Journal of Documentation*, 25 (4), 348–49.

Pruitt, D. G. (1968), Reciprocity and Credit Building in a Laboratory Dyad, in: *Journal of Personality and Social Psychology*, 8 (2), 143–47.

——— **(1981)**, *Negotiation Behavior*, New York: Academic Press.

——— **(1983)**, Integrative agreements: Nature and antecedents, in *Negotiating in Organizations*, M. H. Bazerman and R. J. Lewicki, eds., Beverly Hills: SAGE Publications, 35–50.

——— **(1998)**, Social Conflict, in *The Handbook of Social Psychology*, D. T. Gilbert, S. T. Fiske, and G. Lindzey, eds., New York: McGraw-Hill, 470–503.

——— **and P. J. Carnevale (1993)**, *Negotiation in Social Conflict*, Berkshire: McGraw-Hill.

——— **and J. L. Drews (1969)**, The Effect of Time Pressure, Time Elapsed, and the Opponent's Concession Rate on Behavior in Negotiation, in: *Journal of Experimental Social Psychology*, 5 (1), 43–60.

——— **and D. F. Johnson (1970)**, Mediation As an Aid To Face Saving in Negotiation, in: *Journal of Personality and Social Psychology*, 14 (3), 239–46.

———**, S. H. Kim, and J. Z. Rubin (2004)**, *Social Conflict: Escalation, Stalemate, and Settlement*, 3rd ed., New York: McGraw-Hill.

———**, M. J. Kimmel, S. Britton, P. J. Carnevale, J. M. Magenau, J. Peragallo, and P. Engram (1978)**, The effect of accountability and surveillance on integrative bargaining, in *Contributions to experimental economics*, H. Sauermann, ed., Tübingen: Mohr, 310–43.

——— **and S. A. Lewis (1975)**, Development of Integrative Solutions in Bilateral Negotiation, in: *Journal of Personality and Social Psychology*, 31 (4), 621–33.

——— **and** ——— **(1977)**, The Psychology of Integrative Bargaining, in *Negotiations - Social-Psychological Perspectives*, D. Druckman, ed., Beverly Hills: SAGE Publications, 161–92.

Purdy, J. M. and P. Nye (2000), The impact of communication media on negotiation outcomes, in: *International Journal of Conflict Management*, 11 (2), 162–87.

Putnam, L. L. and T. S. Jones (1982), Reciprocity in Negotiations: An Analysis of Bargaining Interaction, in: *Communication Monographs*, 49 (3), 171–91.

van Raan, A. F. J. (1996), Advanced Bibliometric Methods As Quantitative Core of Peer Review Based Evaluation And Foresight Exercises, in: *Scienctometrics*, 36 (3), 397–420.

Raasch, C., V. Lee, S. Spaeth, and C. Herstatt (2013), The rise and fall of interdisciplinary research: The case of open source innovation, in: *Research Policy*, 42 (5), 1138–51.

Rafaeli, A. and R. I. Sutton (1989), The expression of emotion in organizational life, in: *Research in Organizational Behavior Journal*, 11, 1–42.

Rahim, M. A. (1983), *Rahim organizational conflict inventories: Professional manual*, New York: Consulting Psychologists Press.

——— **(1986)**, Referent Role and Styles of Handling Interpersonal Conflict, in: *The Journal of Social Psychology*, 126 (1), 79–86.

Rahwan, I., S. D. Ramchurn, N. R. Jennings, P. McBurney, S. Parsons, and L. Sonenberg (2004), Argumentation-based negotiation, in: *The Knowledge Engineering Review*, 18 (4), 343–75.

Raiffa, H. (1982), *The Art and Science of Negotiation*, Cambridge: Harvard University Press.

Raudenbush, S. W. and A. S. Bryk (1985), Empirical Bayes Meta-Analysis, in: *Journal of Educational Statistics*, 10 (2), 75–98.

——— **and** ——— **(2002)**, *Hierarchical linear models*, London: SAGE Publications.

Rees, L., S.-C. Steve, R. Friedman, and H.-L. Shih (2019), Anger as a Trigger for Information Search in Integrative Negotiations, in: *Journal of Applied Psychology*, 1–19.

Renwick, P. A. (1975), Perception and Management of Superior-Subordinate Conflict, in: *Organizational Behavior and Human Performance*, 13 (3), 444–56.

Reynolds, N., A. Simintiras, and E. Vlachou (2003), International Business Negotiations: Present Knowledge and Direction for Future Research, in: *International Marketing Review*, 20 (3), 236–61.

Rhoades, J. A., J. Arnold, and C. Jay (2001), The role of affective traits and affective states in disputants' motivation and behavior during episodes of organizational conflict, in: *Journal of Organizational Behavior*, 22 (3), 329–45.

Ritov, I. (1996), Anchoring in Simulated Competitive Market Negotiation, in: *Organizational Behavior and Human Decision Processes*, 67 (1), 16–25.

Robinson, R. J., R. J. Lewicki, and E. M. Donahue (2000), Extending and testing a five factor model of ethical and unethical bargaining tactics: Introducing the SINS scale, in: *Journal of Organizational Behavior*, 21 (6), 649–64.

Rosenschein, J. S. and G. Zlotkin (1994), *Rules of Encounter: Designing Conventions for Automated Negotiation among Computers*, Cambridge: The MIT Press.

Rosenthal, R. (1979), The "File Drawer Problem" and Tolerance for Null Results, in: *Psychological Bulletin*, 86 (3), 638–41.

——— **(1995)**, Writing Meta-Analytic Reviews, in: *Psychological Bulletin*, 118 (2), 183–92.

——— **and M. R. DiMatteo (2001)**, Meta-Analysis: Recent Developments in Quantitative Methods for Literature Reviews, in: *Annual Review of Psychology*, 52 (1), 59–82.

Roth, S. and M. Gmür (2004), Strukturen und Entwicklungslinien der Marketingforschung: Ergebnisse einer Zitations- und Kozitationsanalyse nordamerikanischer Marketing-Journals 1987-2001, in: *Marketing: Zeitschrift für Forschung und Praxis*, 26 (2), 141–58.

Rothman, N. B. (2011), Steering sheep: How expressed emotional ambivalence elicits dominance in interdependent decision making contexts, in: *Organizational Behavior and Human Decision Processes*, 116 (1), 66–82.

——— **and G. B. Northcraft (2015)**, Unlocking integrative potential: Expressed emotional ambivalence and negotiation outcomes, in: *Organizational Behavior and Human Decision Processes*, 126, 65–76.

Rubin, D. C. and J. M. Talarico (2009), A comparison of dimensional models of emotion: Evidence from emotions, prototypical events, autobiographical memories, and words, in: *Memory*, 17 (8), 802–8.

Rubin, J. Z. and B. R. Brown (1975), *The Social Psychology of Bargaining and Negotiation*, New York: Academic Press.

———, **D. G. Pruitt, and S. H. Kim (1994)**, *Social Conflict: Escalation, Stalemate and Settlement*, New York: McGraw-Hill.

Rubinstein, A. (1982), Perfect Equilibrium in a Bargaining Model, in: *Econometrica* , 50 (1), 97–109.

Rudman, L. A. (1998), Self-Promotion as a Risk Factor for Women: The Costs and Benefits of Counterstereotypical Impression Management, in: *Journal of Personality and Social Psychology*, 74 (3), 629–45.

Russell, J. A. (1980), A Circumplex Model of Affect, in: *Journal of Personality and Social Psychology*, 39 (6), 1161–78.

——— **and L. F. Barrett (1999)**, Core Affect, Prototypical Emotional Episodes, and Other Things Called Emotion: Dissecting the Elephant, in: *Journal of Personality and Social Psychology*, 76 (5), 805–19.

Rustenbach, S. J. (2003), *Metaanalyse - Eine anwendungsorientierte Einführung*, Bern: Verlag Hans Huber.

Salacuse, J. W. and J. Z. Rubin (1990), Your Place or Mine? Site Location and Negotiation, in: *Negotiation Journal*, 5–10.

von Salisbury, J. (1159), *Metalogicon*.

Samiee, S. and B. R. Chabowski (2012), Knowledge structure in international marketing: A multi-method bibliometric analysis, in: *Journal of the Academy of Marketing Science*, 40 (2), 364–86.

Sawyer, J. and H. S. Guetzkow (1965), *Bargaining and negotiation in international relations*, New York: Holt, Rinehart and Winston.

Schäffer, U., C. Binder, and M. Gmür (2006), Struktur und Entwicklung der Controllingforschung – Eine Zitations- und Kozitationsanalyse von Controllingbeiträgen in deutschsprachigen wissenschaftlichen Zeitschriften von 1970 bis 2003, in: *Journal of Business Economics*, 76 (4), 395–440.

Schelling, T. C. (1963), *The Strategy of Conflict*, 2nd ed., Cambridge: Harvard University Press.

——— **(1980)**, *The Strategy of Conflict*, 3rd ed., Cambridge: Harvard University Press.

Schewe, A. F., U. R. Hülsheger, and G. W. Maier (2014), Forschungsmethoden: Metaanalyse – praktische Schritte und Entscheidungen im Umsetzungsprozess, in: *Zeitschrift für Arbeits- u. Organisationspsychologie*, 58 (4), 186–205.

Schildt, H. A., S. A. Zahra, and A. Sillanpää (2006), Scholarly Communities in Entrepreneurship Research: A Co-Citation Analysis, in: *Entrepreneurship Theory and Practice*, 30 (3), 399–415.

Schmidt, J. and T. H. A. Bijmolt (2019), Accurately measuring willingness to pay for consumer goods: a meta-analysis of the hypothetical bias, in: *Journal of the Academy of Marketing Science*, 48, 499–518.

Schneider, H. (2010), Determinanten der Kapitalstruktur - Eine meta-analytische Studie der empirischen Literatur, doctoral dissertation, European Business School in Oestrich-Winkel.

Schoop, M., A. Jertila, and T. List (2003), Negoisst: A negotiation support system for electronic business-to-business negotiations in e-commerce, in: *Data and Knowledge Engineering*, 47 (3), 371–401.

Schroth, H. A. (2008), Some Like It Hot: Teaching Strategies for Managing Tactical Versus Genuine Anger in Negotiations, in: *Negotiation and Conflict Management Research*, 1 (4), 315–32.

Schüller, D., S. Imöhl, and G. Berndt (2019), Die Chronologie des Brexits – eine Zusammenfassung des EU-Austritts Großbritanniens, in: *Handelsblatt*, (accessed November 21, 2019), [available at https://www.handelsblatt.com/politik/international/brexit-2019-die-chronologie-des-brexits-eine-zusammenfassung-des-eu-austritts-grossbritanniens/24097616.html?ticket=ST-19212445-PHR9gZZLM9btoVhd5elU-ap1].

Schultz, S. (2009), Porsche und Volkswagen rüsten zum Machtkampf, in: *Spiegel Online*, (accessed November 22, 2019), [available at https://www.spiegel.de/wirtschaft/fusionsverhandlungen-porsche-und-volkswagen-ruesten-zum-macht-kampf-a-623353-7.html].

Schulz, J. W. and D. G. Pruitt (1978), The Effects of Mutual Concern on Joint Welfare, in: *Journal of Experimental Social Psychology*, 14 (5), 480–92.

Schurr, P. H. (1987), Effects of Gain and Loss Decision Frames on Risky Purchase Negotiations, in: *Journal of Applied Psychology*, 72 (3), 351–58.

——— **and J. L. Ozanne (1985)**, Influences on Exchange Processes: Buyers' Preconceptions of a Seller's Trustworthiness and Bargaining Toughness, in: *Journal of Consumer Research*, 11 (4), 939–53.

Schwarz, N. (1990), Feelings as information: Informational and motivational functions of affective states, in *Handbook of motivation and cognition: Foundations of social behavior*, E. T. Higgins and R. M. Sorrentino, eds., New York: The Guilford Press, 527–61.

——— **and G. L. Clore (1983)**, Mood, Misattribution, and Judgments of Well-Being: Informative and Directive Functions of Affective States, in: *Journal of Personality and Social Psychology*, 45 (3), 513–23.

Schwarzer, G., J. R. Carpenter, and G. Rücker (2015), *Meta-Analysis with R*, Heidelberg: Springer.

Schweinsberg, M., G. Ku, C. S. Wang, and M. M. Pillutla (2012), Starting high and ending with nothing : The role of anchors and power in negotiations, in: *Journal of Experimental Social Psychology*, 48 (1), 226–31.

Schweitzer, M. E. and R. Croson (1999), Curtailing deception: The impact of direct questions on lies and omissions, in: *International Journal of Conflict Management*, 10 (3), 225–48.

———**, A. DeChurch, and D. E. Gibson (2005)**, Conflict Frames and the Use of Deception: Are Competitive Negotiators Less Ethical?, in: *Journal of Applied Social Psychology*, 35 (10), 2123–49.

Sebenius, J. K. (1983), Negotiation Arithmetic: Adding and Subtracting Issues and Parties, in: *International Organization*, 37 (2), 281–316.

Shao, B., L. Wang, D. Cheng, and L. Doucet (2015), Anger Suppression in Negotiations: The Roles of Attentional Focus and Anger Source, in: *Journal of Business and Psychology*, 30 (4), 747–58.

Shapiro, D. L. (2004), Emotions in Negotiation: Peril or Promise?, in: *Marquette Law Review*, 87 (4), 737–45.

Shapiro, S. S. and M. B. Wilk (1965), An Analysis of Variance Test for Normality, in: *Biometrika*, 52 (3), 591–611.

Sharma, S., W. P. Bottom, and H. A. Elfenbein (2013), On the Role of Personality, Cognitive Ability, and Emotional Intelligence in Predicting Negotiation Outcomes: A Meta-Analysis, in: *Organizational Psychology Review*, 3 (4), 293–336.

Siegel, S. and L. E. Fouraker (1960), *Bargaining and Group Decision Making: Experiments in Bilateral Monopoly*, New York: McGraw-Hill.

Sinaceur, M., H. Adam, G. A. van Kleef, and A. D. Galinsky (2013), The advantages of being unpredictable: How emotional inconsistency extracts concessions in negotiation, in: *Journal of Experimental Social Psychology*, 49 (3), 498–508.

———, G. A. van Kleef, M. A. Neale, H. Adam, and C. Haag (2011), Hot or Cold: Is Communicating Anger or Threats More Effective in Negotiation?, in: *Journal of Applied Psychology*, 96 (5), 1018–32.

——— and L. Z. Tiedens (2006), Get mad and get more than even: When and why anger expression is effective in negotiations, in: *Journal of Experimental Social Psychology*, 42 (3), 314–22.

Small, D. A., M. J. Gelfand, L. Babcock, and H. Gettman (2007), Who Goes to the Bargaining Table? The Influence of Gender and Framing on the Initiation of Negotiation, in: *Journal of Personality and Social Psychology*, 93 (4), 600–613.

Small, H. (1973), Co-citation in the Scientific Literature: A New Measure of the Relationship Between Two Documents, in: *Journal of the American Society for Information Science*, 24 (4), 265–69.

——— (1980), Co-Citation Context Analysis and the Structure of Paradigms, in: *Journal of Documentation*, 36 (3), 183–96.

Smith, C. A., K. N. Haynes, R. S. Lazarus, and L. K. Pope (1993), In Search of the "Hot" Cognitions: Attributions, Appraisals, and Their Relation to Emotion, in: *Journal of Personality and Social Psychology*, 65 (5), 916–29.

Smith, D. L., D. G. Pruitt, and P. J. Carnevale (1982), Matching and Mismatching: The Effect of Own Limit, Other's Toughness, and Time Pressure on Concession Rate in Negotiation, in: *Journal of Personality and Social Psychology*, 42 (5), 876–83.

Smith, L. C. (1981), Citation Analysis, in: *Library Trends*, 30 (1), 83–106.

Sokolova, M., M. Shah, and S. Szpakowicz (2006), Comparative analysis of text data in successful face-to-face and electronic negotiations, in: *Group Decision and Negotiation*, 15 (2), 127–40.

Solomon, R. C. (1993), *The Passions - Emotions and the Meaning of Life*, Indianapolis: Hackett Publishing Company.

Srivastava, J., D. Chakravarti, and A. Rapoport (2000), Price and Margin Negotiations in Marketing Channels: An Experimental Study of Sequential Bargaining Under One-sided Uncertainty and Opportunity Cost of Delay, in: *Marketing Science*, 19 (2), 163–84.

Stanley, T. D. (2001), Wheat From Chaff: Meta-Analysis As Quantitative Literature Review, in: *Journal of Economic Perspectives*, 15 (3), 131–50.

——— **(2005)**, Beyond Publication Bias, in: *Journal of Economic Surveys*, 19 (3), 309–45.

———, **C. Doucouliagos, and S. B. Jarrell (2008)**, Meta-regression analysis as the socio-economics of economics research, in: *The Journal of Socio-Economics*, 37 (1), 276–92.

Steele, C. M. (1997), A Threat in the Air - How Stereotypes Shape Intellectual Identity and Performance, in: *American Psychologist*, 52 (6), 613–29.

Steinel, W. and C. K. W. De Dreu (2004), Social Motives and Strategic Misrepresentation in Social Decision Making, in: *Journal of Personality and Social Psychology*, 86 (3), 419–34.

———, **G. A. van Kleef, and F. Harinck (2008)**, Are you talking to me?! Separating the people from the problem when expressing emotions in negotiation, in: *Journal of Experimental Social Psychology*, 44 (2), 362–69.

Stevens, C. K., A. G. Bavetta, and M. E. Gist (1993), Gender Differences in the Acquisition of Salary Negotiation Skills: The Role of Goals, Self-Efficacy, and Perceived Control, in: *Journal of Applied Psychology*, 78 (5), 723–35.

Stevens, C. M. (1963), *Strategy and Collective Bargaining Negotiation*, New York: McGraw-Hill.

Stock, B. (1979), Antiqui and Moderni as "Giants" and "Dwarfs": A Reflection of Popular Culture?, in: *Modern Philology*, 76 (4), 370–74.

Stuhlmacher, A. F. and M. Citera (2005), Hostile Behavior and Profit in Virtual Negotiation: A Meta-Analysis, in: *Journal of Business and Psychology*, 20 (1), 69–93.

——— **and T. L. Gillespie (2006)**, Managing Conflict in the Literature: Meta-analysis as a Research Method, in *Methods of Negotiation Research*, P. J. Carnevale and C. K. W. De Dreu, eds., Leiden/Boston: Martinus Nijhoff Publishers, 227–38.

——— **and M. G. Morrissett (2008)**, Men and women as mediators: Disputant perceptions, in: *International Journal of Conflict Management*, 19 (3), 249–61.

——— **and A. E. Walters (1999)**, Gender Differences in Negotiation Outcome: A Meta-Analysis, in: *Personnel Psychology*, 52 (3), 653–78.

Sutton, A. J. (2009), Publication Bias, in *The Handbook of Research Synthesis and Meta-Analysis*, H. Cooper, L. V. Hedges, and J. C. Valentine, eds., New York: Russell Sage Foundation, 435–52.

Swaab, R. I., A. D. Galinsky, V. Medvec, and D. A. Diermeier (2012), The communication orientation model: Explaining the diverse effects of sight, sound, and synchronicity on negotiation and group decision-making outcomes, in: *Personality and Social Psychology Review*, 16 (1), 25–53.

——— **and D. F. Swaab (2009)**, Sex differences in the effects of visual contact and eye contact in negotiations, in: *Journal of Experimental Social Psychology*, 45 (1), 129–36.

Sycara, K. P. (1991), Problem Restructuring in Negotiation, in: *Management Science*, 37 (10), 1248–68.

Tagar, M. R., C. M. Federico, and E. Halperin (2011), The positive effect of negative emotions in protracted conflict: The case of anger, in: *Journal of Experimental Social Psychology*, 47 (1), 157–64.

Tamir, M. and B. Q. Ford (2012), When Feeling Bad Is Expected to Be Good: Emotion Regulation and Outcome Expectancies in Social Conflicts, in: *Emotion*, 12 (4), 807–16.

Tenbrunsel, A. E. (1998), Misrepresentation and Expectations of Misrepresentation in an Ethical Dilemma: The Role of Incentives and Temptation, in: *The Academy of Management Journal*, 41 (3), 330–39.

Thaler, R. H. (1980), Toward a Positive Theory of Consumer Choice, in: *Journal of Economic Behavior and Organization*, 1 (1), 39–60.

——— **(1988)**, Anomalies: The Ultimatum Game, in: *Journal of Economic Perspectives*, 2 (4), 195–206.

Thibaut, J. W. and C. Faucheux (1965), The Development of Contractual Norms in a Bargaining Situation under Two Types of Stress, in: *Journal of Experimental Social Psychology*, 1, 89–102.

——— **and H. H. Kelley (1959)**, *The Social Psychology of Groups*, New York: John Wiley & Sons.

Thompson, L. L. (1990), Negotiation Behavior and Outcomes: Empirical Evidence and Theoretical Issues, in: *Psychological Bulletin*, 108 (3), 515–32.

——— (**1991**), Information Exchange in Negotiation, in: *Journal of Experimental Social Psychology*, 27 (2), 161–79.

——— (**2006**), Negotiation: Overview of Theory and Research, in *Negotiation Theory and Research*, L. L. Thompson, ed., New York: Psychology Press, 1–6.

——— (**2008**), *The Mind and Heart of the Negotiator*, 4th ed., Harlow: Pearson.

——— (**2015**), *The Mind and Heart of the Negotiator*, 6th ed., Harlow: Pearson.

———**, D. Gentner, and J. Loewenstein** (**2000**), Avoiding Missed Opportunities in Managerial Life: Analogical Training More Powerful Than Individual Case Training, in: *Organizational Behavior and Human Decision Processes*, 82 (1), 60–75.

——— **and R. Hastie** (**1990**), Social Perception in Negotiation, in: *Organizational Behavior and Human Decision Processes*, 47 (1), 98–123.

——— **and D. Hrebec** (**1996**), Lose-lose Agreements in Interdependent Decision Making, in: *Psychological Bulletin*, 120 (3), 396–409.

———**, E. A. Mannix, and M. H. Bazerman** (**1988**), Group Negotiation: Effects of Decision Rule, Agenda, and Aspiration, in: *Journal of Personality and Social Psychology*, 54 (1), 86–95.

———**, V. H. Medvec, V. Seiden, and S. Kopelman** (**2001**), Poker Face, Smiley Face, and Rant "n" Rave Myths and Realities about Emotion in Negotiation, in *Blackwell Handbook of Social Psychology: Group Processes*, M. A. Hogg and R. S. Tindale, eds., Malden: Blackwell Publishers, 139–63.

———**, E. Peterson, and S. E. Brodt** (**1996**), Team Negotiation: An Examination of Integrative and Distributive Bargaining, in: *Journal of Personality and Social Psychology*, 70 (1), 66–78.

———**, J. Wang, and B. C. Gunia** (**2010**), Negotiation, in: *Annual Review of Psychology*, 61 (1), 491–515.

Tiedens, L. Z. (**2001**), Anger and Advancement Versus Sadness and Subjugation: The Effect of Negative Emotion Expressions on Social Status Conferral, in: *Journal of personality and social psychology*, 80 (1), 86–94.

Tng, H.-Y. and A. K. C. Au (**2014**), Strategic Display of Anger and Happiness in Negotiation: The Moderating Role of Perceived Authenticity, in: *Negotiation Journal*, 30 (3), 301–27.

Tomlinson, E. C., B. Polin, B. Gray, and B. Barry (2017), Lessons Learned from Working with Roy J. Lewicki, in: *Negotiation and Conflict Management Research*, 10 (2), 126–40.

Topham, G. (2012), Volkswagen swallows Porsche, in: *The Guardian*, (accessed November 22, 2019), [available at https://www.theguardian.com/business/2012/jul/05/volkswagen-buys-porsche].

Totterdell, P. (2000), Catching Moods and Hitting Runs: Mood Linkage and Subjective Performance in Professional Sport Teams, in: *Journal of Applied Psychology*, 85 (6), 848–59.

Tranfield, D., D. Denyer, and P. Smart (2003), Towards a Methodology for Developing Evidence-Informed Management Knowledge by Means of Systematic Review, in: *British Journal of Management*, 14 (3), 207–22.

Tripp, T. M. and H. Sondak (1992), An Evaluation of Dependent Variables in Experimental Negotiation Studies: Impasse Rates and Pareto Efficiency, in: *Organizational Behavior and Human Decision Processes*, 51 (2), 273–95.

Tsay, C.-J. and M. H. Bazerman (2009), A Decision-Making Perspective to Negotiation: A Review of the Past and a Look to the Future, in: *Negotiation Journal*, 25 (4), 467–80.

———, L. L. Shu, and M. H. Bazerman (2011), Naïveté and Cynicism in Negotiations and Other Competitive Contexts, in: *Academy of Management Annals*, 5 (1), 495–518.

Tseng, C.-C. and J.-Y. Shih (2018), Two-stage coalition formation and radio resource allocation with Nash bargaining solution for inband underlaid D2D communications in 5G networks, in: *Journal of Network and Computer Applications*, 111, 64–76.

Tully, S. M. and R. S. Winer (2014), The Role of the Beneficiary in Willingness to Pay for Socially Responsible Products: A Meta-analysis, in: *Journal of Retailing*, 90 (2), 255–74.

Tversky, A. and D. Kahneman (1974), Judgment under Uncertainty: Heuristics and Biases, in: *Science*, 185 (4157), 1124–31.

——— and ——— (1981), The Framing of Decisions and the Psychology of Choice, in: *Science*, 211 (4481), 453–58.

——— and ——— (1991), Loss aversion in riskless choice: a reference-dependent model, in: *The Quarterly Journal of Economics*, 106 (4), 1039–61.

——— and ——— (**1992**), Advances in Prospect Theory: Cumulative Representation of Uncertainty, in: *Journal of Risk and Uncertainty*, 5, 297–323.

Ueno, T., G. M. Fastrich, and K. Murayama (2016), Meta-Analysis to integrate effect sizes within an article: Possible misuse and type I error inflation, in: *Journal of Experimental Psychology: General*, 145 (5), 643–54.

Ulich, D. and P. Mayring (2003), *Psychologie der Emotionen*, Stuttgart: Kohlhammer.

Üsdiken, B. and Y. Pasadeos (1995), Organizational Analysis in North America and Europe: A Comparison of Co-citation Networks, in: *Organization Studies*, 16 (3), 503–36.

Vahidov, R., G. Kersten, and R. Saade (2014), An experimental study of software agent negotiations with humans, in: *Decision Support Systems*, 66, 135–45.

Valente, T. W., K. Coronges, C. Lakon, and E. Costenbader (2008), How Correlated Are Network Centrality Measures?, in: *Connections*, 28 (1), 16–26.

Verbeek, A., K. Debackere, M. Luwel, and E. Zimmermann (2002), Measuring progress and evolution in science and technology – I: The multiple uses of bibliometric indicators, in: *International Journal of Management Reviews*, 4 (2), 179–211.

Vidmar, N. (1971), Effects of Representational Roles and Mediators on Negotiation Effectiveness, in: *Journal of Personality and Social Psychology*, 17 (1), 48–58.

Viechtbauer, W. (2005), Bias and Efficiency of Meta-Analytic Variance Estimators in the Random-Effects Model, in: *Journal of Educational and Behavioral Statistics*, 30 (3), 261–93.

——— (**2010**), Conducting Meta-Analyses in R with the metafor Package, in: *Journal of Statistical Software*, 36 (3), 1–48.

Voeth, M. and U. Herbst (2015a), Preisverhandlungen, in *Handbuch Business-to-Business-Marketing*, K. Backhaus and M. Voeth, eds., 2nd ed., Wiesbaden: Springer Gabler, 537–56.

——— **and** ——— (**2015b**), *Verhandlungsmanagement-Planung, Steuerung und Analyse*, 2nd ed., Stuttgart: Schäffer-Poeschel.

———, ———, **S. Haggenmüller, and M.-C. Weber (2020)**, Wie verhandeln deutsche Manager?, in: *Zeitschrift für Konfliktmanagement*, 23 (1), 21–24.

——— **and A. C. Lenzing (2017)**, Renegotiations - empirical analysis of impacts on business relationships, in: *Journal of Business & Industrial Marketing*, 32 (4), 541–52.

——— **and M. Mayer (2018)**, Ansätze für die Institutionalisierung von Verhandlungsmanagement im Vertrieb, in: *Die Unternehmung*, 72 (1), 87–107.

Vogel, R. (2012), The visible colleges of management and organization studies: A bibliometric analysis of academic journals, in: *Organization Studies*, 33 (8), 1015–43.

——— **and W. H. Güttel (2013)**, The Dynamic Capability View in Strategic Management: A Bibliometric Review, in: *International Journal of Management Reviews*, 15 (4), 426–46.

Volkema, R. J. (2004), Demographic, Cultural, and Economic Predictors of Perceived Ethicality of Negotiation Behavior: A Nine-Country Analysis, in: *Journal of Business Research*, 57 (1), 69–78.

Wahlquist, C. (22019), Volkswagen's $75m fine over diesel emissions scandal rejected by judge, in: *The Guardian*, (accessed November 21, 2019), [available at https://www.theguardian.com/business/2019/oct/16/volkswagens-75m-fine-over-diesel-emissions-scandal-rejected-as-outrageous-by-judge].

Wall, J. A. (1977), Operantly Conditioning a Negotiator's Concession Making, in: *Journal of Experimental Social Psychology*, 13 (5), 431–40.

——— **(1981)**, An Investigation of Reciprocity and Reinforcement Theories of Bargaining Behavior, in: *Organizational Behavior and Human Performance*, 27 (3), 367–85.

Wallace, M. L., Y. Gingras, and R. Duhon (2009), A New Approach for Detecting Scientific Specialties From Raw Cocitation Networks, in: *Journal of the American Society for Information Science and Technology*, 60 (2), 240–46.

Walter, C. and V. Ribière (2013), A citation and co-citation analysis of 10 years of KM theory and practices, in: *Knowledge Management Research and Practice*, 11 (3), 221–29.

Walters, A. E., A. F. Stuhlmacher, and L. L. Meyer (1998), Gender and Negotiator Competitiveness: A Meta-Analysis, in: *Organizational Behavior and Human Decision Processes*, 76 (1), 1–29.

Walton, R. E. and R. B. McKersie (1965), *A Behavioral Theory of Labor Negotiations*, New York: McGraw-Hill.

Wang, L., G. B. Northcraft, and G. A. van Kleef (2012), Beyond negotiated outcomes: The hidden costs of anger expression in dyadic negotiation, in: *Organizational Behavior and Human Decision Processes*, 119 (1), 54–63.

Wang, M., Y. Han, and Y. Su (2017), Social contagion or strategic choice?: The interpersonal effects of emotions during Chinese negotiations, in: *Chinese Management Studies*, 11 (3), 463–78.

Wang, N., H. Liang, Y. Jia, S. Ge, Y. Xue, and Z. Wang (2016), Cloud computing research in the IS discipline: A citation/co-citation analysis, in: *Decision Support Systems*, 86, 35–47.

Wanyama, T. and B. H. Far (2007), A protocol for multi-agent negotiation in a group-choice decision making process, in: *Journal of Network and Computer Applications*, 30 (3), 1173–95.

Webster, D. M., L. Richter, and A. W. Kruglanski (1996), On Leaping to Conclusions When Feeling Tired: Mental Fatigue Effects on Impressional Primacy, in: *Journal of Experimental Social Psychology*, 32 (2), 181–95.

Weingart, L. R., R. J. Bennett, and J. M. Brett (1993), The Impact of Consideration of Issues and Motivational Orientation on Group Negotiation Process and Outcome, in: *Journal of Applied Psychology*, 78 (3), 504–17.

——, **L. L. Thompson, M. H. Bazerman, and J. S. Carroll (1990)**, Tactical Behavior and Negotiation Outcomes, in: *The International Journal of Conflict Management*, 1 (1), 7–31.

Weiss, S. E. (1994), Negotiating with Romans, in: *Sloan Management Review*, 35 (3), 286–301.

Weisz, J. R., B. Weiss, S. S. Han, D. A. Granger, and T. Morton (1995), Effects of Psychotherapy With Children and Adolescents Revisited: A Meta-Analysis of Treatment Outcome Studies, in: *Psychological Bulletin*, 117 (3), 450–68.

Weitz, B. A. (1978), Relationship between Salesperson Performance and Understanding of Customer Decision Making, in: *Journal of Marketing Research*, 15 (4), 501–16.

—— **(1979)**, "A Critical Review of Personal Selling Research: The Need for Contingency Approaches," report, College of Business Administration, University of Oregon.

—— **(1981)**, Effectiveness in Sales Interactions: A Contingency Framework, in: *Journal of Marketing*, 45 (1), 85–103.

White, H. D. (2003), Author cocitation analysis and Pearsons's r, in: *Journal of the American Society for Information Science and Technology*, 54 (13), 1250–59.

——— **and K. W. McCain (1998)**, Visualizing a Discipline: An Author Co-Citation Analysis of Information Science, 1972 – 1995, in: *Journal of the American Society for Information Science*, 49 (4), 327–55.

White, J. B., R. Tynan, A. D. Galinsky, and L. L. Thompson (2004), Face Threat Sensitivity in Negotiation: Roadblock to Agreement and Joint Gain, in: *Organizational Behavior and Human Decision Processes*, 94 (2), 102–24.

White, S. B. and M. A. Neale (1994), The Role of Negotiator Aspirations and Settlement Expectancies in Bargaining Outcomes, in: *Organizational Behavior and Human Decision Processes*, 57 (2), 303–17.

Whittaker, J. (1989), Creativity and Conformity in Science: Titles, Keywords and Co-word Analysis, in: *Social Studies of Science*, 19 (3), 473–96.

Wilken, R., F. Jacob, and N. Prime (2013), The Ambiguous Role of Cultural Moderators in Intercultural Business Negotiations, in: *International Business Review*, 22 (4), 736–53.

Wilson, D. B. (2009), Systematic Coding, in *The Handbook of Research Synthesis and Meta-Analysis*, H. Cooper, L. V. Hedges, and J. C. Valentine, eds., New York: Russell Sage Foundation, 159–76.

Wolfe, R. and K. McGinn (2005), Perceived Relative Power and its affect on Negotiations, in: *Group Decision and Negotiation*, 14 (1), 3–20.

Yang, Y., H. Falcão, N. Delicado, and A. Ortony (2014), Reducing Mistrust in Agent–Human Negotiations, in: *IEEE Intelligent Systems*, 29 (2), 36–43.

———**, S. Singhal, and Y. Xu (2013)**, Alternate Strategies for a Win-Win Seeking Agent in Agent-Human Negotiations, in: *Journal of Management Information Systems*, 29 (3), 223–56.

Yip, J. A. and M. Schweinsberg (2017), Infuriating Impasses: Angry Expressions Increase Exiting Behavior in Negotiations, in: *Social Psychological and Personality Science*, 8 (6), 706–14.

——— **and M. E. Schweitzer (2016)**, Mad and misleading: Incidental anger promotes deception, in: *Organizational Behavior and Human Decision Processes*, 137, 207–17.

Yuan, Y., M. Head, and M. Du (2003), The Effects of Multimedia Communication on Web- Based Negotiation, in: *Group Decision and Negotiation*, 12, 89–109.

Yukl, G. A. (1974a), Effects of Situational Variables and Opponent Concessions on a Bargainer's Perception, Aspirations, and Concessions, in: *Journal of Personality and Social Psychology*, 29 (2), 227–36.

——— **(1974b)**, Effects of the Opponent's Initial Offer, Concession Magnitude and Concession Frequency on Bargaining Behavior, in: *Journal of Personality and Social Psychology*, 30 (3), 323–35.

Zajonc, R. B. (1965), Social Facilitation, in: *Science*, 149 (3681), 269–74.

Zeng, D. and K. Sycara (1998), Bayesian Learning in Negotiation, in: *International Journal of Human-Computer Studies*, 48 (1), 125–41.

Zerres, A. and J. Hüffmeier (2011), Too Many Cooks Spoil the Soup - How Negotiating Teams Deteriorate the Negotiators' Relationship, in: *Die Betriebswirtschaft*, 71 (6), 559–75.

———, ———, P. A. Freund, and K. Backhaus (2013), Does It Take Two to Tango? Longitudinal Effects of Unilateral and Bilateral Integrative Negotiation Training, in: *Journal of Applied Psychology*, 98 (3), 478–91.

Zetik, D. C. and A. F. Stuhlmacher (2002), Goal Setting and Negotiation Performance: A Meta-Analysis, in: *Group Processes & Intergroup Relations*, 5 (1), 35–52.

Zhang, J. and W. P. Bottom (2019), "Blowback: Retaliating against Anger Expressions in Group Decision Making and the Role of Gender," working paper.

Zupic, I. and T. Čater (2015), Bibliometric Methods in Management and Organization, in: *Organizational Research Methods*, 18 (3), 429–72.

Aus unserem Verlagsprogramm:

Caroline Jäckel
Sozioökonomische Effekte durch die Wiederaufbereitung von Informations- und Kommunikationstechnik in Deutschland
Hamburg 2020 / 140 Seiten / ISBN 978-3-339-11688-8

Katarína Remeňová & Nadežda Jankelová
A Descriptive Approach to Decision Making and the Decision-Making Style
Hamburg 2019 / 142 Seiten / ISBN 978-3-339-11218-7

Dirk Förster-Trallo
The Impact of Psychological Occupational Strains on Engineering Managers' Decision Making Behaviour
Hamburg 2019 / 376 Seiten / ISBN 978-3-339-10940-8

Victoria Wojcik
Performanceanalyse mittels Verallgemeinerter Data Envelopment Analysis
Vorgehensmodell und Evaluation
Hamburg 2018 / 452 Seiten / ISBN 978-3-339-10012-2

Maximilian Blum
Der Planungs- und Entscheidungsprozess der wertorientierten Steuerung aus verhaltenswissenschaftlicher Sicht
Hamburg 2017 / 322 Seiten / ISBN 978-3-8300-9517-0

Jan Keidel / Hendrik Jähn / Marco Fischer / Thomas Burghardt / Daniel Neumann (Hrsg.)
Modelle und Methoden zur Entscheidungsunterstützung für betriebswirtschaftliche Wertschöpfungsprozesse
Festschrift für Prof. Dr. Joachim Käschel
Hamburg 2016 / 158 Seiten / ISBN 978-3-8300-9229-2

Désirée Scherer
Kommunikation als integrativer Teil der Umsetzung von Großprojekten im Energiesektor
Analyse eines partizipativen Verfahrens als mögliches Instrument eines gesellschaftlich akzeptierten Kommunikationsprozesses zur Entscheidungsfindung bei einer Großtechnologie am Beispiel der CO_2-Abscheidung und -Speicherung
Hamburg 2016 / 484 Seiten / ISBN 978-3-8300-8826-4

Alexander Smeja
Auswirkungen von Emotionen und deren Regulierung auf entrepreneuriale Prozesse
Hamburg 2014 / 260 Seiten / ISBN 978-3-8300-7844-9